Advances in Science, Technology & Innovation

IEREK Interdisciplinary Series for Sustainable Development

Advances in Science, Technology & Innovation (ASTI) is a series of peer-reviewed books based on the best studies on emerging research that redefines existing disciplinary boundaries in science, technology and innovation (STI) in order to develop integrated concepts for sustainable development. The series is mainly based on the best research papers from various IEREK and other international conferences, and is intended to promote the creation and development of viable solutions for a sustainable future and a positive societal transformation with the help of integrated and innovative science-based approaches. Offering interdisciplinary coverage, the series presents innovative approaches and highlights how they can best support both the economic and sustainable development for the welfare of all societies. In particular, the series includes conceptual and empirical contributions from different interrelated fields of science, technology and innovation that focus on providing practical solutions to ensure food, water and energy security. It also presents new case studies offering concrete examples of how to resolve sustainable urbanization and environmental issues. The series is addressed to professionals in research and teaching, consultancies and industry, and government and international organizations. Published in collaboration with IEREK, the ASTI series will acquaint readers with essential new studies in STI for sustainable development.

More information about this series at http://www.springer.com/series/15883

Noamen Rebai • Mohamed Mastere
Editors

Mapping and Spatial Analysis of Socio-economic and Environmental Indicators for Sustainable Development

Case Studies from North Africa

Springer

Editors
Noamen Rebai
National Engineering School of Tunis
Tunis, Tunisia

Mohamed Mastere
Geopac Research Center
Scientific Institute, Mohammed V University
of Rabat
Rabat, Morocco

ISSN 2522-8714 ISSN 2522-8722 (electronic)
Advances in Science, Technology & Innovation
IEREK Interdisciplinary Series for Sustainable Development
ISBN 978-3-030-21165-3 ISBN 978-3-030-21166-0 (eBook)
https://doi.org/10.1007/978-3-030-21166-0

This Springer imprint is published by the registered company Springer Nature Switzerland AG
The registered company address is: Gewerbestrasse 11, 6330 Cham, Switzerland

Preface

I'm deeply grateful to the Springer Editor in Germany who drove me and my colleague Mohamed Mastere (Scientific Institute, Mohammed V University of Rabat, Morocco) into the adventure of publishing this book, which includes the best works presented in the First International Congress: I SEE GEOMATICS: 'Mapping and Spatial Analysis for Socio-economic and Environmental Indicators for Local and Regional Sustainable Development' which was held on the 25th and 26th of March 2015 in Tataouine, Tunisia.

This Congress was the crowning of a strong collaboration between the Tunisian Association of Geomatics and the Moroccan Association of Young Geomaticists. This partnership dates back to 2013 and culminates in the creation of 'African GEOMatic Association' (AGEOMA).

In my mind, most of the researchers working on topics related to Geomatics in North Africa have definitely been aware of the lack of communication between their community of academicians and the world of other specialists who use Geomatics products and the spatiotemporal data in 'OPEN DATA' or proprietary and cartographic modelling with its different thematic conceptual, treatment and analytical forms from another part.

This situation is mainly related to the absence of strategic development of Geomatics. It is only since the twenty-first century that the teaching of and research on Geomatics has started in Africa in general and in North Africa in particular. It is important to note that till today, this domain is still underrepresented and little engaged in the professional field.

The aim of this work is to provide information and support for regional development. **The Mapping specifies a set of up-to-date indicators that include various domains**: Agriculture, Energy, Industry, Tourism, Transportation, Urban planning, Exploitation of natural resources, Infrastructure, Health, Environment, Education, Information and Communication Technologies, Social affairs and employability, Culture, Sport, Religion, Local authorities, etc.

Environmental and sustainable development indicators have proliferated since the Rio Earth Summit in 1992. The Socio-economic and Environmental Indicators **(SEE I)** allow to observe and analyze the economic data and measure their impact on the environment and their contributions in the social and environmental networks. As examples of the Socio-economic and Environmental Indicators, we can cite:

- The rate of natural resources exploitation (water, oil, useful materials, etc.),
- The strengthening of the adaptation to the climate change,
- The vulnerability of the coastal area to the risk of marine submersion due to the sea-level rise, etc.

Within this framework, a case study of the **Regional Sustainable Development** in Maghreb will be undertaken. A current state of the art of the **SEE I** is necessary to help decision makers in their actions on the basis of the Geomatics approach, the Modeling cartography and the Multi-criteria spatial and temporal analysis of the **SEE I**.

The various contributions confirm the interest of the scientific community in the field and permit the readers to make a synthesis of the research area through the fundamental chapters of this book namely:

1. Modeling and Spatio-Temporal Analysis for the SEE Indicators.
2. GEORISK and Environmental Monitoring.
3. Management and development of the territory (3D GIS, etc.).

Finally, we would like to deeply acknowledge the positive support given by all the participants in the Congress, the experts that gave many conferences as well as the authors for their continual assistance, for their direct contribution, and for sharing their research results.

Tunis, Tunisia

Dr. Noamen Rebai
President of the First International Congress
'ISEE GEOMATICS' Tataouine 2015, Tunisia
For the Steering Committee

# Acknowledgements

This book is the result of a combined effort of many institutions and several research teams working in different scientific research laboratories in Applied Geomatics in Tunisia, Morocco, Benin and Algeria.

On this opportunity, we thank all **Springer staff** for proposing us to write this book. We also appreciate their continual help throughout the different steps of its realization.

This book gathers papers from the First International Congress, dealing with Socio-economic and Environmental Indicators, under the label "I SEE GEOMATICS".

Special thanks for:

- Pr. Guido Zosimi Landalfo (Editorial Director and Asset Manager),
- Dr. Nabil Khélifi, Senior Publishing Editor (Earth Sciences, Geography & Environment and managing: Middle East & North Africa Program—Springer, part of Springer Nature,
- Ms. Reyhaneh Majidi, (Editorial Assistant, Middle East & North Africa Springer, part of Springer Nature Asset Manager).

This work was reviewed by a committee that has enthusiastically agreed to revise chapters and advise authors to follow relevant strategies to improve the quality of their papers. We deeply thank Pr. Michel Boko, Dr. Batita Wided, Pr. Abdelkader El Garouani, Pr. Didier Josselin, Pr. Mohamed Mastere, Dr. Slimene Sedrette and Pr. Noamen Rebai.

We would like to extend our sincere gratitude for the enormous work done by Ms. Zeineb Ben Cheikh and Abdelkader Moussi in formatting the entire document in Latex and for Mrs. Amel Zaouga for the linguistic correction of the manuscript.

All paper authors are thanked for their numerous exchanges for the sake of finalizing this work.

Finally, we are pleased to thank all the members of the Tunisian Association of Geomatics (ATG), the Moroccan Association of Young Geomaticians (AMJG) and the program organizing committee "I SEE GEOMATICS" Congress in its first version.

Dr. Noamen Rebai
President of the First International Congress
"ISEE GEOMATICS" Tataouine 2015, Tunisia
For the Steering Committee

Contents

Contributors

Saâdi Abdeljaouad Laboratoire des Ressources Minérales et Environnement (LRME), Département de Géologie, Faculté des Sciences de Tunis, Université Tunis El Manar, Tunis, Tunisie

Abdelkrim Achbun Laboratoire de géomatique, Institut National d'Aménagement et d'Urbanisme, Rabat, Morocco;
Laboratoire de Géomorphologie et de Cartographie, Geopac Research Center, Scientific Institute, Mohammed V University, Rabat, Morocco

Abdelkrim Arioua Water and Remote Sensing Team "GEVARET", Faculty of Sciences and Techniques, Sultan Moulay Slimane University, Beni-Mellal, Morocco

Wided Batita Department of Applied Geomatics, Université de Sherbrooke, Sherbrooke, QC, Canada

Tarik Benabdelouahab National Institute of Agronomic Research of Morocco, Rabat, Morocco

Zeineb Ben Cheikh RIADI Laboratory, National School of Computer Science, University Campus of Manouba, Manouba, Tunisia

Hafsa Bouamri Water and Remote Sensing Team "GEVARET", Faculty of Sciences and Techniques, Sultan Moulay Slimane University, Beni-Mellal, Morocco

Abdelghani Boudhar Water and Remote Sensing Team "GEVARET", Faculty of Sciences and Techniques, Sultan Moulay Slimane University, Beni-Mellal, Morocco;
Center for Remote Sensing Applications (CRSA), Mohammed VI Polytechnic University, Ben Guerir, Morocco

Claude Caron PRISME–Pôle de recherche en intelligence stratégique et multidimensionnelle d'entreprise, Faculté d'administration, Université de Sherbrooke, Sherbrooke (Québec), QC, Canada

Ali Chaieb LR14ES03 Geotechnical Engineering and Georisk Laboratory, National School of Engineering of Tunis, University of Tunis El Manar, Tunis, Tunisia

Bouchta El Fellah Geopac Research Center, Scientific Institute, Mohammed V University of Rabat, Rabat, Morocco;
Laboratoire de géomatique, Institut National d'Aménagement et d'Urbanisme, Rabat, Morocco

Salma El Fellah Faculty of Sciences, Mohammed V University in Rabat, Rabat, Morocco

Younes El Fellah Energy and Farm Machinery Department, Rural Engineering, Institut Agronomique et Veterinaire Hassan II, Rabat, Morocco

Jean-Michel Follin Ecole Supérieure des Géomètres et Topographes, ESGT, GeF, le Mans, France

Sahar Gaaloul CNCT: Centre National de la Cartographie et de la Télédétection (en Tunisie), Tunis, Tunisie

Sonia Gannouni Laboratoire de Géoressources, CERTE, Technopole de BorjCedria, Tunis, Tunisia

Mohammed Hssaisoune Water and Remote Sensing Team "GEVARET", Faculty of Sciences and Techniques, Sultan Moulay Slimane University, Beni-Mellal, Morocco

Ismail Karaoui Water and Remote Sensing Team "GEVARET", Faculty of Sciences and Techniques, Sultan Moulay Slimane University, Beni-Mellal, Morocco

Salwa Lagdali Faculty of Sciences, Mohammed V University in Rabat, Rabat, Morocco

Youssef Lebrini Water and Remote Sensing Team "GEVARET", Faculty of Sciences and Techniques, Sultan Moulay Slimane University, Beni-Mellal, Morocco

Karim Mansouri ENSI: Ecole Nationale des sciences de l'Informatique Manouba, Manouba, Tunisia

Mohamed Mastere Geopac Research Center, Scientific Institute, Mohammed V University of Rabat, Rabat, Morocco

Abdelkader Moussi LR14ES03 Geotechnical Engineering and Georisk Laboratory, National School of Engineering of Tunis, University of Tunis El Manar, Tunis, Tunisia

Hamza Ouatiki Water and Remote Sensing Team "GEVARET", Faculty of Sciences and Techniques, Sultan Moulay Slimane University, Beni-Mellal, Morocco

Noamen Rebai LR14ES03 Geotechnical Engineering and Georisk Laboratory, National School of Engineering of Tunis, University of Tunis El Manar, Tunis, Tunisia

Stephane Roche Geomatics Research Center, Pavillon Louis-Jacques-Casault, Laval University, Quebec, QC, Canada

Mohammed Rziza Faculty of Sciences, Mohammed V University in Rabat, Rabat, Morocco

François Sabatier Centre Européen de Recherche et d'Enseignement des Géosciences et de l'Environnement, UMR CNRS 6635, Aix-Marseille Université, Aix-en-Provence, France

Murad Salhi CNCT: Centre National de la Cartographie et de la Télédétection (en Tunisie), Tunis, Tunisie

Slimene Sedrette LR14ES03 Geotechnical Engineering and Georisk Laboratory, National School of Engineering of Tunis, University of Tunis El Manar, Tunis, Tunisia

Mohamed Ayet Allah Bilel Soussi Institut Supérieur des Etudes Technologiques de Nabeul, Nabeul, Tunisia

Zouhour Yahyaoui Laboratoire des Ressources Minérales et Environnement (LRME), Département de Géologie, Faculté des Sciences de Tunis, Université Tunis El Manar, Tunis, Tunisie

Chamseddine Zaki American University of Culture and Education in Beirut, Beirut, Lebanon

Part I
Modelling and Spatial-Temporal Analysis

Mapping and Spatial Analysis of Sustainable Development Indicators to Optimize the Quality of Life Using AHP Methods: A Case Study Tataouine, Tunisia

Zeineb Ben Cheikh and Noamen Rebai

Abstract

Many studies have been conducted the implementation of sustainable development (SD) programs by using a set of indicators in various domains such as agriculture, energy, transport, exploitation of natural resources, health, environment, etc. The analysis of those indicators is an important step in improving its relevance. Therefore, to reach an effective result, the correlation between the indicators facilitates in the decision-making process. In this context, it is plausible to refer to the document about the indicators of sustainable development of Tunisia that was edited in 2014 by the Tunisian Observatory center of Sustainable Development: (OTEDD). It contains nine challenges that deal with various socioeconomic and environmental (SEE.I) problems. The endeavor of this study, then, is to determine one of these challenges which is the citizens' quality of life (QoL). Our study area is Tataouine. For reaching our target and prioritizing a set of indicators and geographic representation tools, we have had recourse to the analytical hierarchy process (AHP). We have used the six indicators presented in relation to this challenge: Number of inpatients per basic health center, drinking water supply rate, rate of access to adequate (non-collective) sanitation including rate of connection to the public sewerage network, monitoring of air quality in urban areas, green space areas and urban parks per inpatient and proportion of rudimentary dwellings.

Keywords

Mapping • Spatial analysis • MCDA • AHP • Quality of life • Socioeconomic and environmental indicators (see. I.) • Tunisia • Sustainable development

1 Introduction

With the emergence of environmental problems, the concept of sustainable development has emerged gradually since the late 1960s [1–3]. The first and the most commonly cited definition of sustainable development appeared in 1987 in the Brundtland report [4]. It states that "Sustainable development is a development that meets the needs of the present without compromising the ability of future generations to meet their own needs".

After a wide tour of consultations in several countries of the world, The world commission on environment and development [5] published its report entitled "Notre avenir à tous". The report points out that the main cause of global environmental degradation is the increase of poverty in the south and the sustained economic growth in the north. To meet this real challenge, global actors (communities, associations, individuals) are encouraged to reexamine their practices and behaviors in favor of innovative actions at the "Rio Earth Summit" in Rio de Janeiro in 1992. This conference is characterized by the adoption of a founding text of 27 principles entitled "Rio Declaration on Environment and Development" by 173 Heads of State.

Moreover, in response to these main problems, a program of concrete actions for sustainable development in the twenty-first century has also been adopted. It is known as "Agenda 21". The idea behind it is to propose 21 major actions for the twenty-first century through environmental development. It provides a monumental 800-page plan of action, including the list of major critical problems we face as a rapidly growing global community on a small planet with limited resources [6,7].

Z. Ben Cheikh (✉)
RIADI Laboratory, National School of Computer Science, University Campus of Manouba, Manouba, Tunisia
e-mail: benzeineb@outlook.fr

N. Rebai
LR14ES03 Geotechnical Engineering and Georisk Laboratory, National School of Engineering of Tunis, University of Tunis El Manar, BP. 37, le Belvédère, 1002 Tunis, Tunisia
e-mail: noamen.rebai@enit.utm.tn

N. Rebai and M. Mastere (eds.), *Mapping and Spatial Analysis of Socio-economic and Environmental Indicators for Sustainable Development*, Advances in Science, Technology & Innovation, https://doi.org/10.1007/978-3-030-21166-0_1

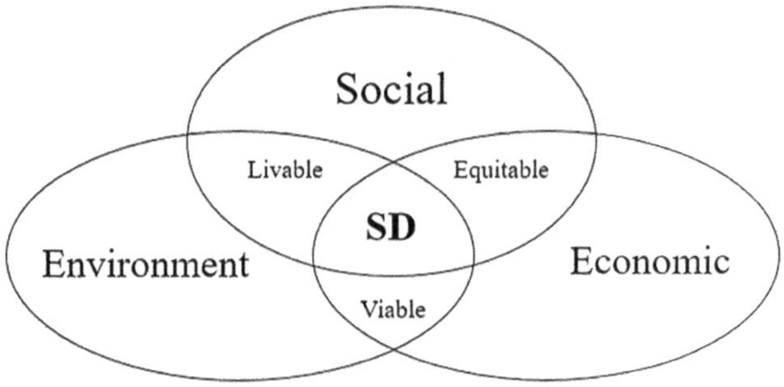

Fig. 1 The three pillars of sustainable development [8]

Sustainable development takes into account three dimensions: economy, society, and environment. Therefore, the development must be equitable, livable, and viable (Fig. 1).

- Environment: The environmental dimension of sustainable development aims to preserve and improve the environment and the natural resources in the long term. Such a goal can be met through maintaining the major ecological balances, reducing risks, and preventing negative environmental impacts.
- Society: The social dimension of sustainable development aims to meet human needs and social equity by promoting all social groups' sharing and enjoyment of health services, housing, consumption, education, employment, culture, etc.
- Economy: The economic dimension of sustainable development seeks to boost efficient economies through sustainable production and consumption patterns.

In this paper, our aim is to improve the quality of life in the case of Tataouine's delegations. We have implemented the analytic hierarchy process (AHP) to weight the different indicators of this challenge, express their relative importance then combine them in order to analyze their behavior in our specific case study. Finally, we have drawn our main findings in a map defining the state of Tatouine in this challenge.

2 Indicators and Challenges

An indicator is an observable element constituting the sign, the trace and the presence of a phenomenon. It allows the measurement of its level of evolution. It is therefore the factual and concrete data that determine a state or an evolution. Also, the outcome indicators demonstrate the achievement of an objective. However, taking only one indicator into account is meaningless since it cannot explain why an objective is only partially achieved. Hence, it is important to link sustainability challenges to other groups of indicators and to acquire (and to be informed about) other indicators.

In this context, the "Tunisian Observatory on Environment and Sustainable Development" [9] has classified the 42 indicators into a number of challenges.

- Sustainable consumption and production
- Promoting an efficient economy, strengthening social equity and tackling regional disparities
- Sustainable management of natural resources
- Promoting more balanced spatial planning based on efficient and sustainable transport
- Promoting a better quality of life for citizens
- Developing energy efficiency and promoting and renewable energies
- Building capacity to adapt to climate change
- Promoting the knowledge of the society
- Adapting governance for better sustainable development promotion

We choose "Promoting a better quality of life for citizens" as a challenge to study in the next sections of our paper.

3 Promoting a Better Quality of Life for Citizens

The concept of the quality of life and its indicators is not a new topic. Bunge [10] has developed this concept by defining its indicators since there are many aspects (health centers, water, green spaces, quality of air, etc.) that have a direct impact on the citizens QoL. However, it is obvious that those indicators can change from one challenge to another according to their importance [11, 12]. In this paper, we are interested in the six indicators defined by OTEDD which are as follows.

3.1 Number of Inpatients Per Basic Health Center (I1)

Improving access to the health center is an important challenge for achieving the social equity. This indicator denotes the population's enjoyment of various health services.

It is presented in several forms such as:

Column chart: The Organization for Economic Cooperation and Development refers to this indicator by a column chart of 35 countries depending on "the resources available for delivering services to inpatients in hospitals in terms of the number of beds that are maintained, staffed and immediately available for use." [13].

Set of values: Eurostat statistics present it by the number of hospital beds in 2002 and 2011 (per 100 000 inpatients) of 35 countries.

3.2 Drinking Water Supply Rate (I2)

The lack of drinking water supply is one of the causes of illness and mortality mainly because of chemical pollutants and poor hygiene. Therefore, the access to drinking water has been one of the objectives of Member States of the United Nations known as The Millennium Development Goals (MDGs). The global target fixed at 88% was reached in 2010. However, statistics in 2015 show that 91% of the population had access to drinking water which exceeded the target. These objectives are fixed differently from one region to another, depending on climate conditions (e.g. 75% for sub-Saharan Africa and 94% for Northern Africa). UNICEF presented those results in different forms, such as a column chart, a world map, etc. [14].

This indicator is calculated as [9] (Eq. 1).

$$\textit{Drinking water supply rate} = \frac{\textit{Population with access to drinking water resource}}{\textit{Total population}} \quad (1)$$

3.3 Rate of Access to Adequate Sanitation Including Rate of Connection to Public Sewerage Network (I3)

This indicator refers to the percentage of the population having a direct access to adequate sanitation in relation to the total population. In urban areas, it corresponds to the connection to a public sanitation system (sewer) or an autonomous one (lost well). In rural areas, basic systems are being deployed with the desire to isolate wastewater from contacting people, animals, crops, and water resources.

This indicator is calculated as follows: [9] (Eq. 2)

$$\textit{Rate of access to adequate sanitation} = \frac{\textit{Population with access to adequate sanitation}}{\textit{Total population}} \quad (2)$$

3.4 Air Quality Monitoring in Urban Areas (I4)

Most air pollution is man-made and comes from the combustion substances such as H2S (Hydrogen Sulfide), NO2 (Nitrogen Dioxide) of biomass fuels used in transport, industry, and a lot of other fields. "Therefore, it is important to continuously monitor the concentration of these substances in industrial environments" [15].

Europe allocated a budget of USD 1.78 billion in 2013 for gas sensors and it is expected to grow by 5.1% in the period between 2014 and 2020 with the aim to monitor Oxygen (O_2), Carbon Monoxide (CO), Carbon Dioxide (CO_2), Nitrogen Oxide (NOx).

3.5 Urban Parks Per Capita (I5)

Urban green areas are major contributors to human health and QoL [16,17]. This indicator captures well measurable and quantifiable physical attributes applied for biodiversity preservation or microclimate regulations [18,19]. Moreover, the green space not only has an effect on the aesthetical side of the cities [20,21] but also contributes to air purification [22].

3.6 The Proportion of Rudimentary Housing (I6)

The housing sector is an area marked by its contradictory flows of development. It mediates between productive and urban development activities and between speculative investment and financing trends. This indicator measures the proportion of urban dwellers living in depriving housing conditions and determines the adequacy of the basic human need for shelter.

This indicator is calculated as follows: [9] (Eq. 3)

$$\textit{Proportion of rudimentary housing} = \frac{\textit{Part of rudimentary housing}}{\textit{Total number of dwellings}} \quad (3)$$

4 Methodology

4.1 Multi-criteria Decision-Making

Multi-criteria decision-making (MCDM) involves the decision-making tools that deal with complex problems for which technological, economic, ecological, and social aspects need to be addressed [23]. In a situation where several criteria are involved, the likelihood of a conflict of opinion is high. Thus, if a logical and well-structured decision-making process is not followed, a multidisciplinary team is designed to make decisions. Using the MCA, members are then not expected to follow neither the same criteria nor ranking alternatives.

Since the 1980s, several multiple-criteria decision analysis (MCDA) methods have been proposed [24]. The most widely applied are using the pairwise comparison [25]. In fact, different MCD methods have been employed in natural resources management, energy planning, water management, etc. The very popular methods are compromise programming (CP), the analytic hierarchy process (AHP), and ELECTRE and PROMETHEE [26].

Table 1 A pairwise comparison matrix

Criteria	C1	C2	C3	...	C_n	W_i
C1	$1/\sum C1$	$W_{21}/\sum C2$	$W_{31}/\sum C3$	...	$W_{n1}/\sum C_n$	$\sum C1/n$
C2	$W_{12}/\sum C1$	$1/\sum C2$	$W_{32}/\sum C3$	...	$W_{n2}/\sum C_n$	$\sum C2/n$
C3	$W_{13}/\sum C1$	$W_{23}/\sum C2$	$1/\sum C3$	...	$W_{n3}/\sum C_n$	$\sum C3/n$
...	...	...	...	...	...	...
C_n	$W_{1n}/\sum C1$	$W_{2n}/\sum C2$	$W_{3n}/\sum C3$	...	$1\sum C_n$	$\sum C_n/n$
	$\sum C1$	$\sum C2$	$\sum C3$	...	$\sum C_n$	

c: criteria
w: weight

Table 2 Scale of pairwise comparison

Intensity of importance	Definition
1	Equal importance
2	Weak
3	Moderate importance
4	Moderate plus
5	Strong importance
6	Strong plus
7	Very strong importance
8	Very very strong
9	Extreme importance

4.2 Essentials of AHP

AHP, proposed by Saaty [27], is a decision-making tool that takes into consideration a set of weighted criteria targeting a better classification of alternatives. It is one of the most commonly used MCDM methods due to its easy implementation, its pairwise comparison and its scalability [28].

Several works have chosen AHP to evaluate economic, ecological and social aspects such as natural resources [29–31] land-use [32–34] energy [35], etc.

4.3 AHP Principles

For measuring the weights of the criteria, AHP calculates in the first step the eigenvalues and the eigenvectors of a square preference matrix (Table 1) that contains the quantified preference of a pairwise comparison of criteria. Saaty [27] proposes the scale shown in Table 2 consisting of values ranging from 1 to 9 where 1 expresses the equality of importance and 9 refers to the extreme importance of one criterion over another.

Figure 2 represents the hierarchical structure of an unstructured problem.

Obviously, the preference values assigned to each pairwise comparison must be well-studied. However, this may not be the case in which expert opinions and the creation of different scenarios may be inconsistent. Such an inconsistency can lead to incorrect results. To resolve this ambiguity, Saaty provided a consistency ratio called CR as the ratio of the CI (Eq. 5) (Consistency Index) to RI (Random Index) (Table 3), so that CR is calculated by Eq. (4).

$$CR = \frac{CI}{RI} \tag{4}$$

$$CI = \frac{\lambda max - n}{n - 1} \tag{5}$$

If CR $\prec$ 0.01 then it is considered as an acceptable limit and the allocation of preferences is coherent. Otherwise, it needs to be revised and adjusted accordingly. the next step in AHP is to calculate the scores of the alternatives. For so doing, we must first multiply each value of the criterion by its weight, then sum up these results according to each alternative. Therefore, we will have a weight assigned to each alternative that defines its priority with respect to the goal.

Spatial multi-criteria decision-making

"Spatial multi-criteria decision-making refers to the application of multicriteria analysis in a spatial context where alternatives, criteria and other elements of the decision prob-

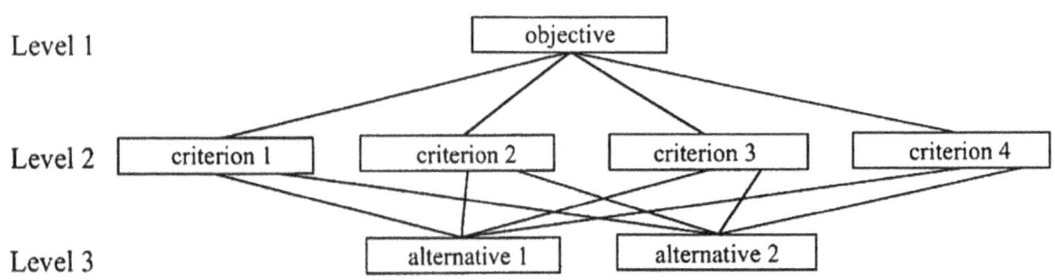

Fig. 2 Structured problem with three different hierarchy levels

Table 3 Random index (RI) of AHP (SAATY, 1977)

Matrix order	1	2	3	4	5	6	7	8	9	10
RI	0.00	0.00	0.58	0.90	1.12	1.24	1.32	1.41	1.45	1.49

	Tataouine N...	Tataouine S...	Smar	Bir Lahmar	Ghomrassen	Dhehiba	Remada
I1	3233.0	2641.0	1479.0	1410.0	1994.0	1431.0	1695.0
I2	84.77	89.82	82.87	90.92	91.68	94.39	77.31
I3	63.12	43.41	0.0	0.0	41.79	0.0	0.0
I4	1.0	1.0	1.0	1.0	1.0	1.0	1.0
I5	0.0	0.0	0.0	0.0	0.0	1.0	0.0
I6	0.3	2.5	0.3	0.4	1.0	3.6	1.9

Fig. 3 Data of Tataouine's delegations

lem have explicit spatial dimensions. Since the late 1980s, multi-criteria analysis has been coupled with geographic information systems (GIS) to improve spatial multi-criteria decision-making." [36]. The use of this couple is becoming increasingly important in land-use planning and environmental planning, as well as for water and agricultural management [37–39].

Therefore, several researchers [40,41] have worked on the combination of AHP with GIS to facilitate the manipulation of these two techniques.

5 Case Study

We have developed a parametric system that allows to personally cross the indicators, divide indicators of a chosen challenge into several intervals, choose the study area according to a need, to deepen the study and contribute to the improvement of the decisions.

5.1 Data

In this paper, we have collected data from the publication that presents the social economic statistics indicators derived from the general population and housing census 2014 for the governorate of Tataouine and its delegations [42,43] (Fig. 3).

5.2 Implementation

We present in the following part a rough summary of the main steps of our program (Fig. 4):

1. Definition of the problem/objective;
On the first step, we need to define our goal by specifying whether it is social, economic, or environmental problem. In our study, our challenge is "Better quality of life".

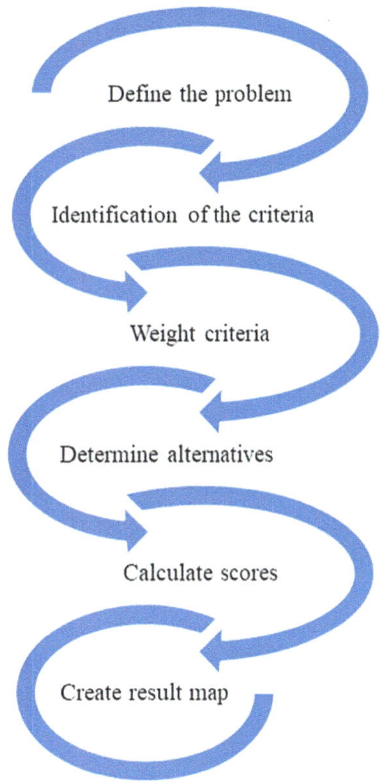

Fig. 4 The iterative steps of our program

2. Determine the Indicators which have an influence on promoting a better quality of life for citizens.
In this step, our program allows to choose a predefined challenge created by OTEED or to create a new analytical interest that meets a specific need.
3. Choose alternatives and collect data (statistical analysis).
We aim to determine the state of the chosen area. In our case, we choose Tataouine's delegations.
4. Weighting of indicators (two scenarios are presented).
This step presents the AHP approach (Decision Matrix, pair-

	Very bad	Bad	meduim	Good	very good
I1		3000 - 5000	1500 - 3000	500 - 1500	
I2		50 - 70	70 - 90	90 - 100	
I3		50 - 70	70 - 90	90 - 100	
I4		0 - 1	1 - 2	2 - 4	
I5		0 - 1	1 - 2	2 - 4	
I6		30 - 100	10 - 30	0 - 10	

Fig. 5 Reclassification of indicators

wise comparison, weight calculation). In our study, we have exposed two different weighting criteria to indicate the influence of the indicators' importance in the final result.

5. Calculate scores.

In this step, we aim to calculate alternatives scores that determine their states. In . in our case, we calculate Tataouine's delegations' scores.

6. Present results in two maps.

Reclassification of indicators.

In our study, the six indicators are divided into three intervals (Bad, Medium, Good). (Fig. 5)

6 Results and Discussion

In order to calculate the values of each alternative, we utilized the Weighted Linear Combination Technique [44]. This technique is used in multi-criteria decision-making with several attributes to be considered. A weight is assigned to each criterion depending on its importance. As result, we have a score for each alternative which is calculated (Eq. 6) by multiplying the weight assigned to each indicator by the scaled value given to that attribute (shown in Table 4) to the alternatives and then summing up the products across all attributes.

$$score = \sum_{k=0}^{n} X_k * V_K \tag{6}$$

n: number of indicators

Xk: Weight of indicator k

Vk: scale value

Tataouine's delegations' data comparison

The final map shows the state of the different delegations from the membership of the scores in the corresponding ranking in the predefined scale values.

The results of prioritization among Tataouine delegations (D1: Tataouine Nord, D2: Tataouine Sud, D3: Smar, D4: Bir Lahmar, D5: Ghomrassen, D6: Dhehiba, D7: Remada) are obtained corresponding the six indicators of the fifth challenge. We can see (Fig. 6) that for the first indicator which is the number of inpatients per basic health center Fig. 6a

Table 4 State scale values

Scale values	Definition
1	Very bad
2	Bad
3	Medium
4	Good
5	Very Good

D1 is very much preferred and closely followed by D2 and then after by D5. On drink water supply rate indicator Fig. 6b, D6 is the most preferred closely followed by D5, D4, and D2 and then after by other alternatives. On rate of access to adequate sanitation including rate of connection to public sewerage network indicator Fig. 6c, D1 is on top priority but closely competing with D2 and D5 and then the other alternatives have no access to adequate sanitation (value equal to 0). About air quality monitoring in urban areas indicator Fig. 6d, it is enough to have a single controller for the governorate to control all delegations. Therefore, this indicator is equal to all alternatives. On urban parks per capital indicator Fig. 6e, there is only one park in the whole governorate on D6 which explains the huge difference between alternatives in this indicator. Concerning the last indicator which is the proportion of rudimentary housing Fig. 6f, D6 is most preferred with a strong lead over D2 and D7 which are already more preferred than D1, D3, D4, and D5.

Scenario 1: a maximum weight assigned to "the number of inpatients per basic health center"

In the first scenario, highest weight is given to the number of inpatients per basic health center (I1: 0.38), followed by the proportion of rudimentary housing (I6: 0.2), the monitoring of air quality in urban areas (I4: 0.16), the access to adequate sanitation rate including rate of connection to public sewerage network (I3: 0.12), the drinking water supply rate (I2: 0.085), urban parks per capita (I5: 0.06). This is shown in Fig. 7.

Considering all indicators simultaneously and processing pairwise comparison in AHP, an overall result obtained in this study is presented in a map (Fig. 8). Each scale of schemes presents the state of delegation in our chosen challenge. Our main findings could best fit in different topics and benefit the country in different ways. In the present context, a ranking

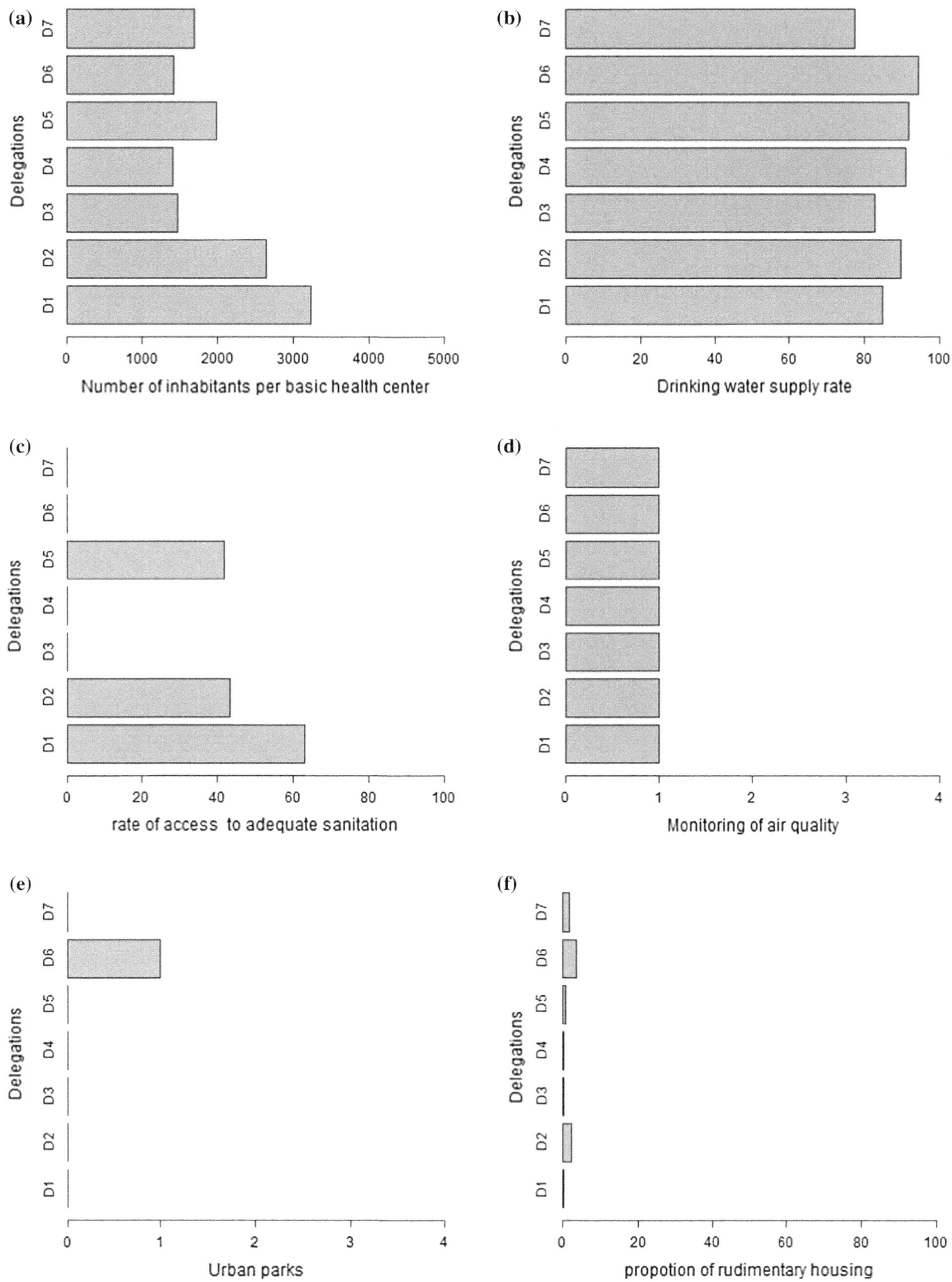

Fig. 6 Priority of alternatives corresponding to the main criteria. **a** Number of inpatients per basic health center, **b** Drinking water supply rate, **c** Rate of access to adequate sanitation, **d** Air quality monitoring, **e** Urban parks, **f** Proportion of rudimentary housing

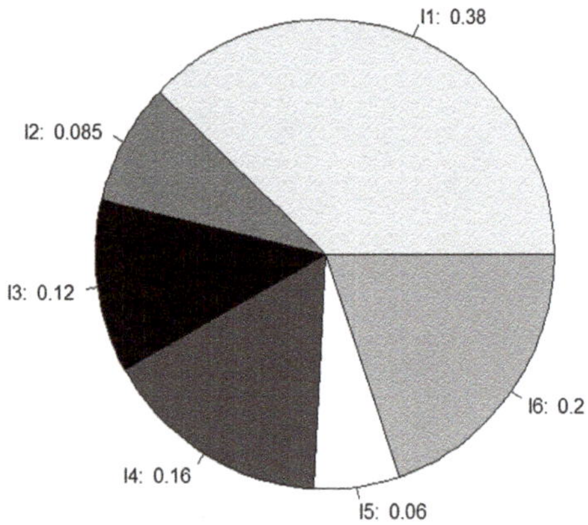

Fig. 7 Weight assigned to criteria in scenario 1

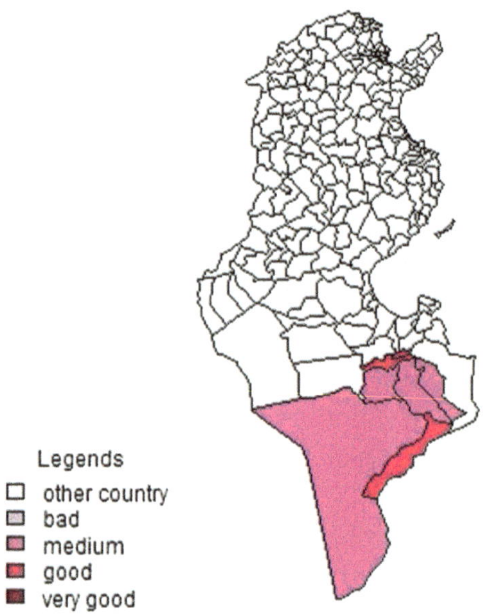

Fig. 8 Map of scenario 1

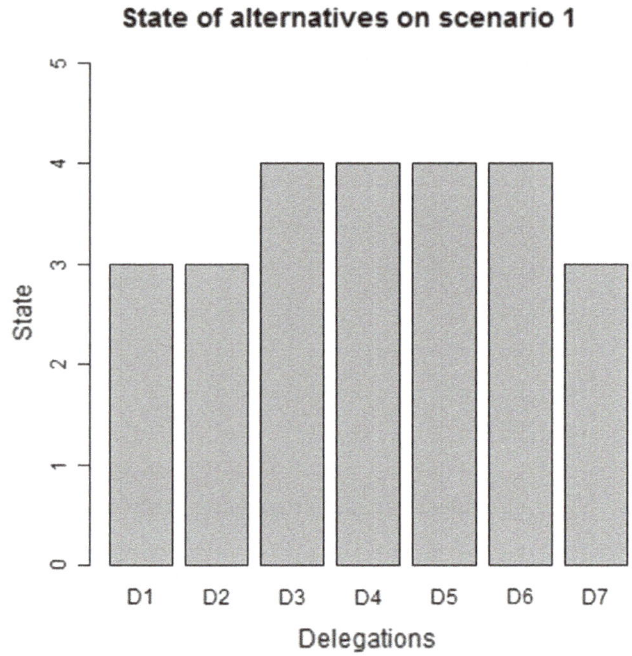

Fig. 9 State of alternatives on scenario 1 (1) Very Bad, (2) Bad, (3) Medium, (4) Good, (5)Very Good

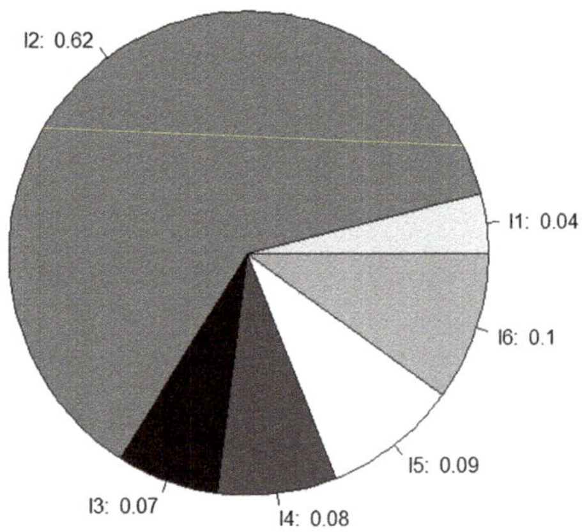

Fig. 10 Weight assigned to criteria in scenario 2

of alternatives is shown in Fig. 9. It appears that they are all important and closely competing with each other but for the overall national benefit, it is important to rank and identify the best fit.

Scenario 2: a maximum weight assigned to "drinking water supply rate"

In the second scenario, the highest weight is attributed to the drinking water supply rate (I2: 0.62), followed by the proportion of rudimentary housing (I6: 0.1), urban parks per capita (I5: 0.09), monitoring of air quality in urban areas (I4: 0.08), access to adequate sanitation rate including rate of connection to public sewerage network (I3: 0.07), number of inpatients per basic health center (I1: 0.04). This ranking is better shown in Fig. 10. While changing the weight assigned to the indicators, the results of the delegation states presented on the map on Fig. 11 changes too and the scheduling of the alternatives also changes as shown in Fig. 12. In this scenario, D4, D5, and D6 are preferred to other alternatives with a "good" state.

7 Conclusion

Indicators are specific quantities used to assign current conditions as well as to predict socioeconomic and environmental changes in a country. Their analysis is an important step to improve their relevance. Sometimes, one indicator is not enough to analyze sustainable development. In this case, it is necessary to go through a set of indicators in order to optimize the results. Moreover, to showcase these results, maps can be a simple means of viewing decisions.

The concept of the decision map is designed primarily for decision-makers in particular and for investors in general. In cartographic modeling, the experts' preferences take the form of a set of weights associated with different indicators. The

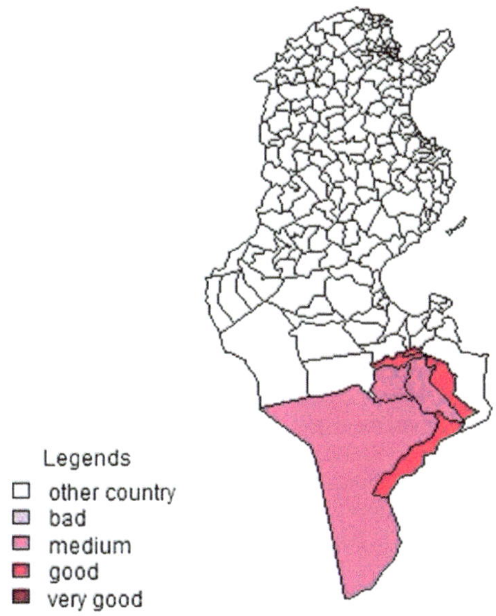

Fig. 11 Map of scenario 2

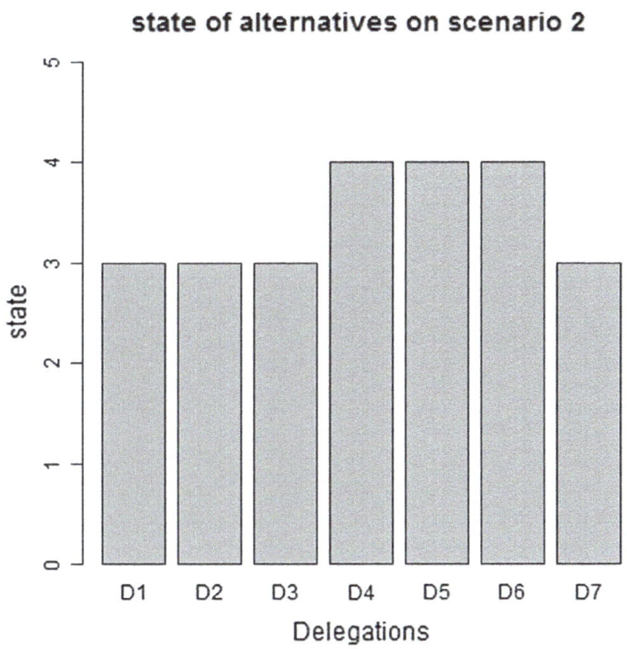

Fig. 12 State of alternatives on scenario 2 (1) Very Bad, (2) Bad, (3) Medium, (4) Good, (5)Very Good

cartographic representation of decisions is a tool for directly defining investments and identifying their geographical positions.

One of the main objectives/challenges of the government is to improve the quality of life of its citizens. For an effective fulfillment of this pledge, several criteria must be taken into account. In this paper, we have zoomed on the case of Tataouine (south of Tunisia). We have recourse to six indicators defined in OTEDD in the fifth challenge that addresses this objective. We have then proposed a model of a multi-criteria evaluation based on the AHP technique. This method allows, by combining the different indicators, to consider and analyze several scenarios. Second, we have classified the various extracted indicators according to a hierarchy which enables us to generate a map highlighting the state of the different delegations of Tataouine. We have also generated two maps presenting two scenarios whose weighting indicators are different. That is to show the influence of choice of the importance of the indicators on the final result. Those maps can be used in different disciplines and by different experts as decision-making facilitators.

References

1. Redclift, M. (2005). Sustainable development (1987–2005): An oxymoron comes of age. *Sustainable Development.*, *13*, 212–227.
2. Hopwood, B., Mellor, M., & O'Brien, G. (2005). Sustainable development: Mapping different approaches. *Sustainable Development.*, *13*, 38–52.
3. Rogers, P. P., Jalal, K. F., & Boyd, J. A. (2008). *An introduction to sustainable development earthscan*. London.
4. WCED. (1987). *Our common future. World commission on environment and development*. Oxford: Oxford University Press.
5. Brundland, G. H. (1998). *Notre avenir à tous* (Fleuv ed.). Québec: Montréal publication.
6. UNCED AGENDA 21 Regency Press, London (1992). Reproduced in an abridged form in The Earth Summit.
7. United Nations Commission on Sustainable Development (2002). Report of the World Summit on Sustainable Development. United Nations publication: Johannesburg, South Africa. 26/08–4/09.
8. Tanguay, G. A., Rojaoson, J., Lefebvre, J. F., & Lanoie, P. (2010). Measuring the sustainability of cities: an analysis of the use of local indicators. *Ecological Indicators.*, *10*, 407–418.
9. OTEDD. (2014). Observatoire Tunisien de l'Environnement et du Développement Durable, les indicateurs de développement durable en Tunisie.
10. Bunge, M. (1975). What is a quality of life indicator? *Social Indicators Research*, *2*, 65–79.
11. Michalos, A. C. (2004). Social indicators research and health-related quality of life research. *Social Indicators Research*, *65*, 27–72.
12. Land, K. C., Michalos, A. C., & Sirgy, M. J. (2012). *Handbook of social indicators and quality-of-life research*. The Netherlands.
13. OECD. (2017). Hospital beds (indicator). https://doi.org/10.1787/0191328e-en. Retrieved 22 Sept 2017.
14. UNICEF. (2015). *Progress on sanitation and drinking water*. New York: United Nations International Children's Emergency Fund.
15. Grand view research (2014). Gas Sensors Market Analysis By Product (Oxygen, Carbon Dioxide, Carbon Monoxide, NOx), By Technology (Electrochemical, Semiconductor, Solid State, PID, Catalytic, Infrared), By End-Use (Medical, Building Automation Domestic Appliances, Environmental, Petrochemical, Automotive, Industrial) And Segment Forecasts To 2020.
16. Lee, A. C. K., & Maheswaran, R. (2011). The health benefits of urban green spaces: a review of the evidence. *Journal of Public Health*, *33*, 212–222.
17. Hartig, T., Mitchell, R., De Vries, S., & Frumkin, H. (2014). Nature and health. *Annual Review of Public Health*, *35*, 207–228.
18. Oliveira, S., Andrade, H., & Vaz, T. (2011). The cooling effect of green spaces as a contribution to the mitigation of urban heat: A case study in Lisbon. *Building and Environment*, *46*, 2186–2194.

19. Takács, Á., Kiss M., & Gulys, Á. (2014). Some aspects of indicator development for mapping microclimate regulation ecosystem service of urban tree stands. *Acta Climatologica Et Chorologica. Tomus* (pp. 47–48, 99–108).
20. James, P., Tzoulas, K., Adams, M. D., Barber, A., Box, J., Breuste, J., et al. (2009). Towards an integrated understanding of green space in the European built environment. *Urban Forestry and Urban Greening*, *8*(2), 65–75.
21. Wolcha, J. R., Byrneb, J., & Newell, J. P. (2014). Urban green space, public health, and environmental justice: The challenge of making cities 'just green enough'. *Landscape and Urban Planning*, *125*, 234–244.
22. Livesley, S. J., McPherson, E. G., & Calfapietra, C. (2016). The urban forest and ecosystem services: Impacts on urban water, heat, and pollution cycles at the tree, street, and city scale. *Journal of Environmental Quality.*, *45*, 119–124.
23. Bogetoft, P., & Pruzan, P. M. (1997). Planning with multiple criteria: investigation, communication and choice. *Journal of Multi-Criteria Decision Analysis*, *12*, 112.
24. Loken, E. (2007). Use of multicriteria decision analysis methods for energy planning problems. *Renewable and Sustainable Energy Reviews.*, *11*, 1584–1595.
25. Feizizadeh, B., & Blaschke, T. (2012). Land suitability analysis for Tabriz County, Iran: A multi-criteria evaluation approach using GIS. *Journal of Environmental Planning and Management.*, *56*, 1–23.
26. Toloie-Eshlaghy, A., & Homayonfar, M. (2011). MCDM methodologies and applications: A literature review from 1999 to 2009.
27. Saaty Thomas. L. (1977). A scaling method for priorities in hierarchical structures.
28. Mark, V., & Patrick, H. (2013). An analysis of multi-criteria decision making methods. *International Journal of Operations Research.*, *10*, 56–66.
29. Fatemi, F., Ardalan, A., Aguirre, B., et al. (2016). Social vulnerability indicators in disasters: Findings from a systematic review. *International Journal of Disaster Risk Reduction.*, *22*, 219–227.
30. Hongwen, Yan Lu, Wang, Xu Yuexiang, & Yang, Yang. (2017). Evaluation of water environmental carrying capacity of city in Huaihe River Basin based on the AHP method: A case in Huai'an City. *Water Resources and Industry.*, *18*, 71–77.
31. Zhang, Z., Lu, W. X., Zhao, Y., & Song, W. B. (2014). Development tendency analysis and evaluation of the water ecological carrying capacity in the Siping area of Jilin Province in China based on system dynamics and analytic hierarchy process. *Ecological Modelling*, *275*, 9–21.
32. Akinci, H., Yavuz zal, A., & Turgut, B. (2013). Agricultural land use suitability analysis using GIS and AHP technique. *Computers and Electronics in Agriculture*, *97*, 71–82.
33. Memarbashi, E., Azadi, H., Barati, A. A., Mohajeri, F., Van Passel, S., & Witlox, F. (2017). Land-Use suitability in Northeast Iran: Application of AHP-GIS hybrid model. *ISPRS International Journal of Geo-Information*, *6*(12), 396.
34. Kundu, S., Khare, D., & Mondal, A. (2017). Landuse change impact on sub-watersheds prioritization by analytical hierarchy process (AHP). *Ecological Informatics*, *42*, 100–113.
35. Ahmad, S., & Tahar, R. M. (2014). Selection of renewable energy sources for sustainable development of electricity generation system using analytic hierarchy process: a case of Malaysia. *Renew Energy*, *63*, 458–466.
36. Chakhar, S., & Mousseau, V. (2008). Spatial multicriteria decision making. In S. Shehkar & H. Xiong (Eds.), *Encyclopedia of GIS* (pp. 747–753). New York: Springer.
37. Ceballos-Silva, A., & Lpez-Blanco, J. (2003). Evaluating biophysical variables to identify suitable areas for oat in Central Mexico: a multi-criteria and GIS approach. *Agriculture, Ecosystems and Environment.*, *95*, 371–377.
38. Sicat, S. R., Carranza, E. J. M., & Nidumolu, U. B. (2005). Fuzzy modeling of farmers' knowledge for land suitability classification. *Agricultural Systems.*, *83*, 49–75.
39. Chen, B., Adimo, O. A., & Bao, Z. (2009). Assessment of aesthetic quality and multiple functions of urban green space from the users' perspective: The case of Hangzhou flower garden, China. *Landscape Urban Planning*, *93*, 76–82.
40. Marinoni, O. (2004). Implementation of the analytical hierarchy process with VBA in arcgis. *Computer Geosciences.*, *30*, 637–646.
41. Chandio, I. A., Matori, A. N. B., WanYusof, K. B., Talpur, M. A. H., Balogun, A.-L., & Lawal, D. U. (2013). GIS-based analytic hierarchy process as a multicriteria decision analysis instrument: A review. *Arabian Journal of Geosciences*, *6*, 3059–3066.
42. INS. (2014). http://www.ins.nat.tn/fr/publication/tataouine-travers-le-recensement-g%C3%A9n%C3%A9ral-de-la-population-et-de-l%E2%80%99habitat-2014.
43. Ods Republic of Tunisia Ministry of Development, Investment and International Cooperation (2015). http://www.ods.nat.tn/upload/CHIFTATAOUINE.pdf.
44. Drobne, S., & Lisec, A. (2009). Multi-attribute decision analysis in GIS: Weighted linear combination and ordered weighted averaging. *Informatica*, *33*, 459–474.

A Qualitative Study of the Relevance of the WikiGIS Functionalities to the Collaborative Dimension of the Geodesign Process

Wided Batita, Stephane Roche and Claude Caron

Abstract

For validating the feasibility and effectiveness of the WikiGIS concept that has been proposed as a solution to better meet the main dimensions and requirements of Geodesign process, we have conducted a qualitative study based on a mixed methodology: questionnaire and interview with experts in various fields such as geomatics, architecture, urban planning, Geodesign, etc. This qualitative study aims initially at ensuring the relevance of developed features, which are managing traceability contributions, navigation in the history of contributions via a time browser and cartographic interface GeoWeb 2.0; and proposed WikiGIS features, which are the parameters of the data quality, the deltification, geoprocessing and sketching tools and the multimedia hyperlinks supporting the argument. Second, it is an opportunity to better understand Geodesign. In fact, thirty (30) experts have been involved. The results show that WikiGIS features are very relevant to support the collaborative dimension of the Geodesign process.

Keywords

WikiGIS • Geodesign • Collaborative dimension • Qualitative study • Mixed methodology

W. Batita (✉)
Department of Applied Geomatics, Université de Sherbrooke, Sherbrooke, QC J1K 2R1, Canada
e-mail: widedbatita@gmail.com; wided.batita@usherbrooke.ca

S. Roche
Geomatics Research Center, Pavillon Louis-Jacques-Casault, Laval University, 1055, avenue du Séminaire, Quebec, QC G1V 0A6, Canada
e-mail: stephane.roche@scg.ulaval.ca

C. Caron
PRISME–Pôle de recherche en intelligence stratégique et multidimensionnelle d'entreprise, Faculté d'administration, Université de Sherbrooke, 2500, boulevard de l'Université, Sherbrooke (Québec), QC J1K 2R1, Canada
e-mail: Claude.Caron@usherbrooke.ca

N. Rebai and M. Mastere (eds.), *Mapping and Spatial Analysis of Socio-economic and Environmental Indicators for Sustainable Development*, Advances in Science, Technology & Innovation, https://doi.org/10.1007/978-3-030-21166-0_2

1 Introduction

Geodesign is an emerging component of larger planning, landscape architecture and design field. It combines geographic information sciences with spatial design. Geodesign is beyond a simple planning approach, but it supports collaborative planning and informs decision-making on critical issues. Geodesign is defined by Shannon McElvaney in ESRI (2013) as: "an iterative design method that uses stakeholder input, geospatial modeling, impact simulations, and real-time feedback to facilitate holistic designs and smart decisions." However, Tess Canfield and Carl Steinitz [1] after Michael Flaxman [2, 3] and Stephen Ervin [4] (2014), have written: "Geodesign applies systems thinking to the creation of proposals for change and impact simulations, in their geographic contexts, usually supported by digital technology."[1]

Recent researches [4,5] show that for an effective achievement of Geodesign process, a smooth management of the historical dynamic semantic and geometric contributions is needed. This allows participants to view, exchange, and collaborate. Indeed, the traceability of geographic components goes hand in hand with the traceability of geometric (location and shape), graphic (iconography) and descriptive components.

In addition to this component, [4] outlined and described fourteen (14) elements of a Geodesign system such as working in synchronous and asynchronous mode, hyperlinks, level of abstraction, simulation tools, dashboards, diagram managers, collaboration tools.

Another important component in the Geodesign process is the deltification (geometric and textual). It denotes the ability to compare two scenarios and display the differences. Until now, few applications and technologies manage this function, despite of its importance in such process. This function helps a lot in decision-making. Decision making

[1] Tess C. and Carl Steinitz, C., after Michael Flaxman and Stephen Ervin, Redlands, California: 4th Geodesign Summit, 2014.

within a multidisciplinary team is a difficult task. In this research, we chose a differentiated consensus to take the best solution. Noucher [6] introduced the term of differentiated consensus in order to looking for consensus based on the identification and recognition of differences and divergences before looking at the convergence in the context of coproduction of geographic data and decision-making by multi-actors.

These functions are important in the Geodesign process but the problem is whether these functions are present or not in the existing tools that could do Geodesign.

To approach this problematic, we have recently made a review of the main existing relevant tools that could meet the Geodesign requirements. This review highlights that these functionalities are not fully present in the studied solutions. However, there are new solutions that answer perfectly the Geodesign requirements such as GeoPlanner, CityEngine of ESRI, but other constraints limit their uses such as license costs. Based on this fact, we have conceptually designed WikiGIS [7]. in which we have detailed the way we managed the updates by versioning in the database and we developed a WikiGIS mock-up to visualize the proposed features. In other words, WikiGIS combines wiki-type content management systems with an interactive map built upon GeoWeb 2.0 technologies. It ensures traceability and history of the user-generated spatial representations, while providing dynamic access to the previous versions of geographic objects, as well as documentation on data quality [9].

In the present work, we precisely propose a validation survey of the developed and proposed WikiGIS functionalities in order to better understand the Geodesign concept and how it is seen by the participants. This survey was carried out through the WikiGIS development cycle from 2012 to 2015. It is based on a mixed methodology: unstructured questionnaire and semi-structured interviews with experts in various fields such as geomatics, architecture, urban planning, architecture, interior design, and Geodesign. This study is quite important at this stage of the research. This is due to the fact that we can adjust the proposed functionalities on basis of mock-up before expensive development effort is done as with the case of AGIL[2] that encourages the prototyping before the development.[3]

Moreover, this paper is a follow-up to three previous papers. Batita Roche, Caron and Bédard [8] drew up the theoretical framework of WikiGIS concept and in their second paper (2014), they worked on the conceptualization of this technology using mainly UML formalism. The third paper [10], elaborated by the same authors, revolves around a computer-based mock-up to test the functionalities and underline their usefulness for the collaborative dimension of the Geodesign process through a simulated use case.

In the next section, we will briefly present the concept of WikiGIS (definition, dimensions, and elements) and its relevance to Geodesign. The third section, however, will be devoted to the methodology used. The qualitative approach as well as the sample and respondents will be more precisely described and justified in the context of this research. Finally, in the fourth section, we will introduce our main findings that back up our validation of the relevance of the proposed WikiGIS functionalities in general and in Geodesign in particular.

2 Context

2.1 Geodesign

Geodesign is a new buzzword but it is not a new concept. Artz [11] [11a,11b] says "We've been doing Geodesign for years." Also, during the first Geodesign summit in 2010, the sentence "Geodesign is both an old idea and a new idea" [12] was repeated several times.

To put it differently, Geodesign describes the integration of GIS and spatial design. The latter is the creation or recreation of entities in geo-scape. Roche [13] highlighted the intersections between GIS and spatial Design. He showed that the basic spatial concepts embedded in GIS such as scale, place, time, multiple presentations are included in spatial design. One of the first GIS was actually designed by and for the community of spatial designers in the sixties at the Harvard Laboratory for Computer Graphic [14]. This concept was updated in December 2008 at the NCGIA special meeting on "Spatial Concepts in GIS and Design" in Santa Barbara. This meeting was extended by the first Geodesign summit that took place in Redlands in January 2010. Since that occasion, many editions have appeared in the same place. The last one took place on January 23–25, 2018.[4] Every summit and every new 3D technology, which become the norm, have emerged to model the future with citizen engagement.

Since Geodesign is a new term, many definitions have been proposed by a group of thought leaders from academia and by a variety of professional experts whom we can call "Geodesigners" such as Dangermond [12], Flaxman [2,3], Ervin [4], Miller [15], Abukhater [16], Walker, Sinton and Lee [33]. However, until today, no consensus has been reached.

In this study, the authors define Geodesign as: "a process that harness the creativity from design and the rational-

[2]http://softwareprototyping.net/requirements-prototyping-agile-methods/.

[3]http://www.esri.com/events/geodesign-summit.

[4]http://www.ccs.neu.edu/course/is4800sp12/resources/qualmethods.pdf.

ity from GIS to improve a multidisciplinary team analysis, making rapid and iterative scenario for rapid feedbacks and decision making within a consensus with public involvement." Although Geodesign as a discipline has recently been launched, it includes at least four basic elements [16]:

- Sketching or drawing sketches and potential plans. In such process, the sketching is often participatory. It provides a collective brainstorming environment.
- The use of spatially informed models to evaluate the usefulness of proposed sketches.
- Rapid feedback gives an idea about the medium-short term,
- Iteration: trying and visualizing different alternatives. The iterations encourage creativity, strengthen group work, and help to simplify complex systems.

Aina and Garba [17] add a fifth element which is 3D visualization. The latter presents design alternatives and scenarios in three dimensions. Pouliot [18] has identified and quantified the value of the 3rd dimension in decision-making which is CityEngine. It constitutes an innovation in GIS 3D and helps the user to make many flexible scenarios that are close to reality.

Based mainly on the Geodesign summits of between 2010 and 2017, we have identified the following characteristics of the Geodesign process: creative, interactive, deliberative, collaborative, participative, iterative, uncertain, multi-scale, multi-actors, and multi-thematic. The Geodesign community is working on supporting and maintaining this process and its dimensions by new tools, technologies, and applications.

2.2 WikiGIS

WikiGIS concept was introduced for the first time in 2006 at the Department of Geomatic Sciences of University Laval [19]. In the current research, Batita, Roche, Caron and Bédard [8] proposed the following definition of WikiGIS: "a 2.0 collaborative platform, supported by wiki as Content Management System to ensure the traceability of geographical contributions and provide a support to the argument and qualification of these contributions, and a dynamic consultation and analysis. The WikiGIS has also powerful editing and creation functions (sketch-mapping) as well as basic GIS analysis functions. It offers a quick and easy access to the ongoing Geodesign process" (translated by the authors).

Indeed, WikiGIS resides on a collaborative platform designed basically on Web 2.0 technologies. It could be an alternative solution to support the collaborative and creative dimensions of Geodesign process. WikiGIS derives its importance from merging the functions of these three subcomponents: the strengths of wiki features, basic GIS features, and design tools. The main concepts of WikiGIS are detailed in [7,8,10].

In order to provide a generic overview of WikiGIS context and highlight its major components, we represent the Web Ontology Language (OWL) by C map Tools in Fig. 1. In fact, the WikiGIS context dimensions have been used to answer for these seven questions: who, when, for what, what, how, where and why?

Those questions have been answered as follows:

1- Who? The profile of the user should be determined. Indeed, his/her physical and professional profiles give us an idea about the contributor and somehow his/her credibility.

2- When? The time in such a spatiotemporal database is very important. Each version is edited in a determined time. WikiGIS is a synchronous and asynchronous platform, where the participants could work together and collaborate in real time.

3- Where? The WikiGIS is multi-scalar. The participant could work in both local and regional contexts.

4- For what? WikiGIS is a platform designed to support a multidisciplinary, contributive and collaborative work. In this way, reaching a decision becomes possible through improving public involvement.

5- What? The participants can edit mainly a point, a line, a polygon and an argument (justification, relevance...).

6- Why? The participants must justify the reason of the editing. They must discuss their contributions. These contributions are supported by a wiki-media through hypertexts. The latter auto-evaluate these arguments on the basis of perceived quality. This process informs us about the credibility of the contribution.

7- How? WikiGIS is elaborated on basis of Web 2.0 technologies. Hence, its user interface is easy, ergonomic and simple (Fig. 2). In other words, the user can have an easy and dynamic access to the different contributions since the synchronization of the different windows of the interface allows the user to see the different details of all the contributions.

Batita, Roche, Caron and Bédard [7] have presented a UML (Unified Modeling Language) based on the WikiGIS conceptual framework. Then, a mock-up is implemented to better fit the collaborative dimension of Geodesign, and a case study is generated to illustrate the usefulness of this concept.

The WikiGIS interface is organized around five areas (Fig. 2):

1. The menu bar: it contains the following drop-down menus:

- Profile users: the administrator who is the manager of the project; the project members who are participants in the project; and the visitors, anyone who can accede to the website (they can only view but in some cases can edit when public participation is permitted); Base map: base map provides a user with context for a map satellite image;

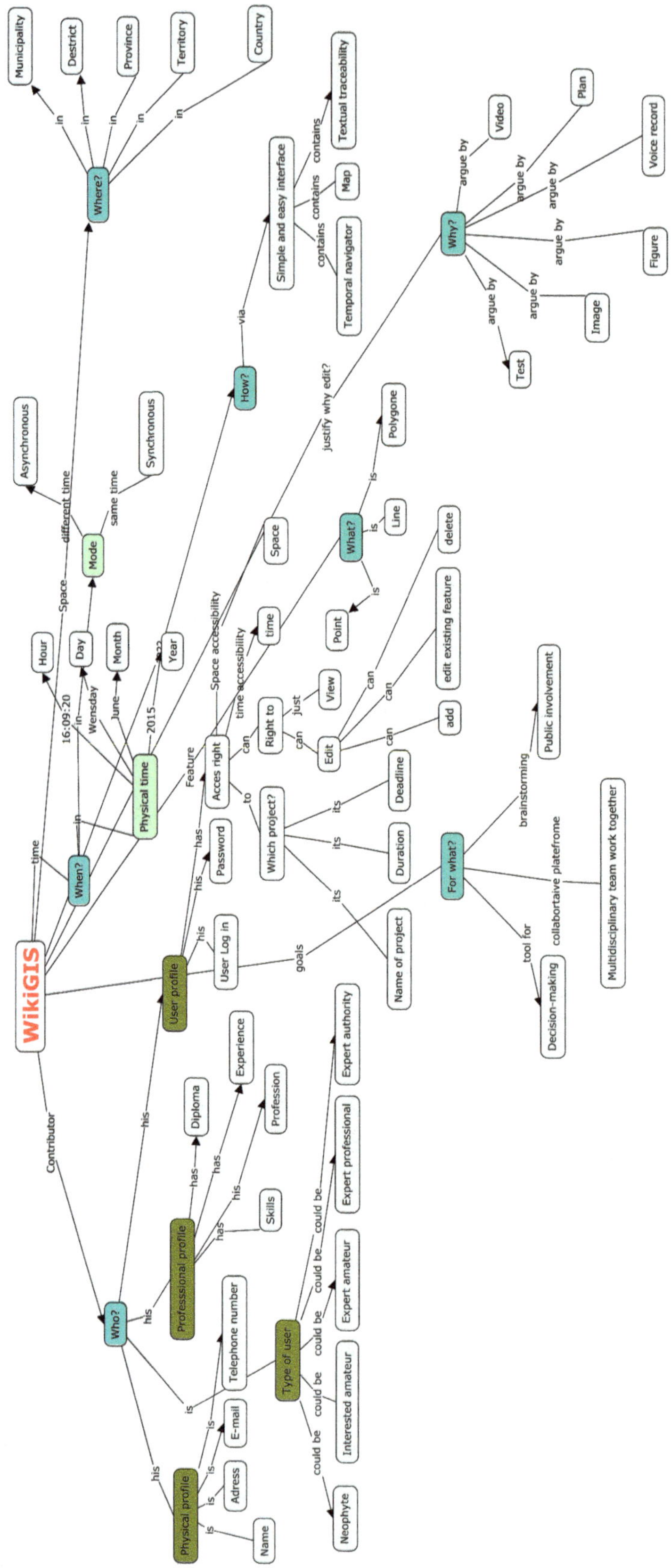

Fig. 1 WikiGIS ontology

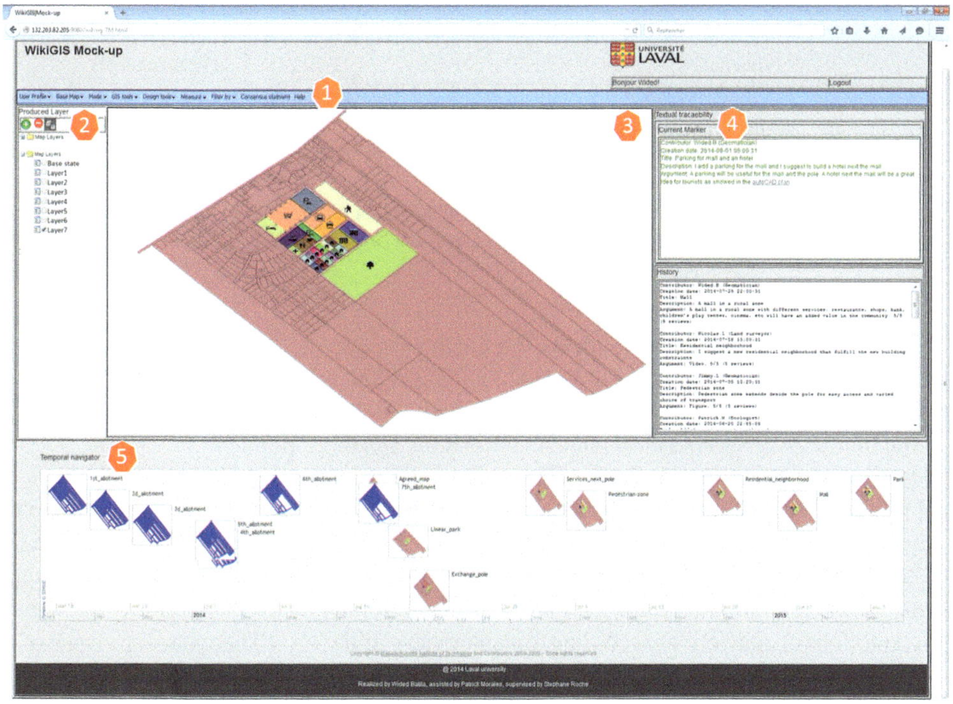

Fig. 2 WikiGIS user's cartographic interface. (1) The menu bar. (2) Layer management. (3) Map. (4) Textual traceability window. (5) Temporal navigator

- Mode: two modes are available: a view mode and an edit mode. The former is available by default for all users;
- GIS tools: a set of GIS operators (union, merge, buffer, intersect diff, etc.) allows not only to manage and transform the created feature, as well as to make its spatiotemporal analysis;
- Design tools: the main drawing tools for sketch-mapping such as draw point, line, polygon, copy, past, cut, erase, resize, color, symbol, etc;
- Measure: to measure distance, area, angle;
- Filter by: the user can filter the contributions by actor, date or argument;
- Consensus statement: when the consensus is reached, the participants will be informed and solicited to fill a survey to take decision;
- Help: a Wikipedia page about WikiGIS;

2. Layer management: When a user edits (creates, modifies, deletes) a feature, he/she has to create a new layer. The layer (s) can be displayed or hidden. During a design process, many layers will be created.

3. Map: shows the context map. A user can add or remove layer (s) on the map. He can also change the scale and extent;

4. Textual traceability window: all the metadata about a contribution appear in a text box on the right of the map. Descriptive data are arranged in two sets: the descriptive data associated with the last updated layer, and the descriptive data associated with former versions of the feature.

5. Temporal navigator: provides the evolution of a feature over time.

The user's WikiGIS mock-up graphic interface illustrates the main developed components. The map panel (Area1), the textual traceability (Area4) and the temporal navigator (Area5) are synchronized in a way that when the user clicks on an object, all the details appear on the three respective windows.

3 Methods: Concepts Validation Through a Qualitative Study

3.1 Qualitative Study

McClure [20] reviewed three effective data collection methods for qualitative study (1) interviews, (2) observation, and (3) questionnaires.

3.1.1 Interviews

There are five types of interviews. The first type is Formal (or structured. They involve a great deal of planning, scheduling, and preparation. The same questions are asked to all the respondents in the same order and with the same way.

Second, there are Focused group interviews (semi-structured interviews. They are conducted with three to six people. Small groups encourage collaboration among individuals, create memorable learning experiences, increase the learners' participation, and limit anxiety. Moreover, semi-structured interviews involve a series of open-ended questions based on the topic areas that the researcher wants to approach [21]. Indeed, this type is often favored in researches aiming at obtaining exploratory qualitative data [22].

Table 1 Advantage and disadvantages of interview and questionnaire

	Advantages	Disadvantages
Questionnaire	– Allows a wider range and distribution of the sample than interviews do., – The collection and analysis of Data are relatively easier and quicker.	– Poorly worded or direct questions might be refused from, the part of the respondents
Interview	– The interviewer can gather other, supplemental information., – Interviews are flexible and more adaptable, to individual situations.	– Certain information may be untold in face-to-face interviews – Interviews pose the problem of recording the information obtained from the respondents

Third, unstructured interviews are those through which the interviewer encourages the respondents to talk freely about a given topic.

The fourth category is Critical incident interviews. During the latter, participants are asked to recall a period of time that had particular meaning and resulted in a memorable change or experience. Finally, there are in-depth interviews which are commonly used in a three-interview series approach. This form of interviews aims at deciphering the respondent's opinions, emotions or convictions according to an interview guide.

3.1.2 Observation

Observation is an intuitive process that enables individuals to collect information about others by viewing their actions and behaviors in their natural surroundings.[5] This method provides researchers the ability to check nonverbal expression of feelings among participants and determine their different interactions and communications.[6]

3.1.3 Questionnaires

Questionnaires are used when it is impossible to interview every respondent individually. they generally consist of open- or closed-ended questions or items that measure facts, attitudes, or values [20]. There are two types of questionnaires: unstructured and structured questionnaires. While the former allow respondents to reply freely, the latter specify their answers. The observation method is eliminated from this study because it is far from its aim. Actually, our focus is on the questionnaires and interviews, though it is crucial to note that both methods have advantages and disadvantages which we have summarized in Table 1.

Clearly, this qualitative study is conducted mainly to prove the usability and effectiveness of WikiGIS and grounded theory is used for collecting and analyzing the data [23,24]. Lawrence and Tar [25] show, in this context, that grounded theory is particularly suitable for dealing with qualitative data of the kind gathered from the participant's observation, from the observation of face-to-face interaction, from semi-structured or unstructured interviews, from case studies or documentary sources.

Accordingly, unstructured questionnaires and semi-structured interviews have been collected. These two types of data collection are often used together in mixed method [26,27].

However, [26] quoted different limitations of the questionnaire-interview study. Indeed, they can engender problems in aligning data, limited and weak evidence of consistency, possible misinterpretation of some questionnaire, lack of variability in the participant's responses.

Though we are aware of those drawbacks, we have recourse to it because of its flexibility and easy implementation, as well as its relevance (openness) for validating the new WikiGIS functionalities. It is also important to note that a Delphi study was launched in 2012 but it has failed. Thus, grounded theory has been used to validate WikiGIS' functionalities. It is an appropriate method for this study since the target is the development of a context—a based description and explanation of WikiGIS' functionalities.

3.2 Presentation of the Sample and the Respondents

This qualitative study is based on questions related to Geodesign and WikiGIS. The questionnaire-interview study is open. This permits the respondent to add comments and suggestions in order to enrich WikiGIS concepts. A set of fifteen (15) questions are distributed as follows:

(1) The first part with open answers is devoted to the new emerging term, Geodesign, where we have tried to validate its main characteristics, propose new features and technologies and finally define it.

(2) The second part with open answers is dedicated for the evaluation of the relevance of the developed WikiGIS functionalities in terms of the traceability of the contributions and navigation in the history, parameters of data quality, deltification, and cartographic user's interface.

A part of open answers seeks to validate, through its characteristics, whether WikiGIS is an efficient solution to support the collaborative dimension of the Geodesign process or not.

The questionnaire took approximately twenty minutes while interviews last almost 1 h. The population targeted by this study is researchers. In total, thirty respondents have been involved, mainly academics: professors, graduate stu-

[5] http://www.qualitative-research.net/index.php/fqs/article/view/466/996.

[6] http://keywordtool.io/.

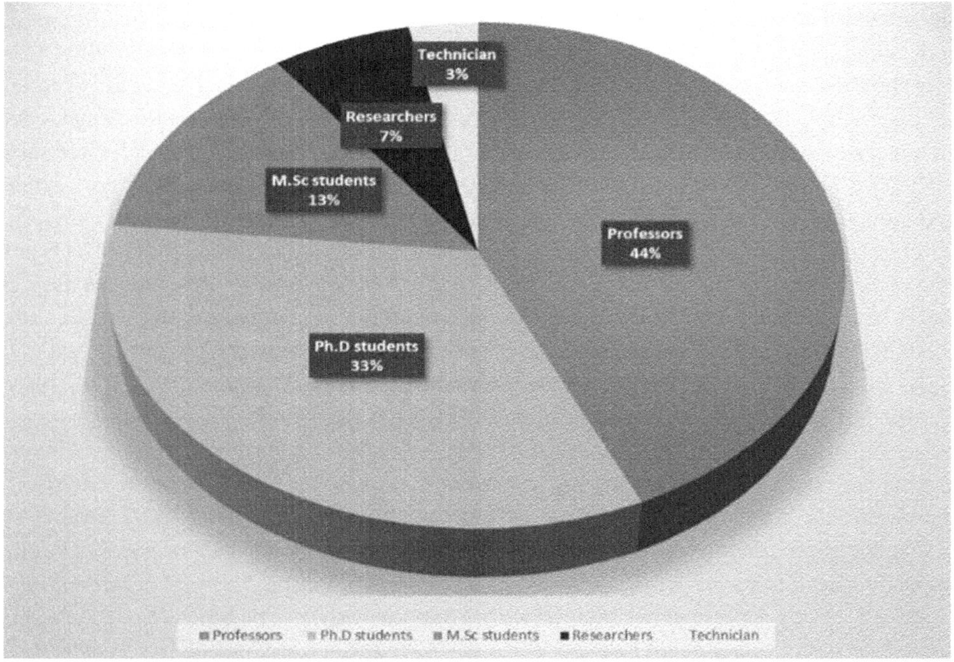

Fig. 3 Respondents' professions

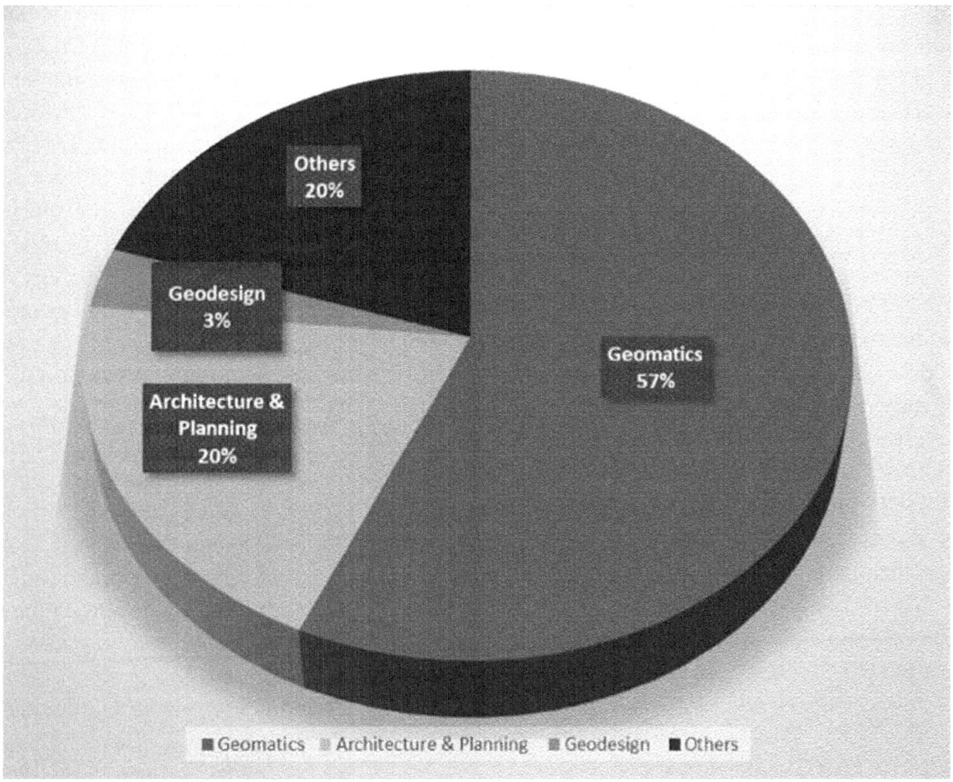

Fig. 4 Research domains of respondents

dents and professional researchers (Fig. 3). Eight professionals have been interviewed and twenty two questionnaires have been filled out. The respondents have been selected according to their knowledge of the subject matter, their legitimacy in relation to the panel of experts that they could represent, and their availabilities. Nearly fifty people were selected and approached at the beginning of this study but only thirty have collaborated.

Thirty (30) respondents were mainly from geomatics (Fig. 4). The geometicians belong to different specific research domains such as geophysics and GIS, 3D modeling from LIDAR data, 3D cadaster, topography, museology, GIS and spatial database, LIDAR points clouds processing, GIS and cartography, GIS, geosciences, geo-decision, simulation, geospatial BI, geospatial data quality, cognitive geomatics, cellular automata, spatial modeling, geo-informatics. The architects, urbanists, and planners have also participated. The category "Others" include Mathematicians, geographers, and computer science specialists. The participants were of different nationalities. In fact, we have dealt with people from USA, Canada, Morocco, Tunisia, Peru, South Africa, Saudi Arabia, France, and United Arab Emirates in order to generalize the potential of the output.

As shown in Fig. 4, the selected respondents vary in terms of expertise and background. We have initially targeted the Geodesigners but only one has finally collaborated, so other experts have been asked to validate the proposed functionalities according to their knowledges.

The question asked at this level of study is **"how many users do I need to have?"**

In such a study, representative samples are not important as in the case of the validation of new video games. In Human–Computer Interaction, five users is an accepted number [28]. Other researches have proved that seven users is an optimal number in a small project and fifteen users is optimal for medium-to-large projects [28].

Therefore, a number of thirty users in this study seems adequate.

4 Qualitative Data Analysis

This section is mainly based on the respondents' answers in the context of unstructured questionnaires and semi-structured interviews conducted in this qualitative study. As already mentioned in Sect. 3.2, the first package of questions revolves around the Geodesign process in order to gather more information about it. The second package is about WikiGIS proposed features and functionalities. A small part is about their utility in the Geodesign process and more specifically in filling its collaborative requirement.

The respondents have been asked to justify their answers, but unfortunately, some of them have answered just by yes or no.

4.1 The Geodesign Part

As Geodesign is a new emerging field, (a recently coined term), and that there is no conventional definition as described in Sect. 3.1, we have explored it via the Keyword Tool[7] and Bing Ads Intelligence.[8] Both are free powerful keyword research tools that allow to build and expand on keyword lists. They enable to easily research keywords and gauge their performance on the Bing Network, and then apply those insights to improve the keyword selection and campaign performance.

In fact, Bing Ads Intelligence provides important data, such as historical traffic, historical performance, geography and demographics.[9] Over the last 30 days, 63.64% males and 36.36% females have generated the keyword Geodesign via their laptops and computers with 50% of the group are between 35 and 49 years old. From September 2013 to June 2014, there were 2262 search queries about Geodesign.

Keyword Tool generates 201 keywords related to Geodesign on google.com; whereas Bing Ads Intelligence generates 87 compound keywords. The different keywords obtained from both tools, from videos of Geodesign summits and from papers and books are illustrated in the tag cloud of Fig. 5.

In Sect. 2.1, we have described largely the main characteristics of Geodesign process such as creativity, collaboration, interaction, participation, iteration, multi-scale, multi-theme and multi-actors and deliberation. As first question, the respondents have been asked to judge the importance of these characteristics in that process (Table 2).

We have done a correspondence analysis of this table and we got a p-value = 0.3234, which asserts that the link between Criteria and level of importance is not significant (no link).

Figure 6 shows that the most important axis (83%) opposes the group of Criteria Collaboration, Multi-scale, Multi-actor, Multi-thematic, Creativity, Interaction (on the left) which is very important to the group Iteration, Participation, Deliberation (right) that is Important or Unimportant.

They judged that the deliberative criterion is less important in the process. However, Roche [13] demonstrated that the spatial design in general and urban design in particular are based on a deliberative approach. This idea is based on the work of Ciobanu [29] whose research is based on the study of Forester [30].

Ciobanu proposes a formal analysis of a process of a deliberative urban design. In the second question, the respondents have been asked to propose other complementary characteristics and features to better define the Geodesign process, so they have suggested the following features:

- multi-versions: have many scenarios of the same feature,
- confrontation: while collaboration convey the meaning of building together atop of the same opinion, confrontation conveys arguing over different opinions, then decision must be made with or without consensus,

[7] http://advertise.bingads.microsoft.com/en-ca/bing-ads-intelligence.

[8] http://advertise.bingads.microsoft.com/en-ca/cl/257/training/bing-ads-intelligence-tool.

[9] http://geogig.org/.

Fig. 5 A tag cloud with terms related to Geodesign

Table 2 Respondents' thoughts on Geodesign characteristics

Criteria	Very important	Important	Unimportant	Not applicable
Creativity	16	10	1	
Collaboration	18	7	2	
Interaction	17	8	2	
Participation	15	10	2	
Deliberation	9	13	5	
Iteration	13	12	2	
Multi-scale	18	9		
Multi-actor	16	11		
Multi-thematic	17	9	1	

- informed by feedback of results from simulations, as real time as possible,
- evolutionary: contextual and temporal process,
- modeling: create models of data and process to solve large, complicated and significant design problems,
- taking into consideration the cognitive aspect for presenting geographical information to people with different backgrounds and sometimes with different abilities,
- voting system. Some of the proposed features have been already quoted by Ervin [4].

The third question of this part is related to the software and technology that could support the process. They suggested tools and software that can improve the Geodesign process such as BIM (Building information modeling); visualization 3D; collaborative software; version management; project management tool, geo-simulation; process and data modeler; spreadsheets; computer-assisted technologies; Object-oriented design; SOLAP, sketchUp, AutoCAD...

As mentioned above, there is no consensus in defining Geodesign. Thus, the participants have been asked to define it from their point of view. Among the definitions provided, just two have been presented because they are representative and consistent. The first one has been given by a Geodesigner and the second by a Professor in Geomatics:

1- "Fusing imaginative and functional creativity in environmental design with analytic geospatial science, geo-simulations, impact analyses, and system thinking, enabled by modern digital technology and collaboration tools. So Geodesign is truly "Computer Aided Design"–but not just 'CAD'!"

2- "Design that harness the power of modern Geo and Web 2.0 technologies to improve group analysis, solution building and decision making."

Those two definitions are almost similar. Both highlight firstly the role of Geodesign in terms of creativity, analysis, geo-simulation, solution building, impact analysis, decision-making. Secondly, they present the technologies that do the Geodesign: the designer presented the CAD and the geomatician presented Web 2.0 technologies.

4.2 The WikiGIS Part

The endeavor of the study is primarily to validate the concepts of WikiGIS and more precisely its functionalities. As noted above, 50 persons who are Geodesigners and urbanists were invited to participate in this study but they did not collaborate. Details about the participants are summarized in Table 3.

Fig. 6 Correspondence analysis

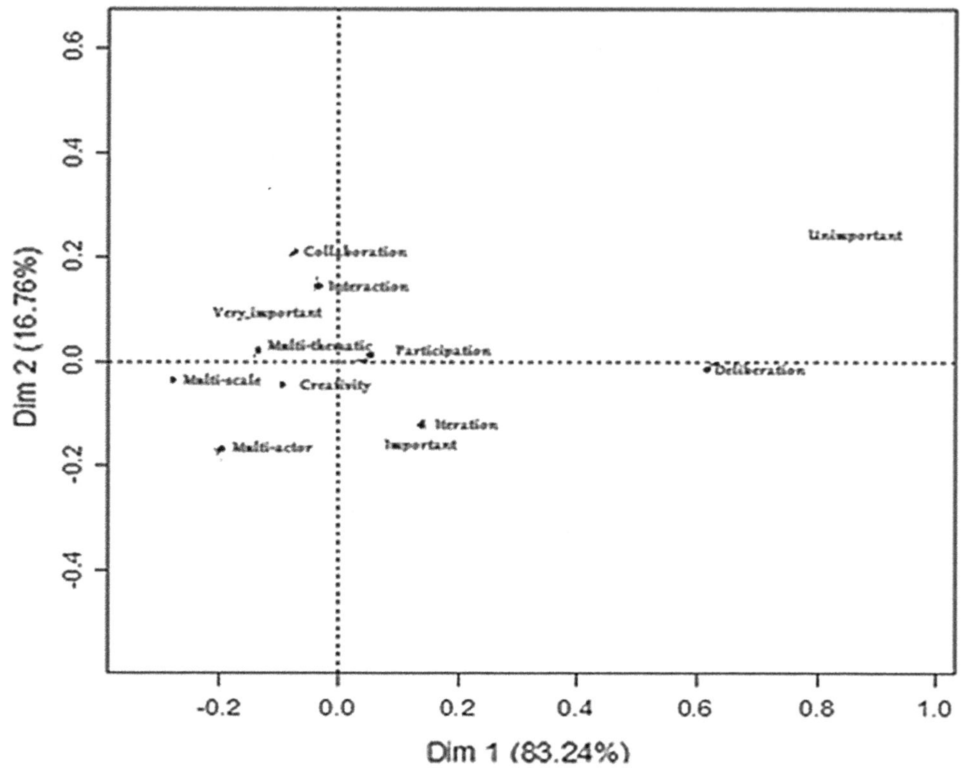

Table 3 Details about the participants

Domain	Research interest	Number
Geomatics	Geophysics and GIS, 3D modeling from LIDAR data, 3D cadaster, topography, museology, GIS and spatial database, LIDAR points clouds processing, GIS and cartography, GIS, geosciences, geo-decision, simulation, geospatial BI, geospatial data quality, cognitive geomatics, cellular automata, spatial modeling, geo-informatics	17
Architecture and Urban planning	Architecture, Urban and territorial planning, Interior design, Landscape	6
Geodesign	Design and GIS	1
Others	Mathematics, Computer sciences, Geography	6

The data gathered from unstructured questionnaires and semi-structured interviews have been examined, coded and categorized into concepts that we aim to highlight throughout this qualitative study. Hence, the codes are about WikiGIS components, history management by versioning, temporal navigator and measures to insure data quality. They are also about how to compare versions of the same feature and about the presentation of the interface. After examination, new codes have been generated. They form a fusion of wiki, GIS, and Design tools; traceability; history navigation; data quality; deltification and cartographic user interface. Following the axial coding [25], a new subcategory and a core category have been formed as described in Table 4.

The technical and organizational categories have been presented to participants. The latter were asked to evaluate them and validate them if they fit their needs. The results uncover the fact that the participants agree almost fully with the proposed features of WikiGIS. Their anwsers and some of their comments are presented in the next section.

Table 4 Axial coding method

Core category	Subcategory	Codes
Technical	Geo-informatics	– Traceability – history's navigation – cartographic user interface
Organizational	Functionalities management	– fusion of wiki, GIS and Design tools – data quality – deltification

About the components of WikiGIS: Wiki, GIS and Design tools, 90% of respondents have assumed that the fusion of wiki, GIS and Design tools is useful and relevant. One respondent answered: "Yes, since they are all part of the Geodesign workflow. The tighter the integration between the tools, the more homogeneous the user interface, the easier and more fluid it is, the better it is for users (Geodesigners)."

Subsequently, the respondents have been asked if they have other suggested components. Eighteen respondents have agreed with the proposed components, whereas nine respondents suggested to add other components such as 3D component; context-awareness component; Simulation for dynamic phenomenon like traffic, pollution; voice recording along mouse movements (with proper tagging and indexing for easier retrieving); modeling the proposals to rapid evaluation and feedbacks; text mining to derive high-quality information from text.

The next question dealt with the traceability. Undeniably, the Geodesign projects, like all design projects, may generate multiple variants and states over the time. Managing multiple versions is a demanding task [4].

Following the logic of traceability of the textual wiki, versioning is the solution to manage the traceability of the geometric component. Versioning is assigned to entire tables (table versioning), to each record of a table (tuple versioning) or to each value taken by a descriptive attribute (attribute versioning and even geometric attributes in some systems). In the case of WikiGIS, "versioning by occurrence" appears to be the most relevant choice because it is the simplest. Every proposition to edit or replace a scenario can be considered as a different and unique action. The page is reloaded instantly and current versions are displayed in a "thread of versioning". By using the versioning by occurrence, the disaggregation is manageable [7].

Regarding the different methods of editing tracking mentioned above, the respondents have been asked if the occurrence versioning is a relevant choice for WikiGIS. 75% of respondents have judged that the versioning by occurrence is the most relevant choice. A participant who has a solid background in spatial Database supported this choice by saying: "Personally, I believe it is the most relevant choice because it is the simpler and, in fact, every proposition to change a scenario or to propose a new one can be considered as a different and unique action (rather than the evolution of a previous action). The relationship with previous actions can be dealt with proper relations built in the database." The interviewer therefore wants to emphasize the relationship between the new and previous actions. Indeed, Batita, Roche, Caron and Bédard, [7] have created the fields of ID current primitive and ID previous primitive that help in making the request in both directions: up and down. They indirectly allow drawing the evolution of an entity. Another interviewer, who is a computer scientist, liked the proposed method of versioning and proposed to explore a new project called Geogig[10] that is based on Git.[11] Indeed, Geogig is an open source tool to handle distributed versioning of geospatial data. However, until now, the last version beta is still unstable.

After validating the best method of versioning, the next question is how the user can navigate through the contributions. So in order to navigate in history, a temporal navigator is proposed as shown in Fig. 2 at the bottom (area 5). 70% of the respondents have appreciated this time browser. The latter is implemented in our interface by adapting the temporal exploration tool of MIT event data.[12] This browser is designed to help user to see all versions over the time.

We move now to data quality and credibility and how to insure the data produced. More and more data, including geospatial data are edited and published on the Web each day. Now in the era of GeoWeb 2.0, everyone even without geospatial knowledge can create maps, edit geospatial data, and make various applications for different reasons and share them with others. However, most of them do not understand the uncertain nature of geospatial data and take the digital data as true without considering their quality and validity for intended application. In the context of anxiety about data quality, Sonnen [31] said: "Data quality is a problem we need to address if we in the geospatial industry expect to be a part of the enterprise IT picture. Our most pressing need is a simple, reliable way to answer: "Are these data fit for this purpose?" each time spatial data are merged or shared in an enterprise system." Bédard [32] highlighted via many situations the problems of uncertainty of geospatial data and invited users to "be assured [that] serious errors bring about legal liability."

To overcome this issue in our study, many parameters have been taken into consideration such as spatiotemporal access restriction, control of contributions, verifying the profile of contributor, filtering of the contributions of the public by the administrator, monitoring the W7 model, statistics testing the user's reputation and the credibility of contribution, arguing the contribution by multimedia [7]. 70% of respondents have found these parameters enough. One respondent who has worked a lot on geospatial data quality has said: "This is a good start considering that it is not possible to include several indicators of credibility. A small group of indicators is enough. Potentially, the addition of the voice recording along mouse movements to show what object one is talking about could help as the quality of the argument could be better conveyed verbally." In question 9, the respondents have been asked about deltification. Hence, to compare and display the differences between the two geometric components, overlaying two layers with different colors (one for each scenario) is used with transparency. This helps to visualize the differences.

Seventeen respondents, who are all geomaticians, have appreciated deltifiation. One of them said: "No one solution is the best for every situation. Sometimes this solution works

[10] https://git-scm.com/.

[11] http://www.simile-widgets.org/timeline/examples/compact-painter/compact-painter.html.

[12] https://en.wikipedia.org/wiki/Markov_decision_process.

well, in other cases it is more efficient to display the delta (i.e. only the difference), and sometimes it is more efficient to display 2 maps (one for each scenario). It depends on the complexity of the differences. I would prefer to have access to the overlay AND to the possibility of having 2 maps side-by-side."

Regarding the cartographic user interface, which was designed upon the GeoWeb 2.0 design, eighteen respondents liked it, while four participants have found it inadequate for a complex project. The majority has found it easy, simple, ergonomic and dynamic and suggested collapsing and expanding windows to focus on the details of modifications in the map panel.

Twenty-one respondents found the WikiGIS useful for their domains: geomatics, design, architecture. Indeed, the track-editing of an entity is important and the way to save and access to all the changes over time is even more important.

For this part of the survey, the use of grounded theory is very successful to get insights from participants about WikiGIS' functionalities.

4.3 The Utility of WikiGIS in Geodesign

In the last part of the questionnaire and interviews, the respondents have been asked about the utility of WikiGIS functionalities in the Geodesign process for answering some of its requirements. Unfortunately, the respondents have not answered all the questions in this section and sometimes their answers have been unjustified. More than half of the respondents found the traceability management, the deltification, the occurrence versioning, and the set of parameters of credibility crucial elements in the Geodesign process. In the same vein, A Geodesigner (one of the leaders of Geodesign Summit in Redlands, USA) states that "Iteration and revision are essential for the design processes."

The majority of respondents couldn't affirm the efficiency of the WikiGIS to support the collaborative dimension of Geodesign. It is clear that they might find this question too fuzzy because this study is based on a simulation via a WikiGIS mock-up instead of freely interactive prototype to test the WikiGIS functionalities. However, if we combine the previous answers in Sects. 4.1 and 4.2, we can say that WikiGIS is a promising tool that supports perfectly the collaborative dimension of Geodesign process in case the suggested features will be taken into consideration. Indeed, the traceability management is a pertinent feature in the Geodesign process. It helps participants to know the state of the studied entity during the process; each version is justified and stocked for the best decision-making and the user can explore the current version as well its previous versions in order to follow its evolution. This traceability is insured by the occurrence versioning because it is the best method to save all the modifications in a spatial database. This versioning allows iteration, fast feedbacks, and mutual revision.

The deltification feature is very interesting, in terms of time, for determining the differences between two scenarios (versions). This feature has been just implemented in new technologies such as GeoPlanner (ESRI product) and Geogig.

The easy and simple WikiGIS interface helps the participants with different qualifications and backgrounds to work together in a collaborative environment.

The proposed parameters to qualify the produced data form a good starting point and are proven important for avoiding problems of VGI. These parameters are the spatio-temporal access restriction, control of contributions, verifying the profile of contributor, filtering of the contributions of the public by the administrator, monitoring the W7 model, statistics testing the user's reputation and the credibility of contribution, arguing the contribution by multimedia.

5 Conclusion

In order to make technical enhancements on developed and proposed WikiGIS functionalities, a validation survey of the conceptual framework on which the WikiGIS has been designed is carried out in this paper. Indeed, unstructured questionnaires and semi-structured interviews have been executed with thirty respondents. However, despite its flexibility and easy implementation, this method presents many drawbacks. For example, it faces problems with aligning data, with the limited and weak evidence of consistency, possible misinterpretation of some questionnaire, lack of variability in participant responses. Fortunately, in our case, none of those problems has been encountered and the goal of the study has been reached since we come to successfully validate the relevance of WikiGIS functionalities.

The result shows that almost all the respondents have agreed with the proposed and developed functionalities and found the tool useful in their domains. The respondents have suggested to add new items to make it more efficient and useful such as: a 3D component; simulation for dynamic phenomena (such as traffic and pollution); voice recording along mouse movements (with proper tagging and indexing for easier retrieving); modeling the proposals to rapid evaluation and feedbacks; text mining to derive high-quality information from text... In fact, WikiGIS in Geodesign can be improved through adding simulation tools to see the extent to which each scenario can develop over time, can be modeled for decision-making such as Markov Decision Process[13] and be designed to study the economic and social impacts of a scenario by Bayesian Network for example. Further work may deal with a real functional WikiGIS prototype in which the features and

[13]https://en.wikipedia.org/wiki/Bayesian_network.

functionalities proposed by respondents will be implemented and tested in a real complex Geodesign use case. In this paper, WikiGIS is presented as a tool for a strict technical use, but the authors will extend that validation to an ethical and sociopolitical use after adding the enhancements mentioned earlier.

WikiGIS is too limited in terms of democracy as the authors have tried to conceptualize a new Geoweb 2.0 tool with insuring data quality that is lost in the majority of GeoWeb 2.0 applications and tools based on a consumer-oriented wiki content management system.

References

1. Steinitz, C. (2014). Which way of designing? *Geodesign by Integrating Design and Geospatial Sciences* (pp. 11–40). Springer: Springer.
2. Flaxman, M. (2010). Fundamentals of geodesign. *Proceedings of Digital Landscape Architecture* (pp. 28–41). Bernburg Dessau: Anhalt Universtiy of Applied Sciences.
3. Flaxman, M. (2010). Geodesign: Fundamental Principles and Routes Forward. Talk at Geodesign Summit 2010. http://video.esri.com/watch/106/geodesign-fundamental-principles. Retrieved 6 Jan 2010.
4. Ervin, S. (2011, May). *A System for Geodesign. Originally presented at Digital Landscape Architecture Conference*. Germany: Dessau.
5. Campagna, M. (2014). Geodesign from theory to practice: From metaplanning to generation of planning support systems. *TEMA: Journal of Land Use, Mobility and Environment*. Retrieved January 17, 2014, from https://www.researchgate.net/publication.
6. Noucher, M. (2009). *La donnée géographique aux frontières des organisations : approche socio-cognitive et systémique de son appropriation*. École Polytechnique Fédérale de Lausanne, Suisse: Thèse de doctorat non publié.
7. Batita, W., Roche, S., Bédard, Y., & Caron, C. (2014). Towards a conceptual framework for WikiGIS. *Future Internet, 6*(4), 640–672.
8. Batita, W., Roche, S., Bédard, Y., & Caron, C. (2012). WikiSIG et Geodesign collaboratif: Proposition d'un cadre logique. *Revue Internationale Géomatique, 22*, 255–285.
9. Roche, S., Mericskay, B., Batita, W., Bach, M., & Rondeau, M. (2012). WikiGIS basic concepts: Web 2.0 for geospatial collaboration. *Future Internet, 4*(1), 265–284. http://doi.org/10.3390/fi4010265.
10. Batita, W., Roche, S., & Claude, C. (2016). *A Simulated Scenario of WikiGIS to Support the Iteration and Traceability Management of the Geodesign Process*. World Academy of Science. *Engineering and Technology International Journal of Computer, Electrical, Automation, Control and Information Engineering, 9*.
11. **11a**. Artz, M. (2010). Geodesign: Changing geography by design. *Directions Magazine*, 3–7. Retrieved October 28, 2010, from http://www.esri.com/library/ebooks/GeoDesign.pdf. **11b**. Artz, M. (2010). Geodesign: An Evolving Field. GIM International: Mapping the world. Retrieved February 24, 2010, from http://www.giminternational.com/issues/articles/id1505-Geodesign_An_Evolving_Field.html.
12. Dangermond, J. (2010). Geodesign and GIS–designing our future. In *Proceedings of Digital Landscape Architecture* (pp. 502–514). Bernburg and Dessau: Anhalt Universtiy of Applied Sciences.
13. Roche, S. (2009). Towards a Leonardo da Vinci approach of GIS for spatial design. Paper presented at Specialist Meeting on Spatial Concepts in GIS and Design. USA: Santa Barbara.
14. Chrisman, N. (2005). Charting the unknown: How computer mapping at Harvard became GIS. ESRI Press.
15. Miller, W. R. (2012). Introducing geodesign: The concept director of geodesign services. Esri Press.
16. Abukhater A. & Walker D. (2010). Making Smart Growth Smarter with Geodesign, Directions Magazine. http://directionsmag.com/articles/making-smart-growthsmarterwith-Geodesign/122336. Retrieved 19 July 2010.
17. Aina, Y. A., Al-Naser, A., & Garba, S. B. (2013). Towards an integrative theory approach to sustainable urban design in Saudi Arabia: The value of geodesign. *Advances in Landscape Architecture*, 531–550.
18. Pouliot, J. (2014, Mai). La valeur ajoutée de la 3é dimension dans des processus décisionnels : représentation cadastrale et découverts de données sur le Web. Conférence de 25ème anniversaire CRG. Canada: Québec.
19. Ciobanu, D., Roche, S., Badard, T., & Caron, C. (2007). Du wiki au wikiSIG. *Geomatica, 61*(4), 455–469.
20. McClure, R. D. (2002). Common data collection strategies effective in qualitative studies using action research in technical/operational training programs. http://evokedevelopment.com/uploads/blog/commonData.pdf.
21. Hancock, B. (1998). *Trent focus for development in primary health care an introduction to qualitative research an introduction to qualitative*. United Kingdom: Trent Focus Group.
22. Rodriguez-Pabon, O. (2005). *Cadre théorique pour l'évaluation des infrastructures d'information géospatiale*. Université Laval, Québec, Canada: Thèse de doctorat non publiée.
23. Glaser, B. G., & Strauss, A. L. (1967). *The discovery of grounded theory: Strategies for qualitative research*. New York: Aldine Transaction.
24. Miles, M. B., & Huberman, A. M. (Eds.). (1994). *Qualitative data analysis: An expanded sourcebook* (2nd ed.). Thousand Oaks: SAGE.
25. Lawrence, J., & Tar, U. (2013). The use of grounded theory technique as a practical tool for qualitative data collection and analysis. *The Electronic Journal of Business Research Methods, 11*(1), 29–40. www.ejbrm.com.
26. Harris, L., & Brown, G. (2010). Mixing interview and questionnaire methods: Practical problems in aligning data. *Practical Assessment Research & Evaluation, 15*(1), 1–19. Retrieved from http://libir1.ied.edu.hk/dspace/handle/2260.2/10032.
27. Lai, E. R., & Waltman, K. (2008). Test preparation: Examining teacher perceptions and practices. *Educational Measurement: Issues and Practice, 27*(2), 28–45. https://doi.org/10.1111/j.1745-3992.2008.00120.x.
28. Lazar, J., Heidi Feng, J., & Hochheiser, H. (2010). *Research Methods in Human-Computer Interaction*. New York: Wiley.
29. Ciobanu, L. D. (2006). *Adaptation des SIG participatifs aux processus de design urbain délibératifs*. Canada: Mémoire de maitrise non publié, Université Laval.
30. Forester, J. (1999). *The deliberative practitioner: encouraging participatory planning processes*. Cambridge: MIT Press.
31. Sonnen, D., (2007). Emerging Issue: Spatial Data Quality. Directions Magazine. http://www.directionsmag.com/article.php?article_id=2372. Retrieved 4 Jan 2007
32. Bédard, Y. (2012). *Geospatial Data Quality Awareness, The Next Challenge: Are We Ready? Invited Keynote Speaker at GEOIDE IV-23 International Workshop on Geospatial Data Quality*. Ethical and Technical Aspects, Quebec City, Canada: Legal.
33. Lee, D. J., Dias, E., & Scholten, H. J. (2014). Introduction to geodesign developments in Europe. In *Geodesign by Integrating Design and Geospatial Sciences* (pp. 3–9). New York: Springer.

Predicting and Assessing Journey Time Through the Application of Itinerary Selection in an Emergency Context

Mohamed Ayet Allah Bilel Soussi, Jean-Michel Follin and Chamseddine Zaki

Abstract

Several studies have recently dealt with the exploitation of data obtained from transiting mobile objects. In this respect, we have formulated the hypothesis that the history of the routes taken by emergency vehicles can provide information that can be exploited in researches on the best itinerary for an intervention. Accordingly, this article proposes an approach for the prediction and evaluation of the speed of a journey based on the technique known as "plotter vehicle." The latter is a dynamic mode with graph per interval retained from route calculation in our study.

Keywords

Travel time estimation • Multi-criteria analysis • Trees and decision rules • Route search algorithm

1 Introduction

Internet has become a crucial element for effectively communicating information concerning transport, traffic conditions and the search for the best route. Information services for route calculation and traffic conditions are available on websites such as sytadyn (www.sytadin.fr) in Paris, coraly (www.coraly.fr) in Lyon or Google Maps (www.maps.google.fr), Mappy (www.mappy.fr), Viamichelin (www.viamichelin.fr) and V-Trafic (www.V-trafic.com). The Geovelo (www.geovelo.fr) portal for cyclists calculates the best itinerary by taking into consideration several criteria such as the total distance, safety, and the cyclist's physical effort. However, these search engines for route calculation are not suitable for use by emergency services. They ground their calculations on established speed limits. Moreover, the widely used research algorithms do not take into account uncertainties about the estimation of journey speed.

Hence, in the following article, we propose a route calculator for emergency services in the region of Le Mansby while exploiting the journey information for interventions. In [1], a general method is presented to enrich a route database (from French National Mapping Agency[1] topographic data) using measurements from the used routes. The method extends from the modeling of routes into the implementation in a time and space database. To implement our approach, a database containing routes matched with the road network in the Sarthe area was created under PostgreSQL/PostGIS.

A strategy is presented in [2] allowing the prediction and evaluation of journey speed. It involves two methods: direct enrichment when there is a sufficient number of observations (number of journeys using a certain route in a defined time slot) and indirect enrichment when the number of recorded observations is insufficient. As a follow-up to this work, we suggest an improvement of the indirect enrichment method. Our proposed methodology will be elaborated through three major steps. First, we will present an existing study on the prediction and evaluation of journey speed. Then, we will highlight our approach for so doing and we will conclude with finalizing our perspectives and main findings.

M. A. A. B. Soussi (✉)
Institut Supérieur des Etudes Technologiques de Nabeul, Nabeul, Tunisia
e-mail: soussi_bilel@yahoo.fr

J.-M. Follin
Ecole Supérieure des Géomètres et Topographes, ESGT, GeF, IRSTV 1boulevard Pythagore, 72000 le Mans, France
e-mail: jmfollin@esgt.cnam.fr

C. Zaki
American University of Culture and Education in Beirut, Beirut, Lebanon
e-mail: chamseddine.zaki@gmail.com

[1] National Institute of Geographic and Forest Information (IGN).

N. Rebai and M. Mastere (eds.), *Mapping and Spatial Analysis of Socio-economic and Environmental Indicators for Sustainable Development*, Advances in Science, Technology & Innovation, https://doi.org/10.1007/978-3-030-21166-0_3

2 The Prediction and Evaluation of an Itinerary Speed Case Study: Montreal, Canada

Throughout this section, we will work on identifying the possible link between the physical characteristics of the road network segments and the traffic flow in Montreal. This is of interest to us as we estimate that the issue of predicting journey time depends essentially on the above-mentioned criteria. The Ministry of Transport in the Province of Quebec collected data from 1998 to 2004 using "plotter vehicles" for measuring traffic congestion [3]. An assessment study of traffic congestion was carried out by [4] using this data. It focuses on frequently-congested road segments.

It should be noted that data collected solely on speed is insufficient for a pertinent assessment of traffic congestion [5]. With the aim of improving the work presented in [4,5] suggests grouping the road segments according to the distribution of the recorded average journey times. This study examines 811 road segments. The authors opted for studying the morning peak period. The road segments were grouped without considering the physical characteristics of the road network (the number of lanes, width etc.).

In order to create a link between the physical characteristics of the road network and the traffic flow in Montreal, [6] proposes an approach that goes through four steps:

1. The first step consists in creating a database for the road segments and their physical characteristics. It is composed of three tables: the first links each of the 811 segments to its group. The second table contains two indicators for each group defined in the first table: the value of the average journey time and its variability. These two tables are from the work of [5]. The last table, however, contains the characteristics of the sample road segments.

2. As for the second step, the author focuses on grouping the road segments according to their physical characteristics (type of road, direction of traffic, authorized speed, number of lanes etc.) The Multiple Correspondence Analysis Method was used for fulfilling this task. The choice of this method can be justified by the existence of more than two qualitative variables (physical factors for the road segments) to be studied. A correlation matrix was generated clarifying the different correlations between the physical factors that may exist in the road segments. Each group is composed of road segments that have a strong correlation between the physical factors. Twelve FP groups were created. For example, the characteristics of group FP1 are the segments of 3-lane freeway with a speed limit of 70 km/h, a hard shoulder and a concrete barrier.

3. The author then links the new group created via the MCA method with the other group based on the distribution of journey time. This step is a way of determining whether the physical factors that characterize the road segments explain the traffic flow or not.

4. The last step through which this method is applied lies in generating a tree and decision rules, which enable us firstly to explain the reason why the segments belong to certain TP groups according to their physical characteristics and to predict the journey time for the remaining road network.

In Fig. 1, we resume the different steps of the method presented by [6] for the analysis and classification of the road segments according to their physical characteristics.

The author specifies the most pertinent physical factors of the road segments that explain the traffic flow in the area of Montreal. Those factors are the type of the road, the authorized speed limit and the number of exits and entrances. In our project, we were inspired by [6] for the prediction of journey speed. In the following section we will present the different stages of our methodology.

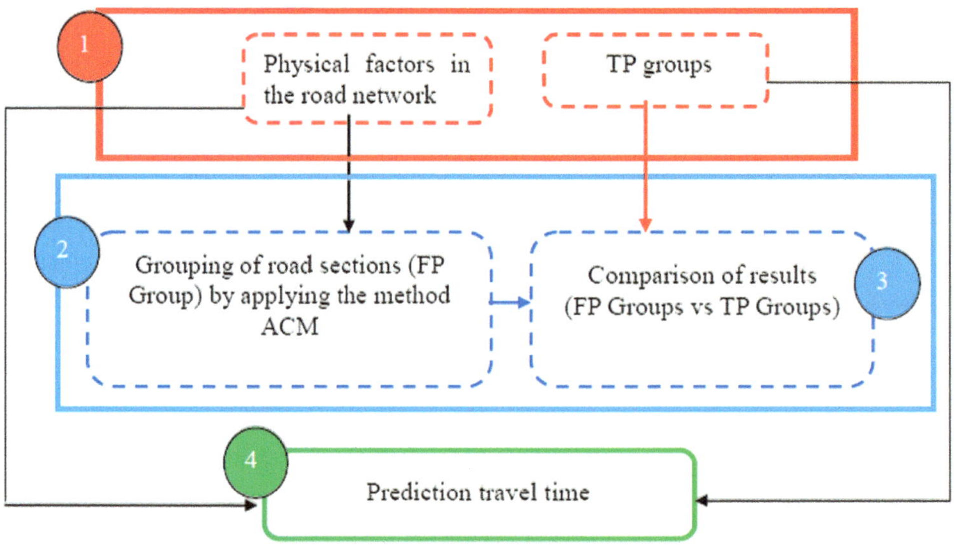

Fig. 1 Analysis methodology and classification of road sections by [6]

3 Proposition for the Prediction and Evaluation of Journey Speed

In this part, we introduce our methodology for predicting route speed. The results obtained from the application of this methodology will be used in the indirect enrichment function. Five stages help elaborate our synthesis.

- The first stage focuses on the analysis of speed variation according to each factor retained in our study (times and the physical characteristics of the road network). The target here is solely to be able to estimate the impact of each factor on speed variation (by studying the distribution of slices speed).
- The second stage is divided into two sub-stages. The first consists in studying the correlation between the different factors retained in the first step of speed observations. Only the factor that is strongly correlated with the speed is retained. In the second sub-stage, we will continue studying the relationship between the retained factor in the first sub-stage and the remaining factors. Only the factors that are of weak correlation will be retained for the following third stage of our methodology. Our idea revolves around minimizing redundancy between the different factors. The AMC method for statistics was used for carrying this task.
- In the third stage, we apply a classifying algorithm in order to generate a tree and decision rules to classify the speed segments according to the factors retained. They serve as tools for the prediction of route speeds on the different road segments. We only use the factors retained in stage two. We propose classifying the speed intervals instead of determinist values. The idea is to take into consideration the uncertainty of the speeds observed.

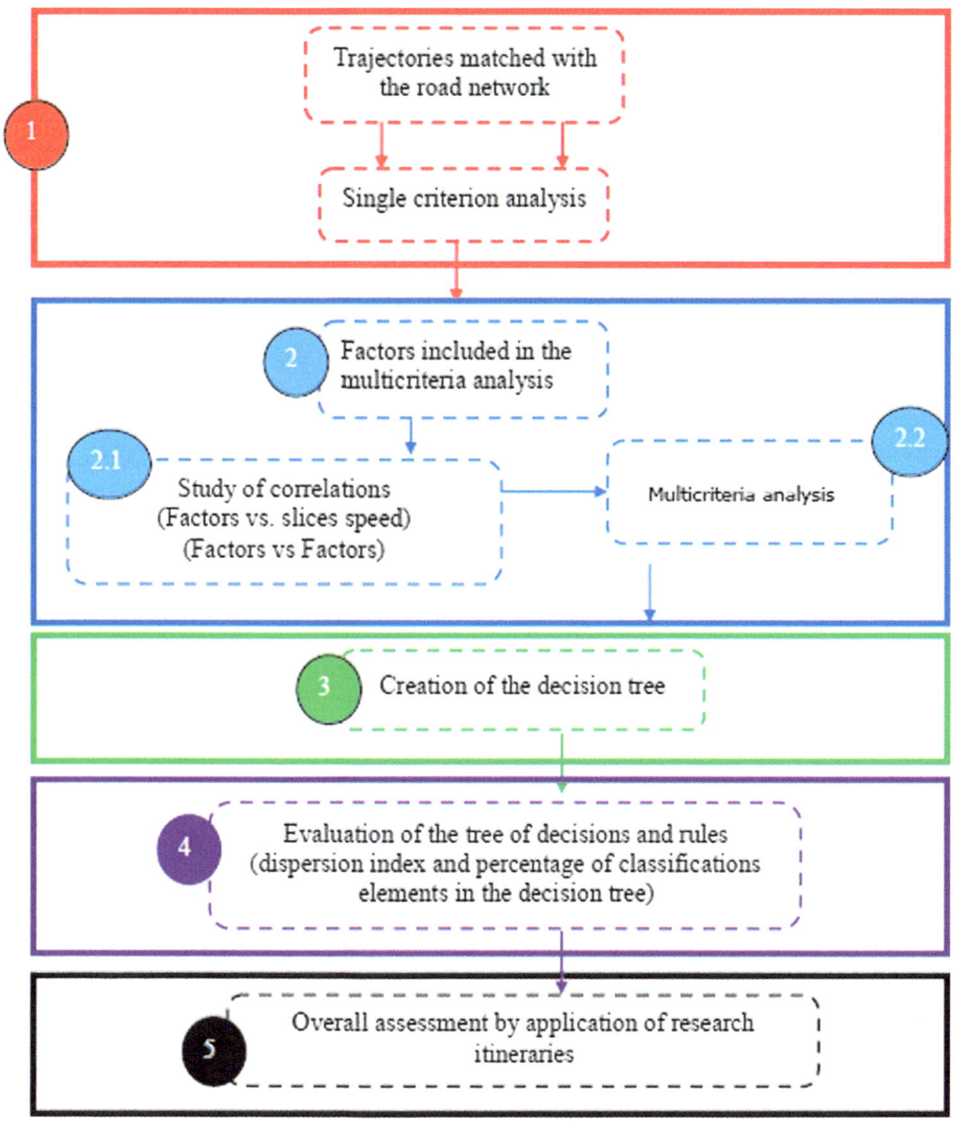

Fig. 2 Methodology for predicting travel speeds

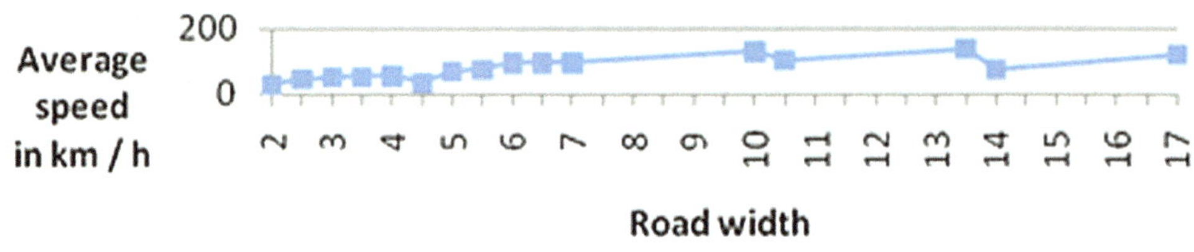

Fig. 3 Average speeds according to the width of the road surfaces

- The fourth stage of our methodology lies in assessing our method of estimation for the speed of a route. In order to do this, we propose the following two sub-stages:

  - The first sub-stage aims at determining the dispersion index for each class linked to a decision rule, designating the relationship between the average speed and its standard deviation, and the percentage of good classification for speed segments.
  - In the second sub-step, we study the histograms of actual speeds as decision rules defined in the third step. The endeavor of this stage is to determine an interval of uncertainty for the speed from the study of speed distributions represented by these histograms. In this respect, the distribution of an empirical distribution of speeds is generated and it is according to which the speed interval is calculated. The probability for this interval is set at 68%.

- In the final stage, a comparison is made between the itineraries proposed by our itinerary calculator and the route histories saved in our database.

In Fig. 2, we resume the different stages of our previously defined methodology enabling the prediction and assessment of the route speeds which will be used by the itinerary calculator.

4 Single-Criterion Analysis

This section elaborates the first stage of our methodology for the prediction of route speeds. It is about studying the impact of each factor on the variation of speed during the itinerary in order to retain only those that have an impact on speeds. The adopted criteria in our analysis are time segments and the physical characteristics of the road network (size of roads, width, number of lanes and classification). The modalities of these factors are defined as follows:

1. The observation period: Four time segments [7h–9h], [11h30–14h], [16h30–19h] and a time segment for the remainder of the day (set for vacation and non-vacation periods from Monday to Friday). These slices are considered only for those days. On Saturday, we have two slices ([13h–21h] and the rest of the day) and one time slot for Sunday.

2. Road width: 4 classes distributed in the following intervals [2 ; 5],] 5 ; 7],] 7 ; 10.5],]10.5 ; 17];

3. Size of the road: 5 groups from 1 to 5;

4. Number of lanes: 4 modalities defined from 1 to 4;

5. The road category: 6 groups for road junction, one road pavement, two road pavement, gravel road, quasi-highway and highway.

The four time segments are determined following a conversation with ambulance drivers. The road widths classes are chosen using an observation of the average speed curve according to road widths. Road size, the number of lanes and the road category are taken directly from the IGN road database. Different histograms, one per factor, are generated and analyzed. In fact, they represent samples for the speeds observed according to the modalities taken for each criterion. Following this, we will only present samples for speeds observed for the road segments corresponding to a given width and size. In order to determine the different categories of the speed observed, we have drawn up a curve of the obtained average speeds in the time period "remainder of the day" according to the width of the road surfaces. This factor indeed appears preponderant, as we will see in what follows. The choice of the time segment is related to the idea of limiting the effects of recurrent traffic congestion as far as possible. It is notable through the observation of the trend in this curve that the road width has a direct influence on speed variation. Using this curve, we are able to determine the four groups for width that are presented above. Moreover, from this same segmentation we have determined the number and the thresholds of the speed groups (Fig. 3).

- v1: speed equal or inferior to 42 km/h,
- v2: speed superior to 42 km/h and equal or inferior to 93 km/h,
- v3: speed superior to 93 km/h and equal or inferior to 113 km/h,
- v4: speed superior to 113 km/h.

5 Analysis According to the Width of the Road Surface

In this type of classification, we present the samples of the observed speeds on roads that have a class 1 width (Fig. 4).

We notice that this histogram presents only one mode with a tight distribution between the segments 20–25 km/h and 75–80 km/h. We also note that the distribution follows a Gaussian

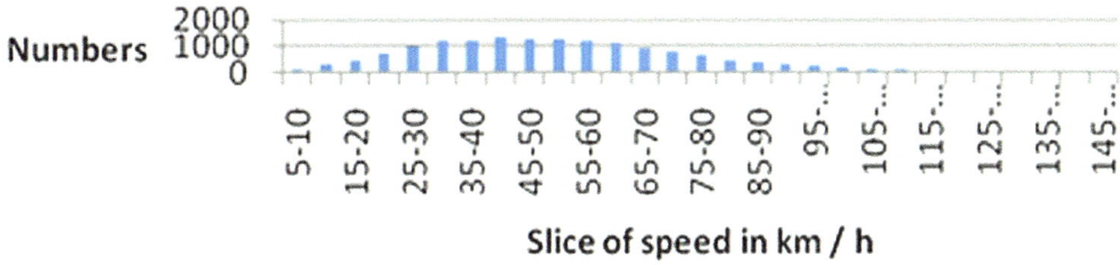

Fig. 4 Numbers of observed speeds on roads whose width pavement belongs to the class 1

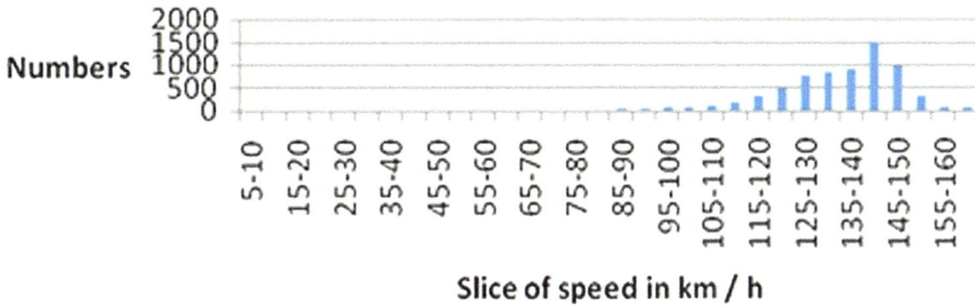

Fig. 5 Numbers of observed speeds on roads whose importance is equal to 1

law. This reveals a relatively homogenous behavior of speeds of around 40–60 km/h on the road segments that are relatively straight.

6 Analysis According to the Size of the Road

In this part, we will shed light on the impact of the size of the road on speed variation.

In Fig. 5, a tight distribution around 125–130 km/h and 145–150 km/h, and in a single mode, can be noted. A peak is observed at 140–145 km/h. This may be explained by the fact that the roads for which the size is equal to 1 are generally motorway-type roads or motorways. Our observation of the two illustrations (Figs. 4 and 5) uncovers the importance of the factors linked to the width of the road surface and the size of the road in explaining speeds. The aim of a single-criterion analysis is to emphasize the impact of the factors previously presented on the variation of the observed speed. Only those factors having an impact on speed variation will be retained for the second step of our methodology (multi-criteria analysis) (Fig. 6).

7 Multi-criteria Analysis

In order to ensure the reliability of our estimations for journey time, we endeavor to find the best combination possible between the different factors that have been retained. Among the different combinations that can be defined, we are searching for the one that will best explain the observed speeds. For

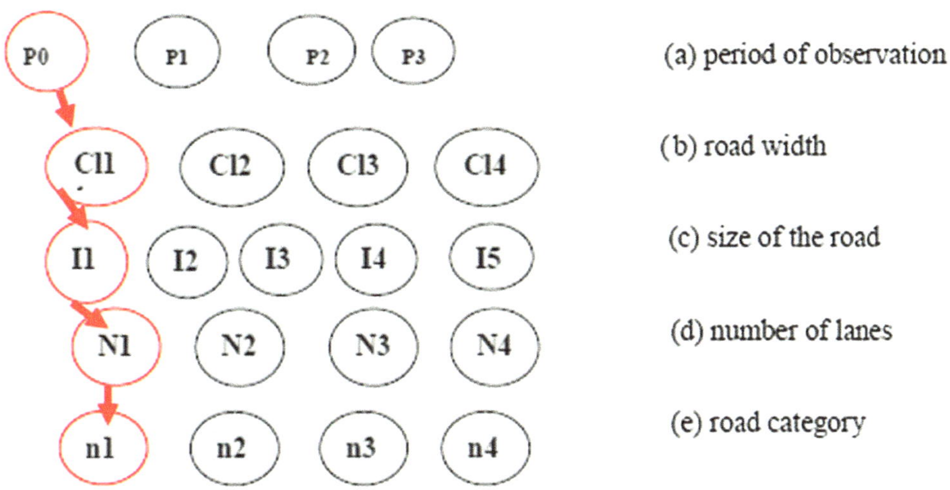

Fig. 6 the factors considered in the multi-criteria analysis

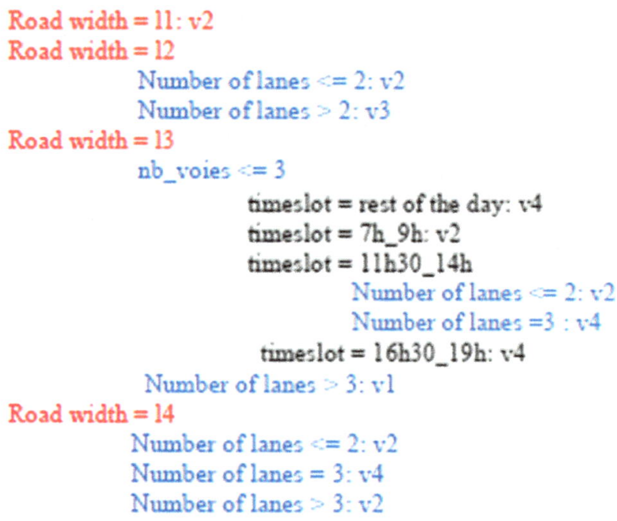

Fig. 7 Decision tree number 1

example, if five factors are used, considering the number of groups defined for each of them, we could come to a total of 1280 groups, including for example the Combination 1, as (P0, CL1, I1, N1, n1) (Fig. 7). Combination 1 can be read in the following way:

- P0 corresponds to the time period "rest of the day",
- Cl1 stands for road widths which are between 2 and 4 m,
- I1 denotes the size of the road that is equal to 1,
- N1 corresponds to a road track number which is equal 1,
- n1 means a highway as way.

It is not possible to analyze all the existing combinations in our study. We risk having histograms and groups that present similarities with others and could therefore be grouped together. To solve this problem, our idea is to carry out a multidimensional descriptive statistic analysis. In this context, we reveal the correlations that can exist between these factors. The objective is to reduce the number of factors to be studied. We apply a classification in order to obtain the best combination between the groups of the retained factors. We generate a tree and decision rules based on the factor groups retained.

In the multidimensional descriptive statistic analysis, we have distinguished two methods: Principal Component Analysis (P.C.A) and Analysis of Multiple Correspondences (A.M.C).
The P.C.A method enables the simultaneous processing of a number of quantitative variables. However, the A.M.C method is used when there is at least two qualitative variables. Hence, the latter is automatically a requirement for using the A.M.C method for the rest of our analysis.

Following the A.M.C method, we determined the correlation matrix between the different factors retained during the single-criterion analysis stage. Thanks to this matrix we can determine the coefficients for the correlations of the variables taken two by two. To study the relationship between the variables we used the following approach. The first step consists in studying the relationship between the speed segments (v1, v2, v3, v4) and the other variables. Only the factors having a strong correlation with the speed segments will be retained.

After the first step, we estimate that the "road width" is the most correlated factor with the speed segments. Consequently, only the width groups will be retained for the second step of our approach. For example, we note that the correlation coefficient between group 1 of road width and the speed v1 is around 0.33. This result is expected as the segmentation of the speeds is based on the curve of average speeds according to the road width. The second step refers to the study of the correlation between the factor retained in the first step and the remaining factors. Only the factors having a weak correlation with road width will be retained for the rest of our study.

At the end of the second step, we note that the less correlated factor with the road width is the number of lanes. After having applied the two previously described steps, the groups linked to the width of the road and the number of lanes will be retained for the prediction of journey time. The algorithm of classification will be based on these groups and the time segments.

The study of our journey database allows us to remark that a significant number of road segments do not have recorded width or numbers of lanes. In order to solve this problem, we carried out another study of the correlation between the groups of speed segments and other factors. We notice that the groups linked to the size of the road have a greater correlation with the speed segment groups. As a result, we decided to apply only the classification algorithm to the time segments and the groups linked to the size of the roads. Indeed, the classification algorithms will be applied twice. Firstly, a decision tree is generated to predict journey time based on time segments, the groups linked to the size of the road and the number of lanes. However, in situations where no information about the road size is available and/or the number of lanes is not specified, a second decision tree is used. It is based on time segments and groups linked to the size of roads.

8 The Creation of Decision Trees

For the creation of decision trees and decision rules, we have recourse to the C4.5 algorithm developed by [2]. In Fig. 7, we present the rules of the first decision.

```
Importance of lanes <= 2
        Importance of lanes = 1: v4
        Importance of lanes =2: v2
Importance of lanes > 2
    Importance of lanes = 3: v2
    Importance of lanes > 3
        Importance of lanes = 4: v2
        Importance of lanes =5: v1
```

Fig. 8 Decision tree number 2

According to the rules presented in the first decision tree, we notice that the time factor only has an influence when the road width belongs to group 13 (road width between 7 and 10.5 m) and when the number of lanes is inferior or equal to 3 m. Concerning the other decision factors, the same speed segment is used for all time segments. In the classification issued from the first decision tree, there is also the presence of unexpected average speeds. For example, the average speed is lower for the I4 class roads with more than three channels that are for the same class roads characterized by three ways. It is explained by the too small sample used for generating these rules. Figure 8 denotes the second decision tree. In relation to the decision rules presented in Fig. 8, we understand that the impact of the time factor is not too significant to be taken into account by the classification algorithm.

9 Evaluation of Trees and Decision Rules

The evaluation of the indirect method for estimating travel speed is based on the following steps:

- determination of the distribution of elements in the two decision trees,
- evaluation of the dispersion index for each rule,
- regeneration of the histograms of speeds according to the decision rules.

By applying the C4.5 algorithm, we conclude that 50 and 48% of items are distributed respectively in the first and second tree. The second evaluation of the indirect method is developed to evaluate the dispersion index for each decision rule. In Figs. 9 and 10, we present the values of the average speeds, standard deviations and dispersion indices for each decision rule of the two trees. By observing the dispersion indices (noted I) presented in the first decision tree, we see that the index has a maximum value of 45% and a minimum of 8%. Regarding rules based on time slots, we note that the dispersion index can reach a maximum 25%. On the second tree, the value of the dispersion index may reach a maximum of 58% and its minimum is 11%. In addition, in the second decision

```
Road width = I1: v2 (16760.0/7027.0)
Speed [29 km/h-77 km/h], I=45%, average speed =53 km/h , standard deviation =24 km/h
Road width = I2
        Number of lanes <= 2: v2 (63959.0/37798.0)
Speed [59 km/h- 117 km/h ], I=31%, average speed = 88 km/h, standard deviation =29 km/h
        Number of lanes > 2: v3 (170.0/55.0)
Speed [80 km/h-104 km/h], I= 12%, average speed =92 km/h, standard deviation =12 km/h
Road width = I3
        Number of lanes <= 3
         timeslot = reste de la journée: v4 (2727.0/726.0)

 Speed [95-145], I =20%, average speed =120 km/h, standard deviation =25 km/h
         timeslot = 7h_9h: v2 (172.0/96.0)

 Speed [73-123], I =25%, average speed =98 km/h, standard deviation =25 km/h
         timeslot = 11h30_14h

        Number of lanes <= 2: v2 (142.0/89.0)
 Speed [98 km/h-146 km/h], I =19%, average speed =122 km/h, standard deviation =24 km/h
        Number of lanes =3 : v4 (202.0/38.0)
 Speed [86 km/h-142 km/h], I =24%, average speed =114 km/h , standard deviation =28 km/h
        timeslot = 16h30_19h: v4 (1073.0/292.0)

 Speed [98 km/h- 146 km/h], I =19%, average speed =122 km/h, standard deviation =24 km/h
        Number of lanes > 3: v1 (27.0)
 Speed [25 km/h-33km/h], I=13%, average speed =29 km/h, standard deviation =4km/h
Road width = I4
        Number of lanes <= 2: v2 (602.0/201.0)
      Speed[57-97], I =25%, average speed =77 km/h, standard deviation =20 km/h
        Number of lanes = 3: v4 (4019.0/137.0)
      Speed [124-148], I =8%, average speed =136 km/h, standard deviation =12 km/h
        Number of lanes > 3: v2 (225.0/90.0)
      Speed [45 km/h-105 km/h], I =40%, average speed =75 km/h , standard deviation =30km/h
```

Fig. 9 The decision rules of the first tree

Importance of lanes <= 2
  Importance of lanes = 1: v4 (6720.0/477.0)
    Speed [119 km/h- 149 km/h], I =11%, average speed =134km/h, standard deviation = 15 km/h
  Importance of lanes=2: v2 (53406.0/33779.0)
    Speed [63 km/h- 119 km/h], I =30%, average speed = 91km/h, standard deviation =28 km/h
Importance of lanes > 2
  Importance of lanes = 3: v2 (19000.0/8393.0)
    Speed [40 km/h- 98 km/h] , I =42% ,average speed =69 km/h, standard deviation =29 km/h
Importance of lanes > 3
  Importance of lanes = 4: v2 (8373.0/2815.0)
    Speed [33 km/h- 77 km/h], I =40%, average speed =55 km/h, standard deviation =22 km/h,
  Importance of lanes =5: v1 (4840.0/2001.0)
    Speed [17 km/h- 65 km/h] , I =58%, average speed =41 km/h, standard deviation =24 km/h,

Fig. 10 The decision rules of the second tree

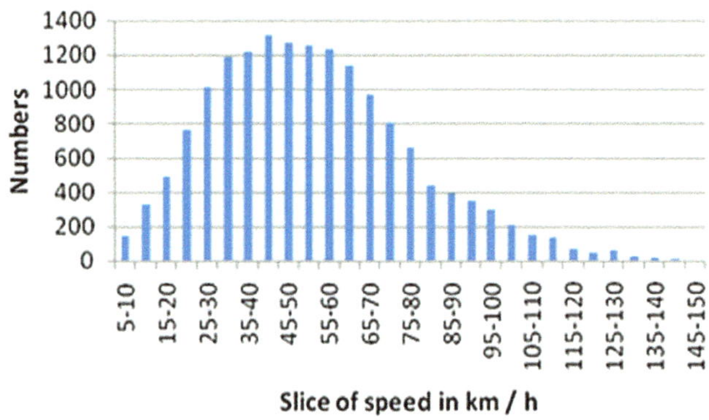

Fig. 11 Actual observed speeds on roads which width belongs to class I1 of the first decision tree

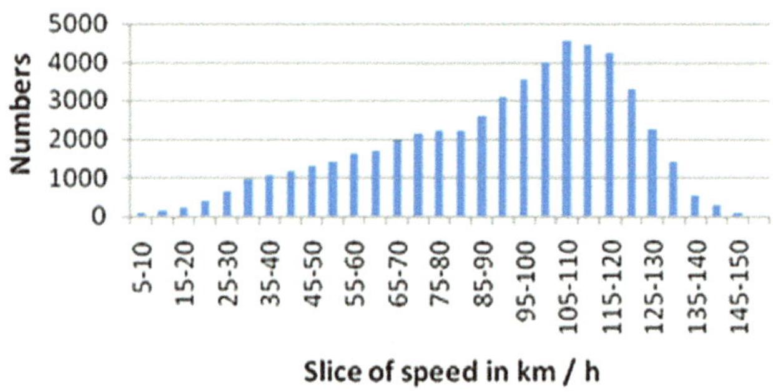

Fig. 12 Numbers of observed speeds on roads which importance is equal to 2

tree, we find that the dispersion index increases relatively to the increase in the degree of importance.

[average speed −standard deviation, average speed +standard deviation]. When the velocity distribution follows the Gaussian one, the probability that the speed included in this range is about 68%, but this is not always the case. To solve this problem, the function draws its empirical distribution. The range of the rate uncertainty shall be adjusted to have a 68% probability. We will take as example the distribution of speeds observed in class I1 in the first decision tree (Fig. 11).

With regard to Fig. 11, we deduce that the distribution is Gaussian. Therefore, it is appropriate to set a range of uncertainty for the average speed based on its standard deviation. We have the uncertainty interval [29–77 km/h]. If we take the other example of the velocity distribution of the class, resulting from the application of the second decision tree for roads with an importance equal to 2 (Fig. 12), we perceive a very wide distribution with one mode. In fact, the average speed calculated from this distribution is equal to 91 km/h, while the interval of uncertainty is [63–119 km/h].

In order to adjust the interval of uncertainty, we draw the function of the empirical distribution of speeds (Fig. 13).

Figure 13 shows that the probability of having a value of velocity in the interval [63–119 km/h] is 65%. Therefore, we expand the range of uncertainty in speed to approximate a probability of about 68%. In our example, the interval is extended to [62–120 km/h] and we have accordingly a probability of 68%. In fact, we perform the same way for modifying intervals uncertainties rates for other classifications.

10 Overall Assessment of Travel Speeds

To evaluate our general method for estimating travel speed, we perform a comparative study between the trajectories of the emergency services and trajectories returned by our calculator [2].

Before doing so, we present in Table 1 a summary of the enrichment of the road database due to the use of the direct and indirect methods and for each time slot percentages. For the direct method, we precise that the number of passes on a road section has been set to two in a time slot.

A close eye on Table 1 allows the observation that the percentage of enrichment by the direct method is very low, compared to the indirect method. In addition, it is clear that the percentage of enrichment based on the decision rules from the first tree is slightly higher than the rules from the second tree.

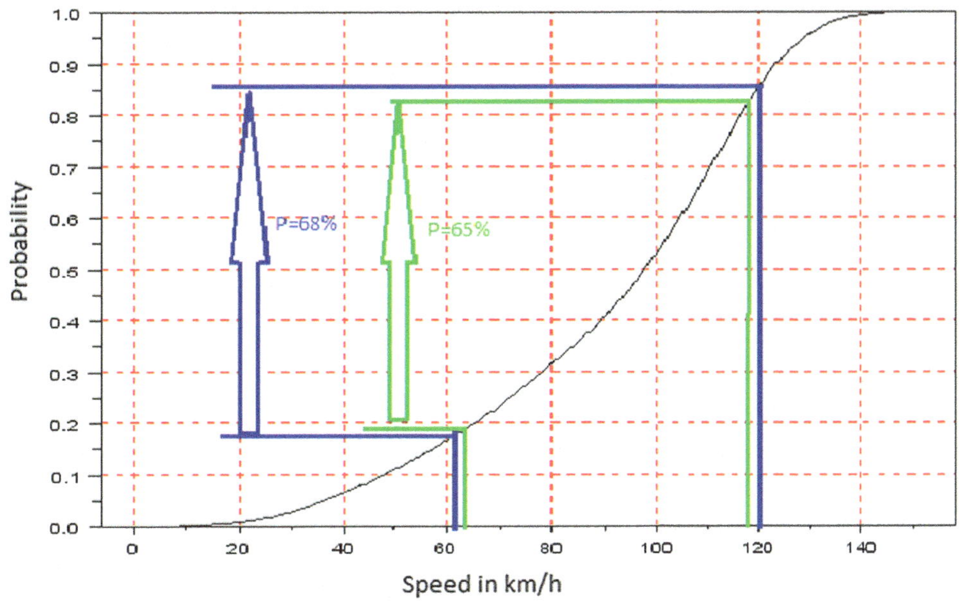

Fig. 13 Empirical distribution of speeds observed on roads whose importance equals to 2

Table 1 Summary of the methods of enrichment percentage for each time-slot

Enrichment method		Enrichment percentage in each time slot			
		[Remainder of the day] (%)	[7h–9h] (%)	[11h30–14h] (%)	[16h30–19h] (%)
Direct method		2	1	0.5	1
Indirect method	First decision tree	52	52.5	52.75	52.5
	Second decision tree	46	46.5	46.75	46.5

Table 2 Results of the comparison between the routes offered by our calculator and trajectories of SMUR

Date and time of interventions	Lenght of the trajectory in km	Percentage of similarity (%)	Difference in travel time (min/s)	Percentage of enrichment by the direct method (%)	Percentage of enrichment by indirect method (%)
2011-04-10 4:44	41	96.6	−2 s	97.9	0.3
2010-12-26 15:18	30	93.4	+16 s	74.6	25.4
2011-02-03 21:04	27	91.7	+13 s	92.9	7.1
2011-04-30 21:53	18	90.1	−42 s	95.7	4.3
2011-02-08 16:20	7	87.9	+2 min,19 s	93	7
2010-12-08 22:06	15	87.6	+22 s	82.1	17.9
2011-04-13 9:20	13	83.4	+42 s	90.2	9.8
2011-03-08 09:47	22	82.7	−2 min,34 s	80	20
2011-04-09 20:24	21	76.3	−2 min,16 s	87.1	12.9
2010-12-01 13:54	49	70.7	+2.22 s	31.6	68.4
2011-03-12 10:05	29	20	−4 min,43 s	41	59
2010-12-11 15:06	31	13.8	−3 min,8 s	78	23

Regarding the evaluation phase of our calculator, we have chosen twelve trajectories that are well-matched with the road network routes. In Table 2, we sum up the main findings of the comparisons done between the routes generated by our calculator and the trajectories of the emergency agents during their interventions. We precise that, the trajectories chosen for this study are not included in the estimated travel time phase.

One of our hypotheses is that paramedics are familiar with the terrain. In other words, we can use the history of the trajectories to validate our methodology for estimating travel time. To get a better estimation, it is necessary to minimize the percentage of dissimilarity between the routes offered by our calculator and the trajectories of the emergency service. Table 2 indicates that 80% of the routes suggested by our calculator have a percentage of similarity greater than 70%. Another important fact is that when the similarity percentage exceeds 75%, the percentage of enrichment by the direct method is higher than 70%. The gain in travel time is on average 35 seconds. The 20% average dissimilarity observed is due to various error sources such as GPS accuracy and matching. Moreover, we show that the dispersion index calculated for the direct and indirect methods of enrichment can reach a maximum of 40 and 58% (Figs. 10 and 11).

11 Conclusion

In this paper, we propose a general method for the prediction and assessment of travel speed based on a technique called "tracer vehicles." Indeed, this technique is best suited to estimate the travel time for a particular category of road users who are responsible for providing intervention services. Our method is divided into the following steps:

- single-criterion and multi-criteria analysis to retain only factors that have an impact on the speed of travel,
- creation and evaluation of trees and decision rules to predict the speed course,
- overall assessment of travel speeds by comparing the trajectories of the emergency services and routes returned by our calculator.

For so doing, we applied a faster time search algorithm. Our idea is to compare the routes suggested by our calculator and routes already borrowed from the emergency agents during their interventions. Our results are promising since we obtained more than 60% of similarity.

For the rest of our project, we suggest to design and implement a system management support of traffic disruption in real time while two types of special events will be taken into consideration: planned events (weather, events, public works and changes in the physical characteristics of the road network), and unexpected events (traffic jams and accidents).

References

1. Soussi, M. A. A. B., Follin, J. M., Moreau, G., Bouju, A., & Polidori, L. (2012a). Enrichissement d'une base de données routière à partir de trajectoires GPS de véhicules d'urgence: Application à l'aide au choix d'itinéraires. *Ingénierie des Systèmes d'information, 17*(1), 35–54.
2. Soussi, M. A. A. B., Follin, J. M., Moreau, G., Bouju, A., & Polidori, L. (2012b). *Une méthode d'estimation du temps de parcours: Application de recherche d'itinéraires dans un contexte d'urgence*. Liège, Belgique: SAGEO.
3. Miro. (2006). Analyse des temps de parcours sur le réseau routier de la grande région de Montréal: Rapport final, projet no. Q94231. Ministère des transports du Québec.
4. Quinlan, J. R. (1993). *Programme for Machine Learning*. USA: Morgan Kaufman.
5. Kalboussi A., (2010). Méthode d'analyse et de classification des segments du réseau routier supérieur de la région de Montréal. Mémoire présenté en vue de l'obtention du diplôme de maitrise des sciences appliquées. Ecole polytechnique de Montréal.
6. Ahmad, S., & Tahar, R. M. (2014). Selection of renewable energy sources for sustainable development of electricity generation system using analytic hierarchy process: A case of Malaysia. *Renew Energy, 63*, 458–66.
7. Loustau, P., Morency, C., Trépanier, M., & Gourvil, L. (2010). Portrait de la fiabilité des temps de parcours sur le réseau autoroutier montréalais. Site internet www.aqtr.qc.ca.

Multi-source Object-Based Approach for Spatio-Temporal Evolution of Land Cover

Mohamed Mastere, Abdelkrim Achbun, Salma El Fellah and Bouchta El Fellah

Abstract

The aim of this study is to apply an object-based approach using multi-time series of high (Spot-5 images) and medium (ETM of Landsat, OLI TIRS of Landast and MSS of Landsat images) spatial resolution to characterize Land Cover of a heterogeneous territory, called Loukkos river basin which is a part of the western area of the Rifean belt. The images segmentation tests regarding several combinations between color levels (0.1–0.9) and scales (0–255) have confirmed a strong relationship among the spectral values of images radiance, with respect to the number of objects. Indeed, the authors have concluded that this relationship, is more strongly related to the values of the Standard Deviation of the images. The "Map difference" used to assess the accuracy of mapping is made less complicated and more accurate to the classical probabilistic methods. Furthermore, the overall accuracy of the Object-Oriented Classification was 80.20%. The study has shown that farmlands have been undergone more changes than urban classes, followed by wetlands and grasslands those have been converted to agricultural lands and lastly the transformation of forests to farmland and natural vegetation.

Keywords

Object-oriented • Mapping • Classification • Land Cover • Shift • Loukkos • Morocco

M. Mastere (✉) · B. El Fellah
Geopac Research Center, Scientific Institute, Mohammed V University of Rabat, Avenue Ibn Battouta, Agdal, Rabat, Morocco
e-mail: mohamed.mastere@gmail.com

B. El Fellah
e-mail: lfellah@israbat.ac.ma

A. Achbun
Laboratoire de géomatique, Institut National d'Aménagement et d'urbanisme, Avenue Allal El Fassi BP 6215 Rabat-Instituts, Rabat, Morocco
e-mail: a.achbun@gmail.com

A. Achbun
Geopac Research Center, Scientific Institute, Mohammed V University, Rabat, Morocco

S. El Fellah
Faculty of Sciences, Mohammed V University, Rabat, Morocco

1 Introduction

Land cover mapping is a very powerful tool in for spatiotemporal monitoring of territorial phenomena and for decision making. Due to advancement in satellite sensors, their analysis techniques are making remote sensing systems fruitful, realistic and attractive for territories mastering and natural resources management.

The description of the pattern and spatial distribution of Land Cover, within driving thematic maps, has traditionally been performed from remote sensing data [1–3]. Furthermore, the use of multi-temporal series of images enables the detection of changes in Land Cover between one or several dates. It can be assisted by the use of different data sources for examples moderate and high-resolution images, thematic and topographic maps, quick Looks of Google Maps, to enhance the quality of images analysis and interpretation. This stage of pre-processing, keeps a great importance in the Land Cover mapping process.

Moreover, the Object-Oriented classification techniques based on image segmentation are gaining interest as methods for generating output maps directly storable into Geographical Information System (GIS) databases. This approach does not only identify Land Cover on a pixel level, but also organizes it such pixels into groups of segments that correspond to real-world objects. Those segments represent basic processing units of Object-Oriented image analysis, called image objects, in replacement of the single pixels [4].

The results of several studies [5–10] indicated that object-based approach provides better results for change detection than traditional pixel-based method because it allows an

N. Rebai and M. Mastere (eds.), *Mapping and Spatial Analysis of Socio-economic and Environmental Indicators for Sustainable Development*, Advances in Science, Technology & Innovation, https://doi.org/10.1007/978-3-030-21166-0_4

effective way to incorporate spatial information and expert knowledge into the change detection process. Furthermore, in the aim to extract Land Cover change between two or several dates, various methods have been employed using remotely sensed data. Two clear categories of those methods may be cited: as pre-classification change detection and post-classification comparison [11]. They are commonly based on techniques such as image differencing, change vector analysis, image regression, and image rationing [12,13]. The pre-classification is a simple "binary" change detection technique, which produces "change" versus "no change" maps, through the creation of difference maps by comparing the Land Cover of two classification maps.

However, the second type which is the post-classification change detection-based object, after being established the segmentation process, the Land Cover classifications produced for several dates (1979, 2001, 2007 and 2013 in our study case), are considered during the generation of detection change maps, as thematic maps, where their objects incorporates spatial and thematic information. The process is done by a filtering stage, which aims to reduce errors of edge [13]. The Loukkos River draws many meanders in the plain [14], is a heterogeneous area.

The specific objective is to experiment the use of multi-sources and multi-temporal data, in order to characterize and assess the spatiotemporal evolution of Land Cover using the performed object-oriented approach. The integration of class maps at different resolutions, helped in developing a methodology based on the use of multi-source available data, to generate Land Cover maps in the Moroccan geographical context, where, the Land Cover is rapidly changing in response to the agricultural crops extension, natural dynamics and other anthropogenic causes. What appeals to consider a methodology that integrates the disposable data to generate Land Cover maps, having a suitable spatial accuracy and responding best possible to the ground truth.

2 Study Area

2.1 Geological and Geomorphological Setting

The study area is a part of the Loukkos basin which has a substantially rectangular shape. It reaches 80 km width. It is mostly formed by heterogeneous outcrops including the Sahel and loamy sandy plateau at northern part, the buttress of the Pre-Rifan belt at the east and south, in the center the alluvial plain. Sandy formations occupy the western and southwestern part of the study area [15].

The Loukkos River draws many meanders in the plain [14]. The basin includes a complex of wetlands classified as a Ramsar site, composed of estuarine waters, Salicornia steppes, salt, wetlands water sweet over grown with water plants and areas of flooding during the low river.

Surrounding this, exist irrigated farmland, including rice, which took place in wetlands that have been drained [16]. The training al-luvial silty or clayey sands, silts and hydromorphic soils are dated from Soltano–Gharbien. It overcomes a sandstone formation with soft shell and blue clay dated of Pliocene.

The whole region constitutes a part of the external domain situated in south west of the Rifean belt, composed by several domains [17,18]. From inside to outside of the chain, we find: internal domain, flyschs domain and external domain (Figs. 1 and 2). This last, is a large structural domain, which represents the Tethyan margin of the African plate, it is covered by flysch thrust sheets and partly by the internal field. The internal domain is divided into three main parts which are from the inside to the outside that is to say from north to south: The Intrarif the Mesorif, and Prerif. In Loukkos unit [19,20]. The predominate facies are calcareous-marl of the Albo-Cenomanian and Senomien. The facies age fluctuates between the Albo-Aptian and the Eocene [21,22].

3 Materials and Methods

3.1 Materials

A datasets of three Spot scenes of images, projected in Lambert Conic Conform system (Clarke 1880 spheroid, datum Merchich), were used with a scene of Landsat 7 and Landsat 2, Landsat 8, images projected in the UTM system (spheroid WGS 84), those where download from the Global Land Cover Facility Database: GLCF, [23] more descriptive characteristics of the imagery are developed in Table 1. Then a subset of the specific site, were done.

3.2 Methods

As it appears in the Table 1, the imagery was acquired from different sensors, with various spectral and spatial characteristics. Prompting to seek, techniques that harmonize the use of these data; and this by ensuring the maximum possible preservation of the information contained and detected during the realization of the difference maps. In this sense, the Land Cover change detection based on the multisources data, requires to bring the maps which will being compared, to the same level of generalization process, defined by Weibel [30] as a process intended to drive from a detailed source of spatial database, a map or database with a reduced complexity, while retaining the spatial and thematic characteristics of the source data to satisfy to a required purpose, this process was tested on the imagery used in this study, specially to assess the accuracy.

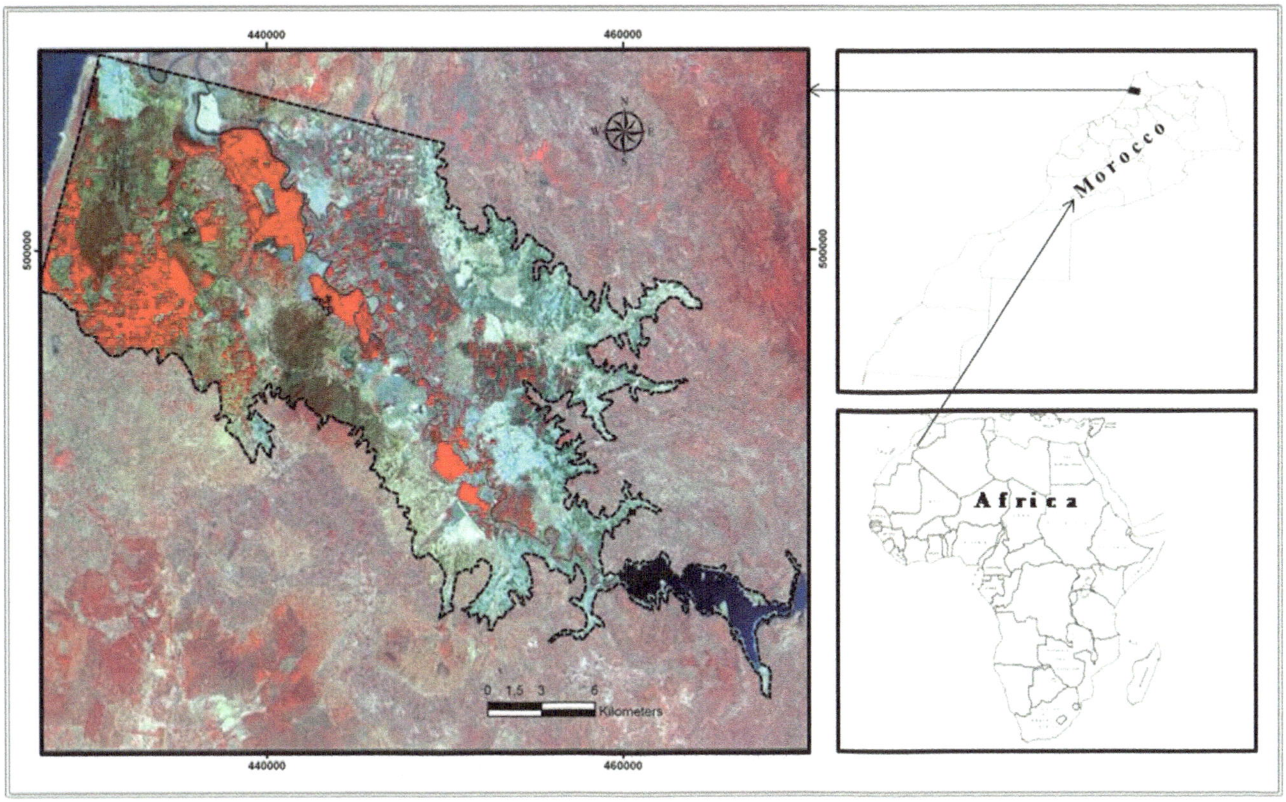

Fig. 1 Study area

Accordingly, and the preprocessing, images were interpreted on the basis of multi-source out the key of interpretation. The latter was used to assess changes in Land Cover through "Map Difference" preclassification process [11] and the old file date that is subtracted from the recent, to quantify the change detection occurred between two periods (1979–2001), (2001–2007) and (2007–2013). Another process of Object-Oriented Classification was conducted, using the shapefiles of the interpreted images and those derived from Object-Oriented classification, and in order to assess the accuracy, interpreted files were converted to the raster format according to the analogue resolution of each image, and Map Difference process was used to calculate accuracy cell-by-cell.

4 Pre-processing

The Landsat imagery was firstly re-projected to the Lambert Conic Conform projection system and georeferenced using the topographic maps of (El Ksar Kbir and Larache at the scale of 1:100 000) and corrected applying a first-order polynomial model and a nearest neighbor algorithm. A subset covering the study area was deed using a DEM, considering the average altitude varying from −6 to 33 m above sea level. While it is a generally flat area, the ortho-correction of the imagery has been neglected. The two images acquired from Landsat 8 were appended on a mosaic to cover the study area.

5 Results

5.1 Segmentation Results

Segmentation algorithms are used to subdivide entire images at a pixel level, or specific image objects from other domains into smaller image objects [13]. Many different approaches have been followed. A Multi-resolution segmentation algorithm was selected to be applied during the present study.

It is one basic procedure in the software eCognition for object-oriented image. It is used here to produce image object primitives as a first step for a further classification and other processing procedures. Objects of interest typically appear on different scales in an image simultaneously. The extraction of meaningful image objects needs to take into account the scale of the problem to be solved [24,25]. Being as an optimization procedure which, for a given number of image objects, minimizes the average heterogeneity and maximizes their respective homogeneity; in this circumstance, homogeneity is used as a synonym for minimized heterogeneity. Internally, three criteria are computed: color, smoothness, and compactness [26].

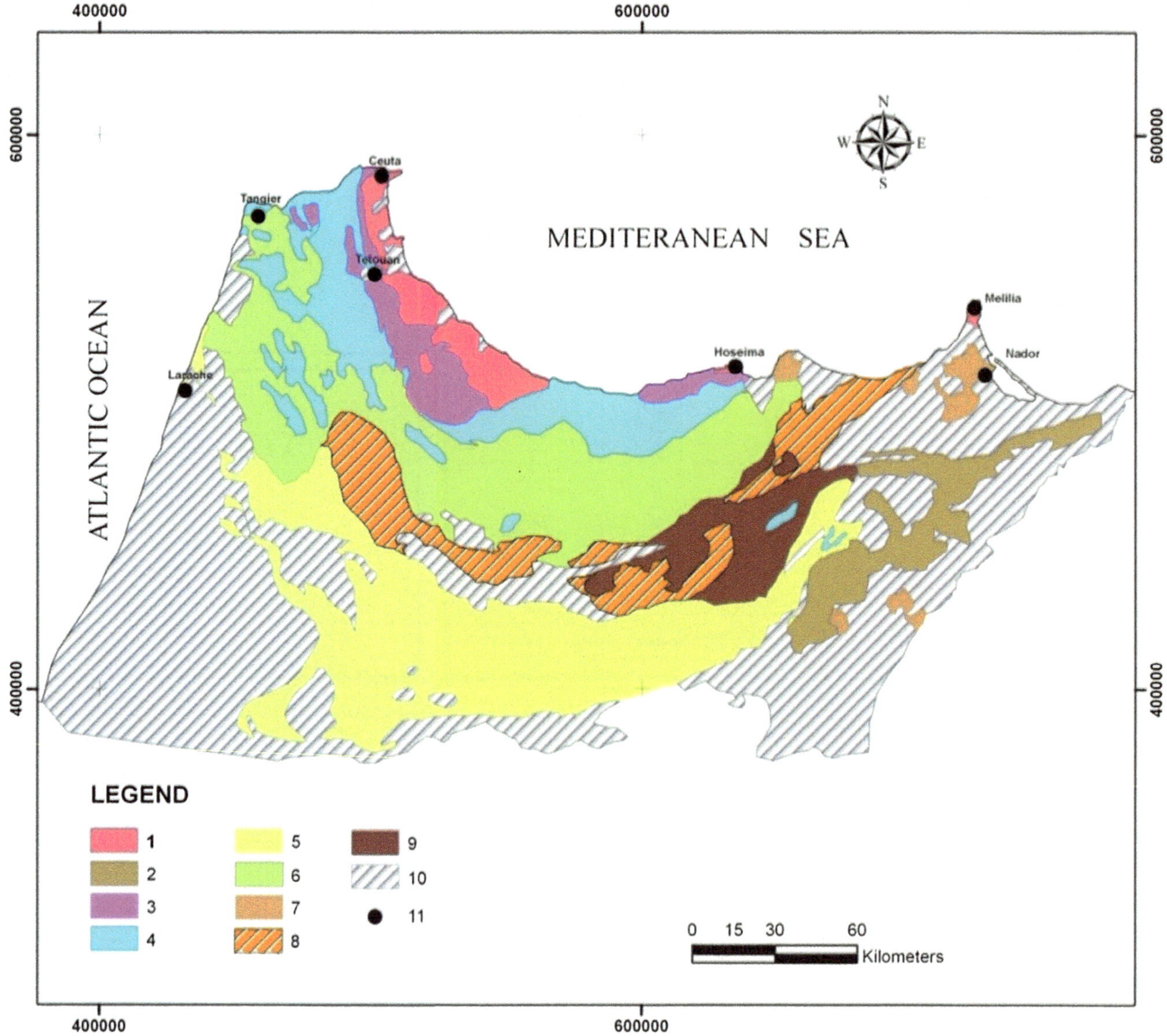

Fig. 2 Geological setting of Rifian belt. 1: Sebtides and Ghomarides; 2: foreland; 3: limestone ridge; 4: flysch thrusts; 5: pre-Rif area; 6: Tangier–Ketama unit; 7: Neogene to quaternary

For each one of the tree images used for the tests, a segmentation process was conducted using different combinations between the two criteria: color defined as digital values (radiance) of the image objects and shape (smoothness, and compactness) as the textural homogeneity of the resulting objects, measured by the scale variable. Table 2 presents the spatial and spectral characteristics of the images and the results of the best significant combination of the scale and the level of color used. For the given data, a level of color 0.9 was the most effective to represents all the five classes interpreted, and best scales are successively for the sensors MSS, ETM and SPOT images with 5, 25, and 30 was used. To define the determinants those moderates the number of objects during the segmentation and classification processes, the coefficients of correlation (R^2) between the image characteristics and objects number produced, was calculated. The Table 3 presents the results. According to the above data and considering the values of the correlation coefficient (R^2) exceeding 0.8, the number of objects is strongly linked to firstly; the Standard Deviation value of the image (0.99), and secondly, it is inversely linked to the spatial resolution of the image (−0.82).

Furthermore, and as defined in the multi-resolution algorithm, the value of the coefficient of correlation is equal to 1.00, confirms the linear relationship between the scale and the mean spectral values of the image, and for the spatial resolution of which the correlation coefficient is very high; recording a value of −0.98.

Table 1 Satellite images details

Date	Satellite	Instrument	Pixel size	Number of bands
15/12/2013	Landsat 8	OLI TIRS	30 m; 15 m for panchromatic band	11
08/12/2013	Landsat 8	OLI TIRS	30 m; 15 m for panchromatic band	11
22/08/2007	SPOT 5	HRG 2	10 m; 20 m – 5 m for panchromatic band	4
06/06/2001	Landsat 7	ETM+	28.5 m – 14.25 m for panchromatic band	8
15/05/1979	Landsat 2	MSS	62 m	4

Table 2 Segmentation results

Image info				Segmentation		Classification
Sensor	Date	Spectral mean values	Standard deviation	Color	Scale	Objects number
MSS	1979	27	4	0.9	5	531
ETM	2001	102	9	0.9	25	723
SPOT	2007	122	32	0.9	30	4417

Table 3 Correlation coefficients (R^2)

	Resolution	Spectral mean value	Standard deviation	Scale	Objects number
Resolution	1	−0.98	−0.88	−0.98	−0.82
Spectral mean value		1	0.78	1.00	0.70
Standard deviation			1	0.77	0.99
Scale				1	0.69
Objects number					1

5.2 Classification Results

Two parallel processes of classification were triggered to allow a subsequent comparison of Land Cover maps accuracy and quality.

The available thematic maps and a knowledge expert system, were explored, with aim to improve the image interpretation and the definition of the interpretation key, which will be used as a reference of the Object-Oriented classification. Each of the images was interpreted visually referring to many sources of information (Google Maps—Geoeye Quick looks-, MedWet maps, Land Cover maps, Spot 5 image with 2.5 m resolution, and topographic maps) and simultaneously digitalized, the maps having shape format, represented a referential data base of the change detection process, the digitalized features representing the interpreted Land Cover classes for each date of data (1979, 2001, 2007 and 2013); were converted to raster images with the corresponding resolution of the images interpreted.

In the object-based classification, each classification task addresses a certain scale. The image information can be represented in different scales based on the average size of image objects and the same imagery can be segmented into smaller or larger objects. Separately on the data collected for the 4 years, an object-based classification was conducted using the eCognition software. The classification process was deed on the base of the key interpretation, which describe five classes: "Water", "Agriculture and natural vegetation", "Forest", "Wetlands, Grasslands and herbaceous", and "Urban" (Fig. 3).

The Hierarchical Classification algorithm (Fig. 4), evaluates the membership value of an image object against a list of selected classes. Classes without a class description are assumed to have a membership value of 0. Class-related features are considered only if explicitly enabled by the according parameter Definiens Developer (2011).

5.2.1 Accuracy Assessment of Object Oriented Classification

Estimating the accuracy of maps derived from remotely sensed data requires statistical sampling [27], estimated by the Kappa statistic (K) [28]. Differently, we have tried to use a determinant accuracy assessment method, based on map difference using the raster calculator implemented in

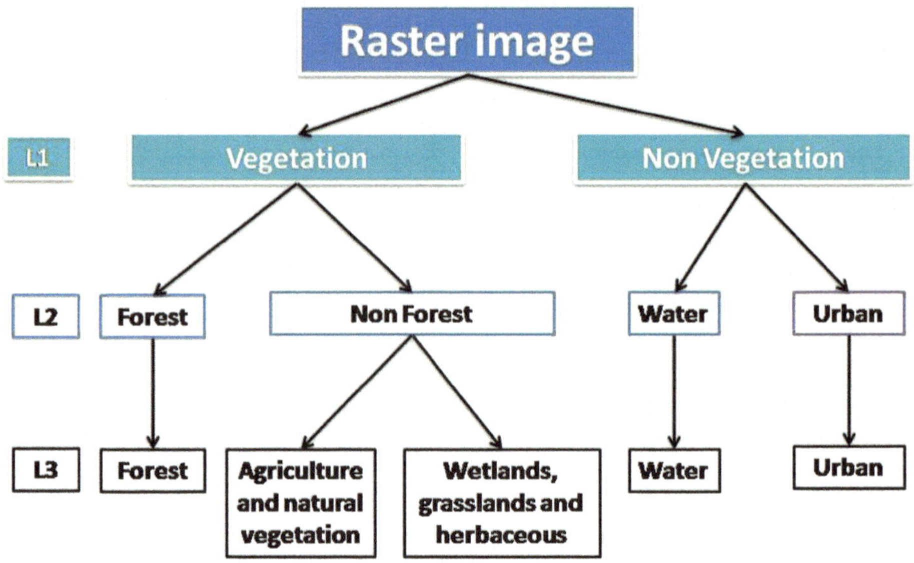

Fig. 3 Hierarchical scheme of the classification processing

ArcGIS Software, which allows to create and execute a "Map Algebra" expression; that will output a resulted raster. Subtraction operator was used to extract the 'compatible' areas, between Object-Oriented classification raster maps and interpreted maps converted to the raster format; the process uses the pixel count, as variable. The accuracy is given by a ratio of compatible pixel count, divided by total pixel count of the image (Table4). The classification accuracies derived from ratio calculation are listed in table under.

5.2.2 Multi-temporal Results

The Tables5, 6 and 7 show the areas and the percentage of multi-temporal Land Cover change successively occurred between the first period 1979–2001, the second 2001–2007 and the third (2007–2013). The process was deed on the bases of the image interpretation to estimate the Land Cover change from 1979 to 2013 in the area of study. The intersected no changed classes areas was calculated, and the changed areas were deducted, using the Intersect tool implemented in the ArcGis (ArcInfo) software, that creates a new coverage by overlaying the features from the input coverage and intersect polygon coverage ArcGIS 10, help (2010).

17% (9074 ha) of the Land Cover classes were changed, from 1979 to 2001; composed principally of the change occurred on 42.3% of the class "Agriculture and natural vegetation" to "Urban" class, 23.8% of the "Wetlands, Grasslands and herbaceous" class to the "Agriculture and natural vegetation" and 11.2% of the class "Forests" was transformed to the "Agriculture and natural vegetation" class.

Furthermore, between the years 2001 and 2007; 3.72% (2029 ha) of the Land Cover classes have been transformed to the news. Where the most significant changes are: 43.02% (877 ha) of "Agriculture and natural vegetation" class has been changed to the "Urban" class, 20.9% from the "Wetlands, Grasslands and herbaceous" to the "Agriculture and natural vegetation", 16% (323.9 ha) of the class "Water" has been changed to the "Agriculture and natural vegetation" Land Cover class. Between the years 2007 and 2013, 63.3% the "Agriculture and natural vegetation" class, was shifted to the "Urban" class. Important rate (9.6%) of the "Agriculture and natural vegetation" class was converted to the "water" class, 7% of the "Wetlands, Grasslands and herbaceous" was converted to the "Agriculture and natural vegetation", and 6.1% of the forests was converted to agricultural areas. The studied area represented approximately 54464 ha.

5.3 Discussion

5.3.1 Segmentation

To explore the results of the Object-Oriented classification, we began by surrounding the fundamental process of image segmentation; this by evaluating the correlations between intrinsic image information (StdDev, Mean, Color level and scale) and the number of objects produced. To do this, we have established several combinations tests of color levels (0.1–0.9), with levels of scale (5–250). According to the results above, the number of objects is more strongly related to the values of the Standard Deviation of the images ($R^2 = 0.99$)

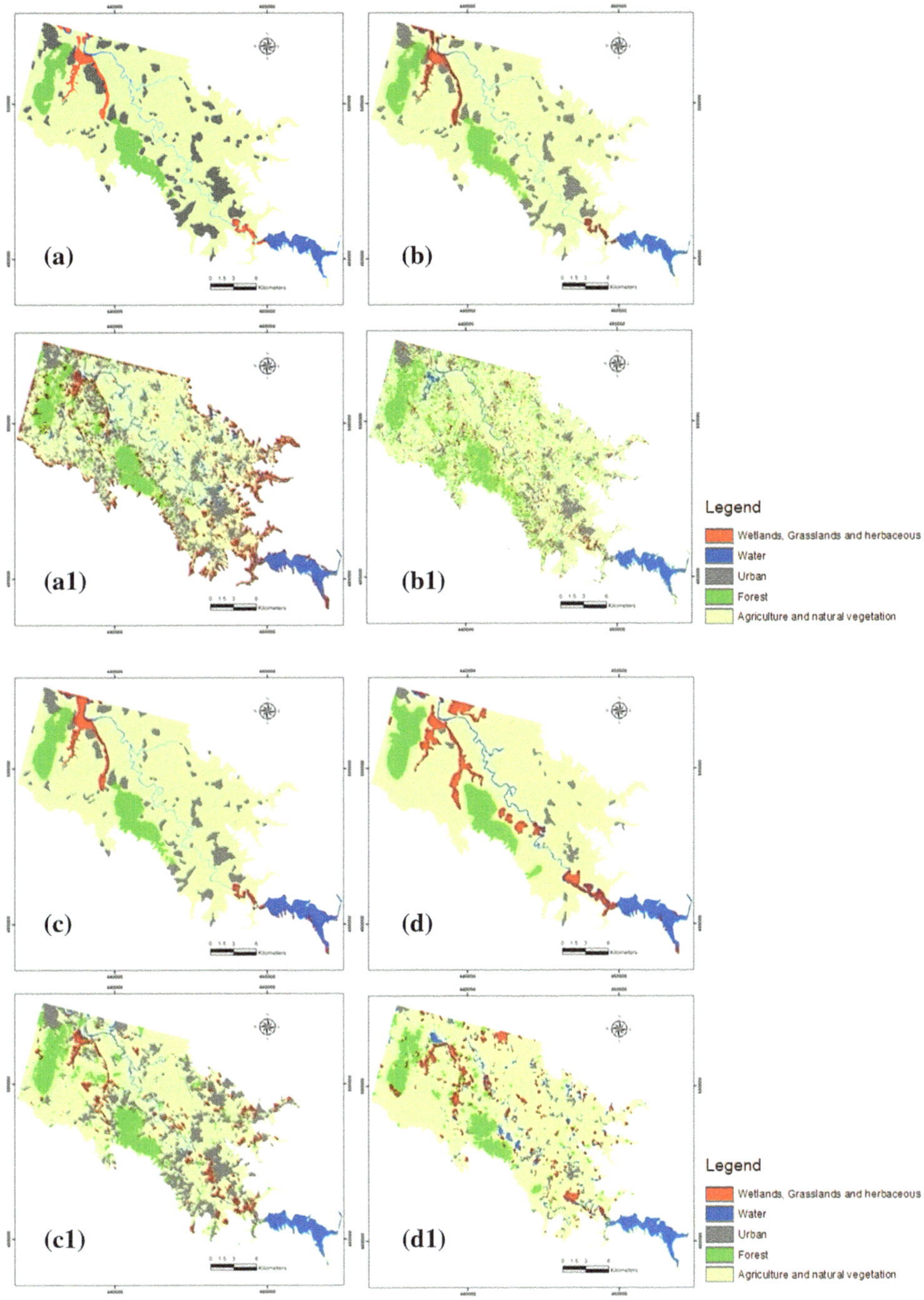

Fig. 4 Classification comparative results: (**a**: O.O. classification map of OLI TIRS image, **a1**: interpretation map of OLI TIRS image, **b**: O.O. classification map of SPOT image, **b1**: interpretation map of SPOT image, **c**: O.O. classification map of ETM image, **c1**: interpretation map of ETM 2001 image), **d**: O.O. classification map of MSS image, **d1**: interpretation map of MSS image)

Table 4 Object oriented classification accuracy results

Image	Total pixel count	Compatible pixel count	Ratio % (accuracy)	Pixel resolution (m)
SPOT 2007	7782724	5688904	73.10	8
ETM 2001	2772367	2103335	75.87	14
MSS 1979	141714	121471	85.72	62
ETM 2013	558044	480666	86.13	30
Global mean accuracy	80.20			

Table 5 Land Cover changes between 2001 and 2007 derived from the interpreted images SPOT and ETM

Changed classes		Area (ha)	% of change
Agriculture and natural vegetation	To urban	3838.8	42.3
Agriculture and natural vegetation	To wetlands, grasslands and herbaceous	392.4	4.3
Agriculture and natural vegetation	To water	63.0	0.7
Agriculture and natural vegetation	To forests	167.4	1.8
Forests	To agriculture and natural vegetation	1050.5	11.6
Forests	To urban	121.4	11.2
Urban	To agriculture and natural vegetation	332.2	3.7
Water	To agriculture and natural vegetation	472.7	5.2
Water	To urban	1.3	0.0
Water	To wetlands, grasslands and herbaceous	219.7920217	2.4
Wetlands, grasslands and herbaceous	To agriculture and natural vegetation	2159.8	23.8
Wetlands, grasslands and herbaceous	To urban	212.1	2.3
Wetlands, grasslands and herbaceous	To water	42.7	0.5
	Total	9074	100
No changed classes			
Wetlands, grasslands and herbaceous		1180.2	2.6
Forests		4337.0	9.6
Agriculture and natural vegetation		36617.1	80.7
Urban		1088.4	2.4
Water		2165.9	4.8
Total		45388	100
54464			

which represents the distribution of the brightness and not only fixed values of the image (pixel value); it describes the way this brightness is repartees on the pixels. The second factor is the image resolution, the correlation between the values of this variable is significantly stronger with the number of objects ($R^2 = -0.82$).

5.4 Accuracy Assessment of the Object-Oriented Classification

To calculate the accuracy of the classification and differently to the classical probabilistic; Monte Carlo methods we have attempted to test the method of "Map Difference" to estimate more precisely pixel by pixel in a deterministic way, precision, we used the raster calculator implemented in ArcGIS Software, Which Allows to create and execute a "Map Algebra " – (Minus) expression. The – (Minus) operator is an overloaded operator, it subtracts the values of the second input raster from the values of the first input raster cell -by -cell. Therefore, the overall accuracy of the compatible raster pixels between Interpreted and Object-Oriented Classification raster was 80.20%, which can be considered as important, since the classification process had been conducted thoroughly, and the precision depends on pixel resolution that increases for low resolution (Table 4 and Fig. 3).

5.4.1 Multi-temporal Results

This work aimed to estimate the Land Cover change in the study area. For reasons of simplifying the Object-Oriented classification and get overall results of the changes of classes in the study area. We used the shape files of the interpreted classes, and we grouped these classes, into five major categories (Fig. 5) as mentioned above. "Intersect" tool Implemented in the ArcGIS (Arc Info) software was used to assess the changes occurred between years 1979 and 2013 (Table 5). Thus, among the 16.7% (9074 ha) of the Land Cover classes were changed; this change is composed by the shift of 42.3% of agricultural lands was converted into "Urban" class; which we define as urban or rural agglomeration. Followed by the

Table 6 Land Cover changes between 2001 and 2007 derived from the interpreted images SPOT and ETM

Changed classes (2001–2007)		AREA (ha)	% of change
Agriculture and natural vegetation	To urban	877.0	43.2
Agriculture and natural vegetation	To wetlands, grasslands and herbaceous	8.7	0.4
Agriculture and natural vegetation	To water	0.1	0.0
Agriculture and natural vegetation	To forests	77.1	3.8
Forests	To agriculture and natural vegetation	287.2	14.2
Forests	To urban	3.0	0.0
Urban	To forests	1.0	
Urban	To agriculture and natural vegetation	22.1	1.1
Water	To agriculture and natural vegetation	323.9	16.0
Water	To urban	0.0	0.0
Water	To wetlands, grasslands and herbaceous	4.2	0.2
Wetlands, grasslands and herbaceous	To agriculture and natural vegetation	423.9	20.9
Wetlands, grasslands and herbaceous	To urban		0.0
Wetlands, grasslands and herbaceous	To water	1.0	1.0
	Total	2029	100
No changed classes			
Wetlands, grasslands and herbaceous		1369.9	2.6
Forests		4213.7	8.0
Agriculture and natural vegetation		39728.1	75.8
Urban		5179.7	9.9
Water		1944.2	3.7
Total		52435	100
54464			

change occurred on 23.8% of the wetlands those were converted to agriculture's. Finally, a deforestation process has caused 11.6% of change in favor to agricultural activities.

It should be noted that the change occurred on farmland has been notably increased between 2007 and 2013, reflecting the development of the agricultural activities, that the Loukkos basin is characterized. This development was accompanied by urban expansion of the towns and rural villages (Fig. 6).

By cons, it was noted that the accuracy increases for lower resolutions (MSS 73.10%). rather, in our opinion, it is that the hierarchical classification scheme (key interpretation), which had a significant effects on the results and therefore not the pixel resolution which is directly has influenced this accuracy, also, other parameters related to classes previously defined, those have; so the class of "Wetlands, Grasslands and Herbaceous" is a complex one, consisting of lakes, mudflats, salt marshes and wetland vegetation, this class has been confused with other classes such as "Agriculture and natural vegetation" and "water", during the classification process.

The class of "Agriculture and Natural vegetation" had intrusions with "forests" class and with "Wetlands, Grasslands and Herbaceous", this confusion is probably due (Fig. 3) to the complexity of the classes; where some classes included heterogeneous patches having different spectral and spatial reaction, could be the most favorable factor that explains the heterogeneity of those results. Furthermore, and since presented in the Table 8, the rate of annual Land Cover change, have decreased during the three periods.

The occurred transformations in the study area between 1979 and 2013 appear geographically coherent, if we take the fact that the population of the area has intensified with +0.7% [29] causing the expansion of the urban space, in this direction we have noted the proliferation of villages with wider areas, and the expansion of the two biggest cities included in the study area called Larache and El Ksar Kbir (Fig. 5) On the other hand; and in consideration of the situation that the perimeter of the Loukkos area, was fitted in 1970 as part of the national hydro-agricultural policy . This dynamic has boosted agricultural activity and therefore has resulted transgressions

Table 7 Land Cover changes between 2007 and 2013 derived from the interpreted images SPOT and OLI TIRS Land Cover changes between 2007 and 2013 derived from the interpreted images SPOT and OLI TIRS

Changed classes (2007–2013)		AREA (ha)	% of change
Agriculture and natural vegetation	To urban	1031.1	63.3
Agriculture and natural vegetation	To wetlands, grasslands and herbaceous	52.1	3.2
Agriculture and natural vegetation	To water	156.7	9.6
Agriculture and natural vegetation	To forests	30.2	1.9
Forests	To agriculture and natural vegetation	99.6	6.1
Forests	To urban	51.3	0.1
Urban	To agriculture and natural vegetation	23.1	1.4
Water	To agriculture and natural vegetation	32.5	2.0
Water	To urban	0.0	0.0
Water	To wetlands, grasslands and herbaceous	6.6	0.4
Wetlands, grasslands and herbaceous	To agriculture and natural vegetation	114.2	7.0
Wetlands, grasslands and herbaceous	To urban	20.6	1.3
Wetlands, grasslands and herbaceous	To water	11.5	0.7
	Total	1629	100
No changed classes			
Wetlands, grasslands and herbaceous		1236.5	2.3
Forests		4140.9	7.8
Agriculture and natural vegetation		39515.1	74.8
Urban		6036.6	11.4
Water		1906.3	3.6
Total		52835	
54464			

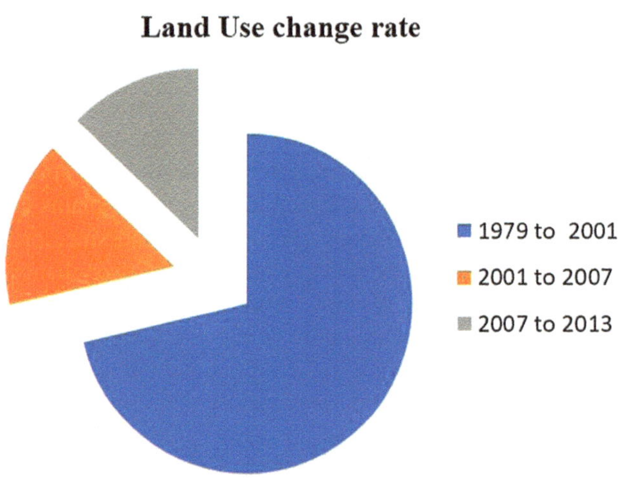

Fig. 5 Proportions of Land Use change during the tree periods

on wetlands and natural areas. A deforestation is considered third determinant of change in Land Cover, probably because the main human activity is agriculture and not grazing and timber harvesting and also probably because the forests in the area have been under protection.

The occurred transformations in the study area between 1979 and 2013 appear geographically coherent, if we take the fact that the population of the area has intensified with +0.7% [29] causing the expansion of the urban space, in this direction we have noted the proliferation of villages with wider areas, and the expansion of the two biggest cities included in the study area called Larache and El Ksar Kbir (Fig. 5) On the other hand; and in consideration of the situation that the perimeter of the Loukkos area, was fitted in 1970 as part of the national hydro-agricultural policy . This dynamic has boosted agricultural activity and therefore has resulted transgressions on wetlands and natural areas. A deforestation is considered third determinant of change in Land Cover, probably because the main human activity is agriculture and not grazing and timber harvesting and also probably because the forests in the area have been under protection.

Table 8 Rates of Land Cover changes between 1979 and 2013

Period	Years	Land Cover change rate	Annual Land Cover change rate
1979–2001	22	16.7	0.8
2001–2007	6	3.7	0.6
2007–2013	6	3.0	0.5

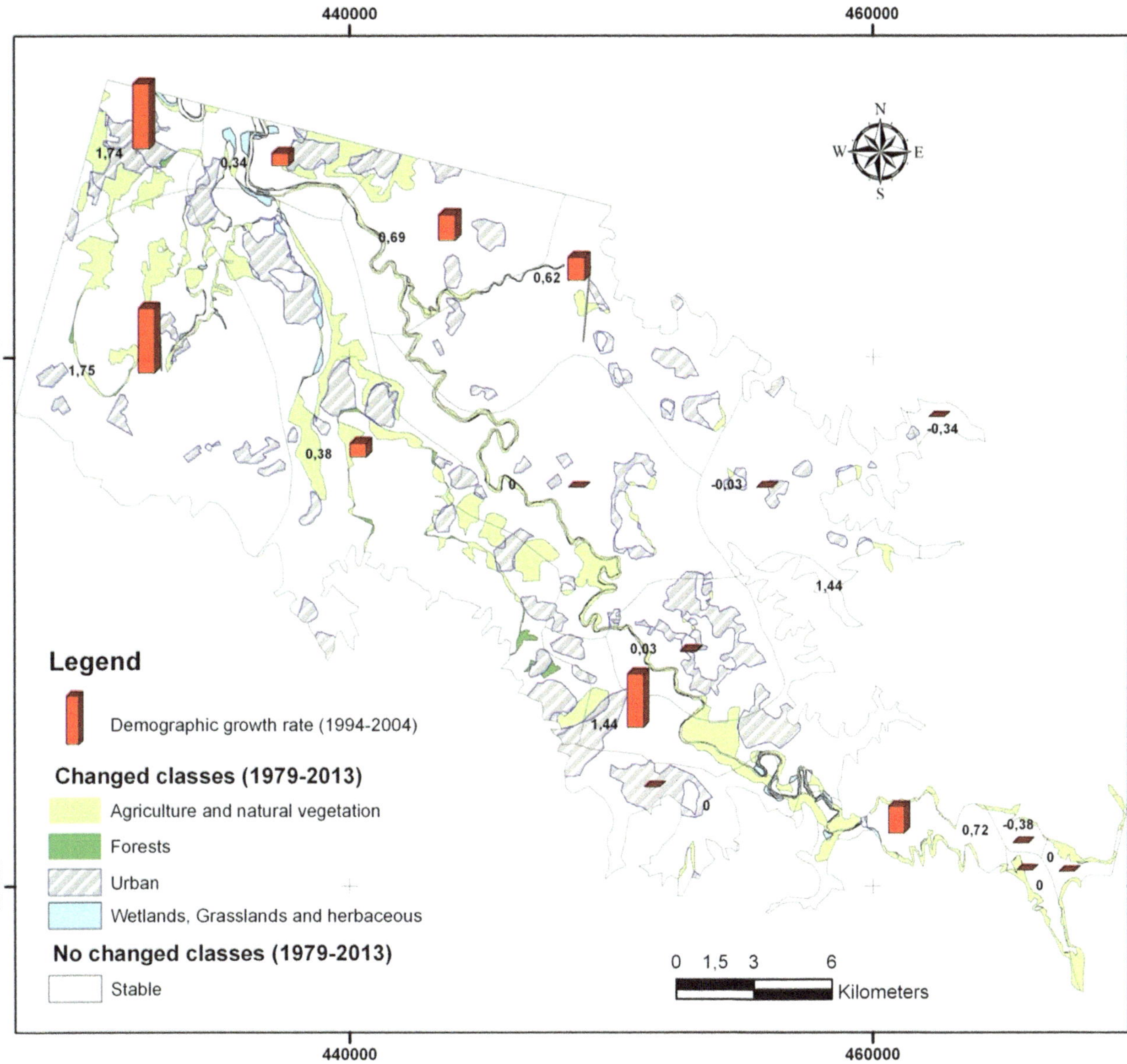

Fig. 6 Land Use change from 1979 to 2013 relatively to the demographic grows trends by towns [29]

6 Conclusions

The segmentation tests established on the images; based on several combinations of color levels (0.1–0.9) and scales, have confirmed a strong relationship between the Spectral values of images, specially the Standard Deviation and the number of objects is more strongly related to the values of the Standard Deviation of the images ($R^2 = 0.99$) which represents the distribution of the brightness (radiance) and not only fixed values of the image (pixel value); it means of how this brightness is distributed on the pixels. The second factor is the image resolution, the correlation between the values of this variable is significantly stronger with the number of objects ($R^2 = -0.82$). Classification tests performed on the selected images in order to best approximate the accuracy of this process, the assessment was established on the base of a deterministic approach, towards the classical probabilistic methods of Monte Carlo. Which method is based on the "Map Difference" cell -by -cell between raster images classified and those interpreted, after unified the resolution of the input and output images. The accuracy is given by the ratio of compatible pixels between inputs raster. We believe that the Map difference is made less complicated and more accurate. For this type of calculation, we do not need to look for samples. Furthermore, the overall accuracy of the Object-Oriented Classification was 80.20%, in our case it was more significant for the low resolutions, probably as explained above, the result was not determined by the purpose of pixel effect (high resolution more accurate), but by the complexity of the key interpretation of the classes, where some classes included heterogeneous patches having different spectral and spatial reaction. The study of the Land Cover evolution, has shown that 16.7% of Land Cover was changed between 1979 and 2001, 3.7% between 2001 and 2007 and 3% between 2007 and 2013. Between 42 and 63% of agricultural and vegetation natural lands, have been converted into urban space, between 7 and 24% of wetlands, grasslands and herbaceous was changed to agricultural lands, and between 1 and 11% of forest lands have been changed to agriculture and natural vegetation class. The development of hydro-agricultural infrastructures implemented in the Loukkos basin since 1970, under the national policy plan, and the global population growth in this area; were the two main important reasons for this evolution. This is said to conclude that the interpreted and classified Land Cover classes, have to be more clear and distinct on both the spatial and spectral levels that confirms the conclusion of [30]; who has said that the rules used to label the map must be rigorous and well defined. The use of the post-classification [11] approach using multi-temporal and multi-source data, in the study of the Land Cover must be accompanied by a good pre-processing, the right establishment of the interpretation key, that should reflect the reality on the one hand and on the other hand take into account the spectral and spatial harmony classes and categories of land cover interpreted on the base images. Furthermore, the segmentation must be conducted properly and carefully while choosing the best combinations possible between the levels of color and spatial scales and to enhance the accuracy of the maps produced.

References

1. Achbun, A., Mansour, M., Layelman, M., & Smiej, F. (2011). Etude comparative de la classification orientée objet d'une image spot5 pour la cartographie de l'occupation du sol via eCognition® 9, Géoobservateur, 19. ArcGis 10, help. http://help.arcgis.com/en/arcgisdestop/10.0/help/.
2. Mastere, M., Van Vliet Lanoë, B., & Brahim, L. A. (2013). Land Cover mapping and its relation to mass wasting and gullying in North-Western Rif (Morocco). *Geomorphology: Relief, Process, Environment*, *3*, 335–352.
3. Chintan, A. S., Arora, M. K., & Pramod, K. V. (2004). Unsupervised classification of hyperspectral data: An ICA mixture model-based approach. *International Journal of Remote Sensing*, *25*, 481–487.
4. Benz, U. C. (2004). Multi-resolution object-oriented fuzzy analysis of remote sensing data for GIS-ready information. *ISPRS Journal of Photogrammetry and Remote Sensing*, *58*(3–4), 239–258.
5. Achbun, A., Mastere, M., & El Fellah, B. (2015). Up-scaling approach for a multi-series object based classification using CORINE typology. *European Journal of Scientific Research*, *132*(2), 107–117. Pau et des pays de l'Adour, 308 p. (1991).
6. Cleve, C., Kelly, M., Kearns, F. R., & Moritz, M. (2008). Classification of the wildland urban interface: A comparison of pixel- and object-based classifications using high-resolution aerial photography. *Computers, Environment and Urban Systems*, *32*(4), 317–326.
7. Corcoran, P., & Winstanley, A. (2008). Using texture to tackle the problem of scale in land-cover classification. In T. Blaschke, S. Lang, & G. Hay (Eds.), *Object-based image analysis* (113–132). Berlin: Springer.
8. Zhou, W., Troy, A., & Grove, M. (2008). Object-based Land Cover classification and change analysis in the Baltimore metropolitan area using multi-temporal high-resolution remote sensing data. *Sensors*, *8*(3), 161–163.
9. George, P. P., Chariton, K., & Krishna, P. V. (2012). Support vector machines and object-based classification for obtaining Land-Use/cover cartography from Hyperion hyperspectral imagery. *Computers and Geosciences*, *41*, 99–107.
10. Hájek, F. (2008). Process-based approach to automated classification of forest structures using medium format digital aerial photos and ancillary GIS information. *European Journal of Forest Research*, *127*(2), 115–124.
11. Singh, A. (1989). Digital change detection techniques using remotely-sensed data. *International Journal of Remote Sensing*, *10*, 989–1003.
12. Lunetta, R., & Elvidge, C. (1999). *Remote sensing change detection*. London: Taylor & Francis, 320 p.
13. Berberoglu, S., & Akin, A. (2009). Assessing different remote sensing techniques to detect Land Cover/cover changes in the eastern Mediterranean. *International Journal of Applied Earth Observation and Geoinformation*, *11*(1), 46–53.
14. Thauvin, J. P. (1979). Le basin du bassin Loukkos (Extrait de ressources en eau du maroc. TomeI: Domaine du rif et du Maroc oriental). *Notes.et Mémoires. Service. Géooogiquel. Maroc*, *231*, 113–125.
15. El Morhit, M. (2009). *Hydrochimie, éléments traces métalliques et incidences Ecotoxicologiques sur les différentes composantes d'un écosystème estuarien (Bas Loukkos)*. Thèse Doctorat ès science, Université Mohammed V de Rabat, 232 p.
16. El Kellouti, M. (2004). Modernisation de la gestion de l'irrigation dans le périmètre du Loukkos (Maroc). Projet INCO-WADEMED Actes du Séminaire Modernisation de l'Agriculture Irriguée (pp. 2–6). Rabat.
17. Durand-Delga, M. (1963). Essai sur les structures des domaines émergés autour de la Méditerranée occidentale. *Geologische Rundschau Stuttgart*, *5-3-2*, 534–535.
18. Kornprobst, J. G. (1974). Contribution à l'étude pétrographique et structurale de la zone interne du Rif (Maroc septentrional). *Notes et Mémoires. Service. Géologique. Maroc*, 251–256.
19. Maychou, S. (2009). *Étude morphostructurale et cartographie SIG du Rharb Septentrional et du Prérif (Maroc). Analyse sismotectonique et modélisation de la déformation de la région de Moulay Bousselham*. Thèse Doctorat ès science, Université Chouab Doukkali, Al Jadida (pp. 20–26).

20. Rampnoux, J. P., Angelier, J., Colleta, B., Fudral, S., Guillemin, M., & Pierre, G. (1979). Sur l'évolution néotectonique du Maroc septentrional. *Géologie méditerranéenne*, *6*, 439–464.
21. Lespinasse, P. (1975). *Géologie des zones externes et des flyschs entre Chaouen et Zoumi (Centre de la chaine rifaine, Maroc)*. Thèse. Université (Vol. VI). Paris, 248 p.
22. Ben Yaich, A. (2011). *Evolution tectono-sédimentaire du Rif externe Centro-Occidental (région de M'sila et Ouazzane, Maroc): La marge africaine du Jurassique au Crétacé, les bassins néogènes d'avant-fosse*. Thèse Doctorat dEtat, Université de.
23. Global Land Use facility database. GLCF. http://glcfapp.glcf.umd.edu:8080/esdi/.
24. Weibel, R., & Jones, C. B. (1998). Computational Perspectives On Map Generalisation. *Geoinformatica*, 301–304.
25. Baatz, M., & Schape, A. (2000). Multiresolution segmentation: An optimization approach for high quality multi-scale image segmentation. In T. Strobl, T. Blaschke, G. Griesebner (Eds.), *Angewandte Geographische Informationsverabeitung. Beitragezum AGIT -Symp. Salzburg, Karlsruhe* (Vol. XII, pp. 11–22).
26. Definiens developer, software (2014). http://www.definiens.com/.
27. Card, D. H. (1980). Using known map categorical marginal frequencies to improve map accuracy. *Photogrammetric Engineering and Remote Sensing*, *48*(43), 1–439.
28. Cohen, J. A. (1960). Coefficient of agreement of nominal scales. *Educational and Psychological Measurement*, *20*, 37–46.
29. Recensement Général de la Population et de l'Habitat (RGPH), Haut-Commissariat au Plan, Maroc (1994–2004).
30. Congalton, R. G. (1991). A review of assessing the accuracy of classifications of remotely sensed data. *Remote Sensing of Environment*, *37*, 35–46.

Comparative Analysis of the Classification of Maximum Reality (MVS) and the Spectral Angle Mapper (SAM) of an Aster Image. Case Study: Soil Occupancy in the Main Area (Tunisia)

Sonia Gannouni and Noamen Rebai

Abstract

The vegetation cover of the area of Mateur (Tunisia) is characterized by the heterogeneity of its settlements. Such a heterogeneity is caused by the interaction of anthropic, pedological, and climatic factors. In addition, these space and spectral heterogeneities limit the reliability of the conventional methods of classification related to the satellite imagery. Thus, in the present study, we propose the recourse to the methods based on the spectral similarity to chart the dominant vegetable species of the ecosystem of Mateur, that is to say the Spectral angle mapper (SAM) and the classification of maximum of probabilities (MVS). We also aim at not only comparing procedures of extraction of the "pure" spectral signatures prototypes, known as endmembers for the SAM approach but also identifying the pieces of drives for the classification MVS in terms of the cartography of the dominant vegetable species of this area. For so doing, we have used images acquired by the sensor thermal (Advanced ASTER spaceborne emission and reflection radiometer. The results obtained show that the use of the methods of SAM and MVS led to similar results in terms of distribution of the species charted, but with differences in the plan of the surfaces affected by these species. The comparison between the results obtained using MVS and those of classification by maximum of probability indicates that SAM allows to classify the dominant vegetable cover with a better precision than MVS.

Keywords

Mapping • Endmembers • MVS and SAM

1 Introduction

Internet In the present work, our primary endeavor is to identify and map the dominant plant species in the region of Mateur using spectral similarity approaches, namely the Spectral angle map (SAM) and the maximum likelihood classification (MVS). Then, we will compare these methods in terms of distinguishing dominant plant species. Finally, we will compare the methods of selection of prototypical signatures, called endmembers, and the training plots and identify the one that is most suitable for the classification of satellite imagery of semi-arid environments. Clearly, the great variety of the physical environment of Tunisia, its geographical and climatic contrasts and its Mediterranean and Saharan double nature explain in large part the diversity and the richness of its ecosystems. In fact, from the point of view of remote sensing, this region is characterized by a double heterogeneity: one is spatial while the other is spectral. This represents a major challenge for the application of remote sensing methods to land cover classification and mapping. The reasons behind that are the effects of the optical properties of soils, atmospheric disturbances, and the effect of the environment and information at the subpixel level. Traditionally, the estimation of the vegetation cover rate from satellite imagery has either been through using conventional classification or vegetation indices [1–4]. The results obtained by these indices are often tainted with errors due to contamination from the atmosphere and underlying soils, [5]. The effect of the latter dominates especially in environments where vegetation is more or less sparse [6,7]. Furthermore, new developments in remote sensing tools have been an asset for quantifying and identifying different plant cover species [8]. Thus, the mapping of plant species has benefited from the development of new

S. Gannouni (✉)
Laboratoire de Géoressources, CERTE, Technopole de BorjCedria, Tunis, Tunisia
e-mail: gannounisonia2017@gmail.com

N. Rebai
LR14ES03 Geotechnical Engineering and Georisk Laboratory, National School of Engineering of Tunis, University of Tunis El Manar, BP. 37, le Belvédère 1002, Tunis, Tunisia
e-mail: noamen.rebai@enit.utm.tn

N. Rebai and M. Mastere (eds.), *Mapping and Spatial Analysis of Socio-economic and Environmental Indicators for Sustainable Development*, Advances in Science, Technology & Innovation, https://doi.org/10.1007/978-3-030-21166-0_5

approaches, some of which are based on artificial intelligence such as neural networks and cellular automata, while others are based on spectral similarities such as MVS and SAM [9]. The former, which are founded upon the concept of training and learning, have shown a high accuracy compared to the classification methods that are built around classical statistics [10–13]. As for the second type of approaches, they exploit the spectral similarity and the information at the subpixel level for the relative contribution of the elements. SAM has been widely used for the study of soil degradation and desertification [14–16]. The approach has also been implemented in the study of the degradation of arid natural environments and has yielded satisfactory results [17–23].

These methods rest on the spectral similarity between the different homogeneous components of the observed environment. The effectiveness of their application to the analysis and classification of images depends on the quality and representativeness of the prototype signatures selected to represent these homogeneous components. Thus, the most important step is the extraction of a limited number of prototype signatures that can show and explain the spectral variability in the image. For this purpose, several extraction methods may be employed:

(1) manual selection by the user, either from a spectra bank measured in the laboratory or in the field [24], or from the image itself [25,26];

(2) automatic selection based on one of the following approaches:

(a) radiometric indices PIR-R/PIR + R [27,28];

(b) Principal Component Analysis (PCA) [29];

(c) iterative error analysis (AEI) [30];

(d) PPI purity index (Pixel purity index) [31] and

(e) the PPI average (APPLE Average Purity) [32].

2 Presentation of the Study Area

The region of Mateur is 50 km away from Bizerte and 83 km from Tunis. It is located in North West Tunisia (Fig. 1). It corresponds to the first reliefs of the Hedils; located SW of the plain of Ichkeul. To the west and to the NW are WSW–ENE oriented limestone chains slid down to the massive limestone chain that leans on the northern edge of the Hédils plateau. The central part of Mateur comprises a succession of soft hills, interrupted SW, by the plateau of Jefna-Bordj El Arab. In the East, the wadi Ed Debab and El Jeladi valley bypass the rounded mass of Jebel Ballouta [33]. The hydrographic network of the basin of Mateur is formed by several wadis which cross the landscape such as wadi El Jeladi, wadi El Gouss, wadi EL Melah, and wadi Gzenni which drains the dike of the Jalta mine and empties in Lake Ichkeul. These different streams are heading toward the NE. They allow the irrigation of mainly cereal crops [34]. The region is provided with Mateur–Tabarka railway with Jalta and Jefna stations. In addition, the GP7 road, from Mateur to Tabarka, follows the valley of the El Gouss wadi. The mining sector itself is very well served by a regional road linking the mining village with the national road 7 for two kilometers.

3 Materials and Methods

3.1 Materials

During this Aster multispectral work, satellite data have been used. Their utility in plant or lithological mapping of large ensembles has been proven despite their low spectral resolution that is close to 0.1 m compared to hyperspectral data which resolution is close to 0.01 m. This high potential is due to the very good quality of the current multispectral data (S/N close to 500:1) compared to HYPERION hyperspectral data, which currently offer low quality data (S/N close to 50:1) combined with a spatial resolution of 15 m in the VNIR of the ASTER sensor. Its strong signal-to-noise ratio and its 6 spectral bands in the SWIR are special features that help increase the potential of lithological discrimination in comparison to LANDSAT (http://eo1.usgs.gov/ASTER.php).

3.2 Methods

The following methodological approach has been followed throughout this study (Fig. 2):

- application of different methods of training plots extraction (prototype parcels);
- classification of land use via the MVS;
- application of different methods of prototype signatures extractions;
- classification of land use via SAM.

Pretreatment of Satellite Images (Radiometric Correction)

ASTER images have been transformed into reflectances through following the classical image processing chain, i.e., the correction of not only the geometric distortions but also the atmospheric and radiometric effects. Satellite data are always provided as digital accounts (DN). However, the latter have no physical meaning. Hence, they need to be transformed into physical values such as luminance and reflectance, which are recorded at the level of the satellite sensors and which are called respectively the exo-atmospheric luminance the (L^*) and exo-atmospheric reflectance (ρ^*).

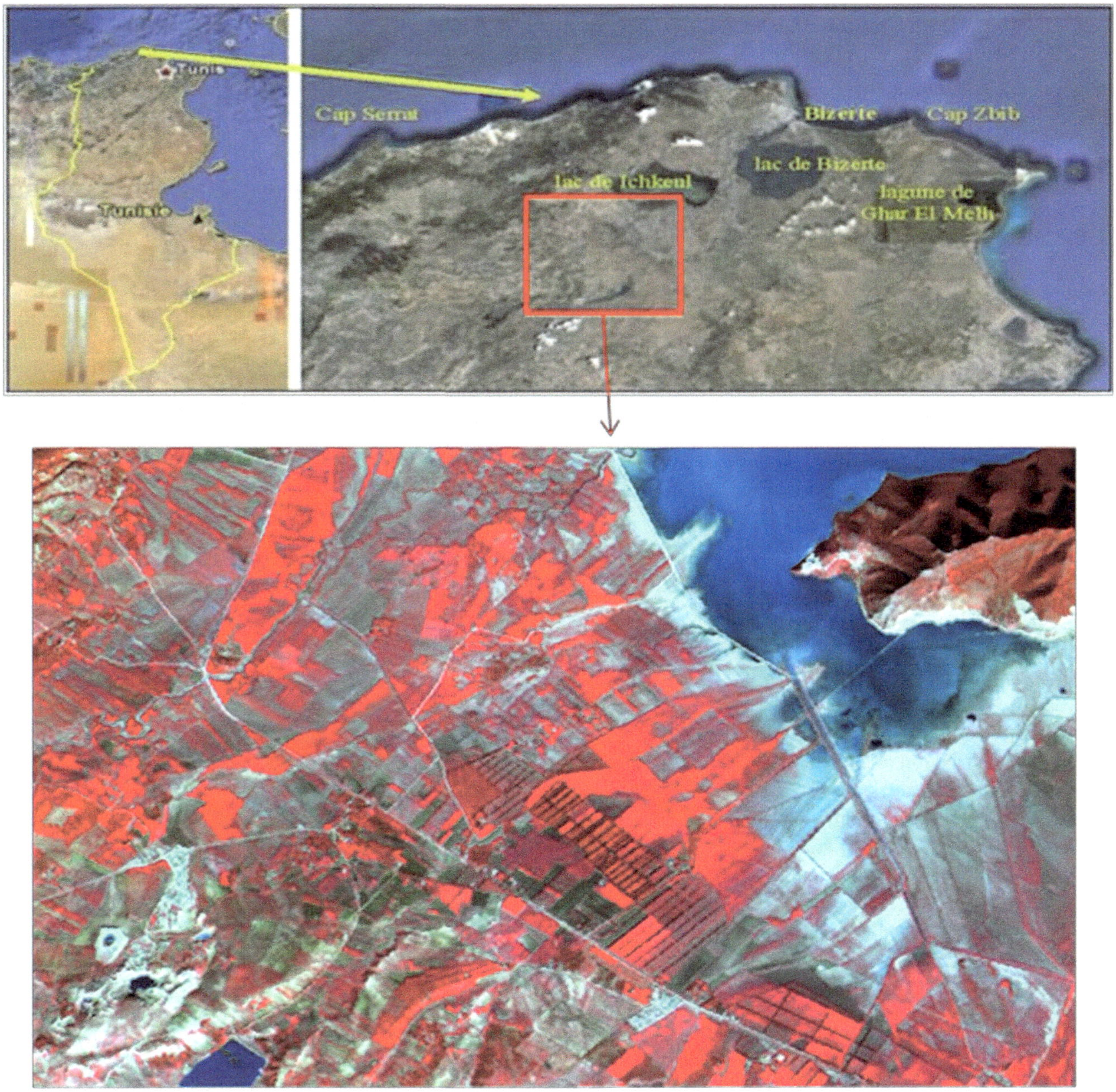

Fig. 1 The map of situation of the zone of study (Aster image of Mateur (Tunisia) taken in February, 2004)

- **Calculation of the exo-atmospheric luminance L***

The linear equation used to calculate the exo-atmospheric luminance L* of a band visible, near or medium infrared of an ASTER image, is:
L * = (DN value − 1) * Unit Conversion Coefficient
ASTER Unit Conversion Coefficients (watts/meter2/steradian/micrometer)/DN.

- **Calculation of the exo-atmospheric reflectance $\rho*$**

The values of the exo-atmospheric luminance are transformed into exo-atmospheric reflectance values. Indeed, we are more interested in reflectance than in luminance, because our objective is to carry out a comparison with the reflectance data acquired in the field. The relation that connects the two physical quantities is linear and expressed by the following equation:
$\rho * = (\pi, d2, L *)/(\text{Esun}\lambda \cos\theta s)$
ρ *: the exo-atmospheric reflectance
L *: spectral radiance
d: the earth–sun distance on the day of acquisition of the image
Esunλ: the average exo-atmospheric illumination. The unit is the w.m–2.μm–1
θ s: solar Zenithal angle in degrees.

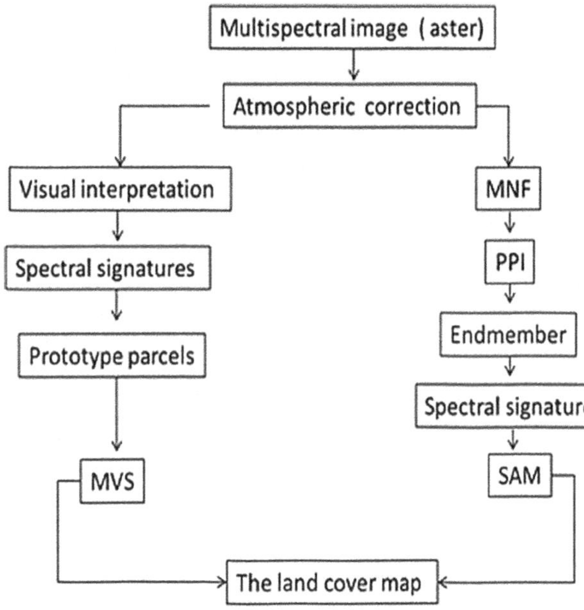

Fig. 2 The methodology of realization of the land cover map

- Methods of extracting information from imagery
- Maximum likelihood classification.

This classification makes possible the evaluation and more precisely the distribution of the reflectance values in a multidimensional space of values. However, it requires more time on the computer. The algorithm is a probabilistic technique based on Bayes' theoreme. It utilizes the statistical data of the different classes (variance, covariance, mean.) and assumes that the pixel values of each class are represented by a Gaussian distribution law. The algorithm introduces an equiprobability model, ie an "a priori" probability equal for each class [35,36]. According to Bayes' theoreme, the decision rule of assigning a pixel to a class minimizes the cost of the error:
P (C_i/X) = P (C_i/X) × P (C_i)/(P (X/C_i) × P (C_i)) with,
X = vector representing the radiometric values of a pixel.
P (C_i/X) = probability of appearance of the value X to the class C_i.
P (X/C_i) = probability density of class C_i.
P (C_i) = probability "a priori" of appearance at class C_i.

Classification Steps (MVS)
They incorporate:

- Visual examination of the colored composition highlighting different zones of homogeneous hues which describe the main lithological units. However, while some of them are similar, the others mix and merge. This is especially the case of populations of pixels with similar spectral responses and therefore very similar hues. The knowledge gained from the field helps us to differentiate between them.
- The establishment of the signature classes amounts to splitting the spectral space into several regions. Two fundamental conceptualizations loom up: [37].

- all objects belonging to the same class are characterized by identical signatures;
- all object class signatures are perfectly distinct from each other.

- Classification: We used the maximum likelihood algorithm.
- Estimation of the precision: This step is done by the evaluation of the matrix of confusion which allows to check the effective separability between the constituted classes, as well as to account for the quality of their description. The confusion matrix records, for each learning zone, the number of pixels assigned respectively in each of the classes, which makes possible the examination of their confusion.

Spectral Angle Mapping Method (SAM) Minimizing Remote Sensing Data
The amount of information contained in remote sensing images is considerably big. It is 120 MB for multispectral data covering an area of 60 × 60 km. The calculation time required to process these data is very important. Therefore, reducing the dimension of the study is an important step in image processing. Several dimensional reduction methods exist, namely, the Principal Component Analysis method and the Maximum Noise Fraction Transform method applied to hyperspectral data. The latter method will be applied because our study is based on hyperspectral analysis techniques for ASTER images.

Transformation into Maximum Noise Fraction
Principal Component Analysis applied to multispectral and hyperspectral images may lead to the relegation of useful information-bearing components to higher order principal components while noisy components are found in the first major ones. From this observation, [38] constructed a data reduction method called "Maximum Noise Fraction Transformation". This method arranges the new components in a new space, according to their degree of noise. It has been shown that a Maximum Noise Fraction Transformation is equivalent to a Principal Component Analysis when the noise variance is identical regardless of the spectral band.

The Index of Pure Pixels PPI
It is a technique of spatial data reduction. All remote sensing images contain a phenomenon known as mixed pixels. These are pixels that contain a mixture of materials (e.g., minerals, plants, water) [39]. In multispectral analysis, it is useful to separate the purest pixels from the mixed pixels in order to reduce the number of pixels to be analyzed and to guarantee the identification of endmembers. In fact, the pixel purity

index is a way of finding the most spectrally pure pixels in the images [29,31].

The PPI specifies how many times the pixel exists in extremes in the simplex [31,40,41]. The vertices of this simplex are the pure poles. Pure pixels typically correspond to spectrally unique materials. The PPI is the product of an image in which the pixel values correspond to the number of times that pixel has been recorded as an endpoint. In this way, the threshold of the PPI image can specify the most extreme pixel results.

Identification of Endmembers

Endmembers can be considered as points in a scattering dimensional parcel of n, where n is the number of bands [31]. The coordinates of the points in the n-space consist of the values of "n" which are simply the spectral values of radiance or reflectivity in each band for a given pixel. The distribution of these points in n-space can be used to estimate the number of spectral endmembers and their pure spectral signatures [42].

Spectral Angle Mapping (SAM)

The SAM method is a classification approach based on the angular similarity between the spectra of the image and reference spectra. Extracted from the image itself, the latter are extreme spectra, called spectral poles, with respect to which the SAM method measures the spectral angle of the pixels [9]. The latter and the spectral poles are considered as vectors having values in a space of dimension equal to the number of bands of the used image. Their representation in a multidimensional space permits the measurement of the angular difference that separates them (Fig. 3).

This spectral angle α is calculated using the following equation [9].:

$$\alpha = \arccos\left(\frac{\vec{s}\,.\,\vec{t}}{\|\vec{s}\|.\|\vec{t}\|}\right)$$

S: is the reference specter;
t: is the evaluated specter.

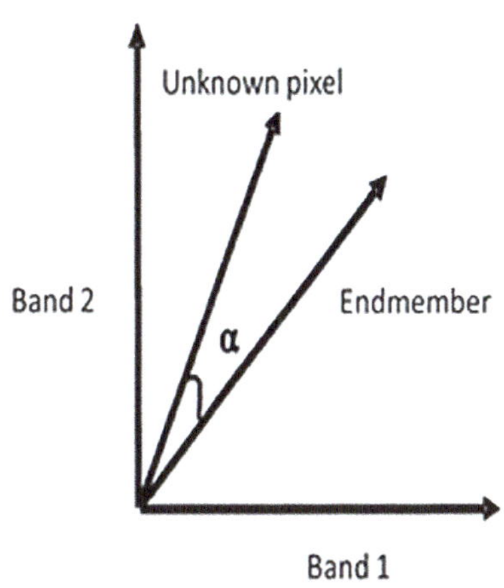

Fig. 3 The concept of SAM method [9]

The attribution of a pixel to a given theme is based upon the measurement of angle α, that is to say the angle existing between the reference specter of the vector and that of each vector in the image.

The smaller is the angle, the bigger is the similarity between the specter of the evaluated pixel and the reference. Moreover, the pixel is to be assigned to the class of the spectral pole with which it has the smallest angular difference. This measure is less sensitive to variations in illumination which increase or decrease does not affect the direction of the vector, but rather its amplitude [8]. A wrong choice of the angle a can generate a minimal number of classes or an undesirable confusion and a maximum number of unclassified pixels [43].

4 Results

MVS Classification

The classes that have been selected are seven: cereals, rangelands, plowed soil, forest, vegetation + water, bare soil, and water (Figs. 4 and 5). The examination of the confusion matrix reveals for each class the percentage of well-ranked pixels (Table 1). The diagonal of this matrix expresses the number of well-ranked pixels compared to other classes. For cereals

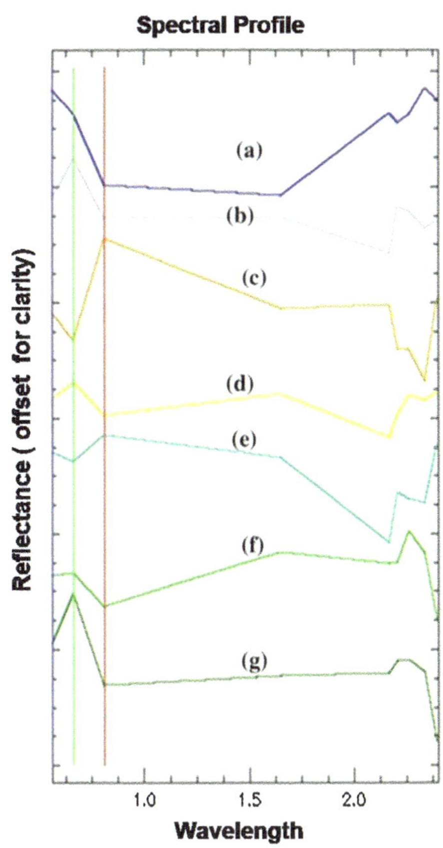

Fig. 4 The spectral signatures and the choice of classes. **a** Water, **b** Bare soil, **c** Cereal, **d** Plowed soil, **e** Vegetation + water, **f** Rangelands, **g** Forest

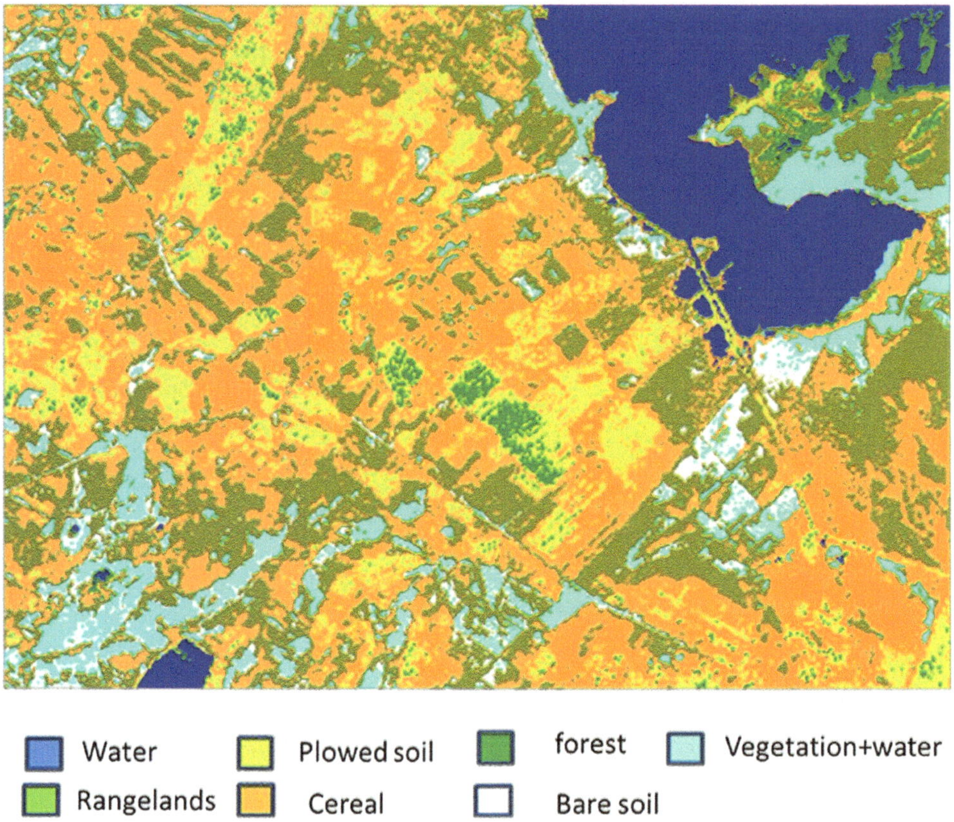

Fig. 5 The map covers land from the MVS method

on 37130 pixels classified at the level of the image, there are 37121 pixels that are well classified, 5 are confused with the courses and 4 with the forests. However, water has no confusion. We also note a little confusion between forests and rangelands. This is due to the fact that some courses are represented by acacia and cactus that have a spectral response close to that of the forests.

SAM Classification

- Preparation for the selection of prototype signatures

– Minimum Noise Fraction MNF

For extracting the prototype signatures, the PPI index requires the calculation of the Minimum Moise Fraction (MNF) components through a procedure similar to the principal component analysis [38]. It is a transformation that consists in minimizing noise by compressing information into a finite number of MNF bands [38,41]. For the image used in this study, the first five components are considered significant (Fig. 6).

– Spatial Reduction by the PPI Pure Pixel Index

The Pixel Purity Index is used to identify the most spectrally pure pixels in an image (and it usually does not reach 1%). The value of the threshold is used to define some pixels that are apparent as ends in the executions of the projected vector. The threshold value would be approximately 2–3 times the noise level in the data (while this value is 1 when the MNF transformed data is used). In this study, the PPI is executed via the recourse to the five MNF bands. It is processed with 100 iterations and produced using threshold 3. Brighter pixels more spectrally represent extreme "shots" and indicate pixels that are more spectrally pure. Darker pixels, however, are less pure. The pixel value is the number of times the pixel was recorded as an endpoint during the PPI process (Fig. 7).

– Selection of Endmembers Prototype Signatures

Endmembers can be considered as points in a scattering dimensional parcel of n, where n is the number of bands [31]. The coordinates of the points in the n-space consist of the values of "n" which are simply the spectral values of radiance or reflectance in each band for a given pixel (Fig. 8). The distribution of these points in n-space can be used to estimate the number of spectral endmembers and their pure spectral signatures. Using the dimensional viewer of n, there are endmembers located by the group of extreme pixels in the dimensional space of n.

Table 1 The confusion matrix for image classification

Classes	C1	C2	C3	C4	C5	C6	C7
C1	37121	0	5	0	0	0	4
C2	0	72503	1	2	0	0	0
C3	0	10	2618	2	0	0	4
C4	0	2	1	55405	0	0	0
C5	0	0	0	0	188903	2	0
C6	0	0	0	0	1	42136	0
C7	0	0	0	0	0	0	81280

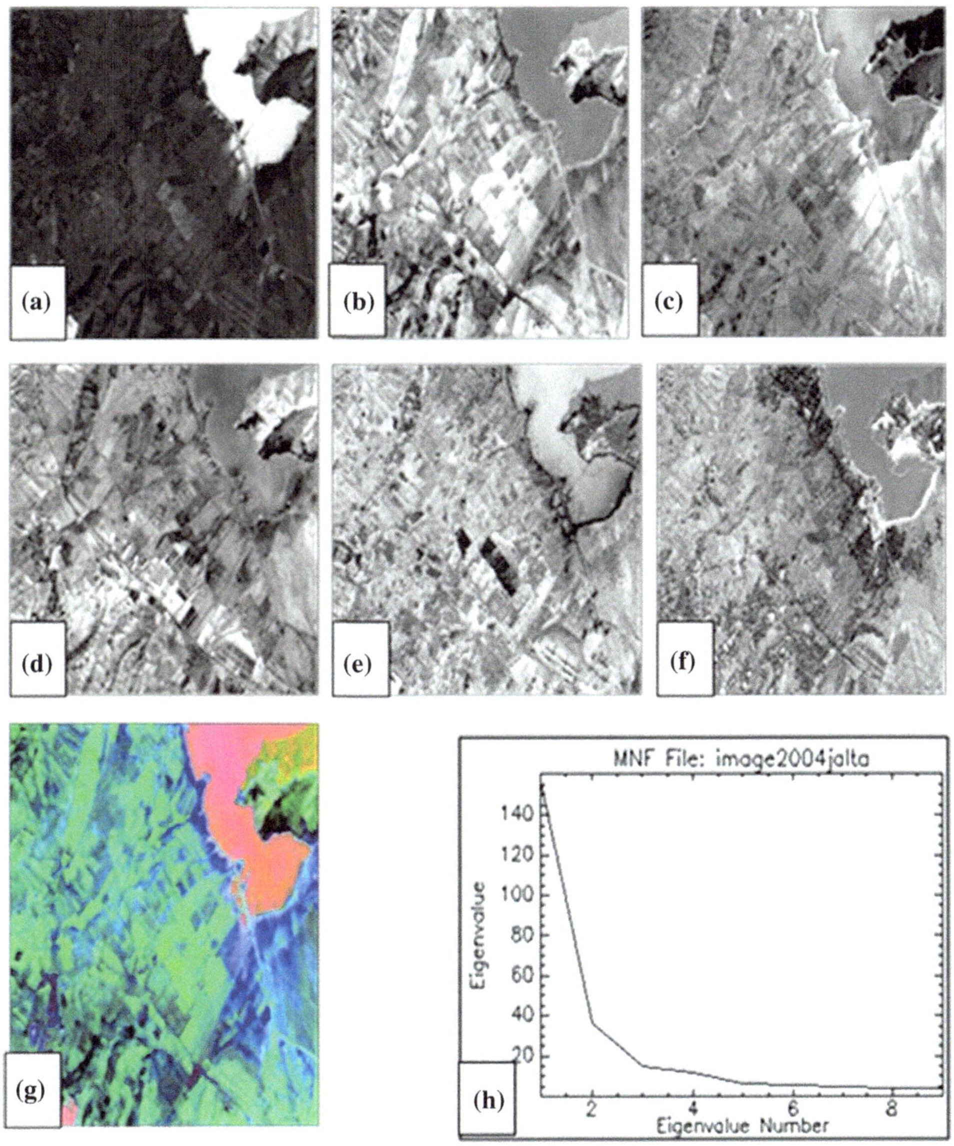

Fig. 6 MNF (Minimum noise fraction) :**a f** 6 bands MNF obtained for 9 bands of aster image; **g** The colored compound (RGB) in the three first bands MNF; **h** Eigenvalue numbers of MNF

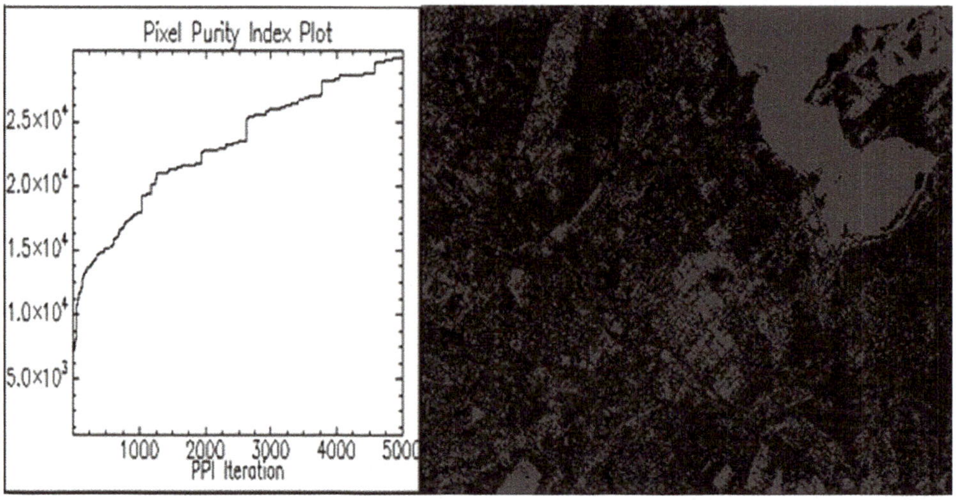

Fig. 7 Pure Pixel Index (PPI)

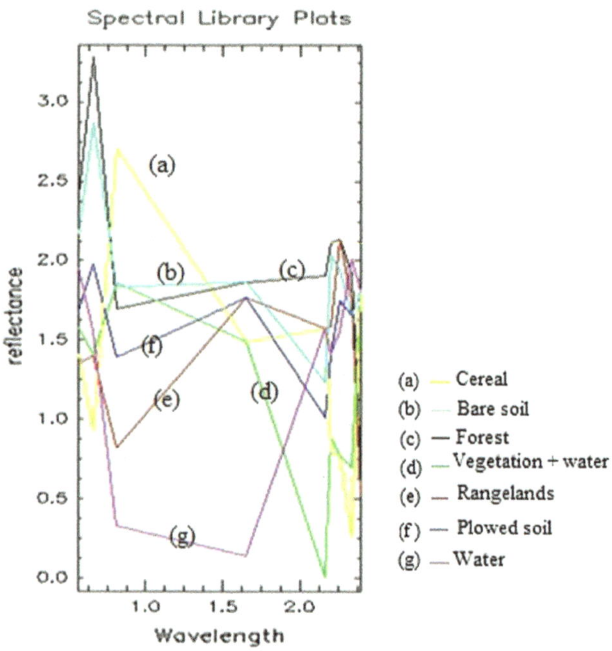

Fig. 8 The spectral signatures of the selected endmembers

- Spectral mapping using the spectral mapping method

Spectral Angle Mapping (SAM) is applied to compare image spectra with the different pure spectra of a spectral library. The algorithm determines the similarity between two spectra by calculating the "spectral angle" between them [41].

For the implementation of the SAM approach, several attempts have been made to identify the combination of the optimal spectral angle and the appropriate prototype signatures to model the image used. The best result is obtained for an angle of 0.18 radians combined with the prototype signatures generated by the PPI index method, which makes possible the distinction between these seven classes (Fig. 9).

5 Discussion

– **Spectral signatures**

The prototype signatures extracted directly from the image and by the PPI method are presented in Fig. 10. They are spectra representing the seven classes. The results of the two methods are different: Signal amplitude: the spectra derived by the PPI method show a relatively higher level of reflectance than the spectra obtained directly from the image, in particular in the visible and near-infrared bands and in the first band of the infrared way;

– **Spectra of "bare ground":**

The spectra produced using PPI and those produced directly from the image differ in their shape and level of reflectance.

– **Spectra of forest vegetation:**

Although quite similar, the spectra obtained by the two methods show some differences in details; the spectra from the PPI method better express the spectral behavior of the vegetation, in particular bands 1, 2, 3, and 4, respectively corresponding to green, red, near infrared, and the mean infrared.

The two sets of prototype signatures are used separately to perform the SAM and MVS. The results are in the form of images of relative fractions of the themes selected. Each image of fractions represents the relative abundance of a given homogeneous component. It expresses its distribution in proportion to the other components involved in the model. It therefore reflects its land occupation rate. Thus, each homogeneous component can be reclassified and edited in the form of a thematic map that can facilitate the analysis and interpretation of the surface conditions of the studied territory.

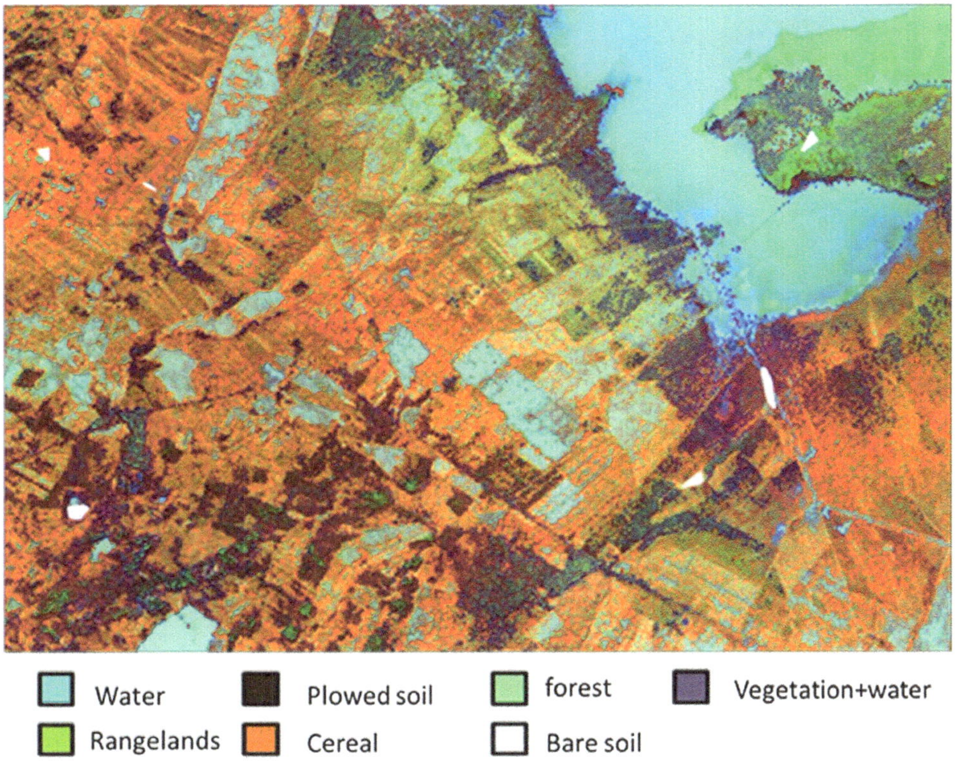

Fig. 9 The map covers land from the SAM method

In order to evaluate the performance of the PPI index, the images obtained from the prototype signatures of the PPI index and those provided directly from the image are compared using linear regression. The results obtained generated a different relative abundance for each of the plant species studied (Table 2). This is illustrated by a relatively low correlation for forests, with a correlation coefficient in the order of 0.59 and a residual error of 0.76, Significant correlations for cereals and rangelands. The correlation coefficients for both classes is 0.98. For the "tilled soil and bare soil" themes, the prototype signatures of the PPI index and those obtained directly from the image have weak correlations.

Thus, the spectra obtained by PPI and directly from the image lead to similar qualitative results in terms of distribution mapping, but they generate different quantitative results in terms of the value of the relative abundance of the classes. This has an impact on the combination procedures of class abundance.

- Distinction of themes using SAM and MVS methods

In general, forest cover classifications produced using the MVS and SAM approaches show some similarity in terms of the overall distribution of land cover (Figs. 5 and 9). More specifically, they show appreciable differences (Table 3):

- The area occupied by the class of the vegetation cover (forest, cereals and rangelands) seems much larger in the case of SAM than MVS;
- The area representing the bare soils is overestimated by the SAM method which has, in fact, attributed the very sparse vegetation to this class.

6 Conclusion

At the core of this paper is a focus on the distinction of the dominant zoning regulations land use in the area of Mateur and their distribution mapping. This allows us to compare the methods developed in order to take into account the problems of spatial heterogeneity and spectral complexity which characterize semi-arid regions and degraded natural environments. These are methods for extracting pure signatures from remote sensing images (pixel purity index and approaches measuring the resemblance of spectral vectors (SAM and MVS)). In the light of the main findings, it is proven that the choice of prototype signatures is a decisive step in the implementation of the information extraction methods, based on the spectral similarity from the ASTER multiband data. This is due to the fact that the method of selecting prototype signatures via the PPI index method generated relatively different spectral signatures, which had a significant influence on the

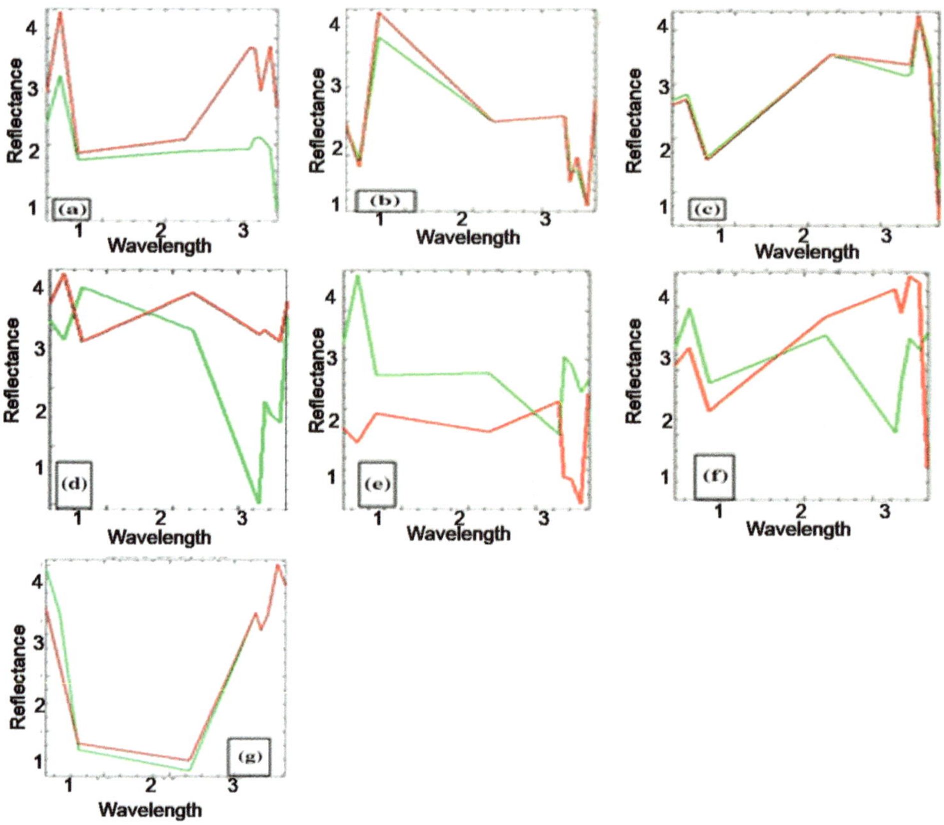

Fig. 10 The spectral signatures of every class following treatment with PPI (red) and collected directly by the image (green). **a** Forest, **b** Cereal, **c** Rangelands, **d** Vegetation + water, **e** Plowed soil, **f** Bare soil and **g** Water

Table 2 The coefficients of the correlations between specters obtained by PPI and directly of the image

Classes	Coefficients of the correlations (R)	Residual error (RMSE)
Forest	0.59	0.76
Rangelands	0.98	0.09
Vegetation + water	0.5	0.2
Plowed soil	−0.24	0.57
Cereal	0.98	0.13
Bare soil	−0.17	0.43
Water	0.95	0.18

Table 3 Comparison between the results of classification by the MVS method and SAM method

Classes	Pixels (%)	
	MVS	SAM
Cereals	19.2	39.3
Bare soil	15.1	4.1
Range lands	0.5	0.7
Vegetation + water	11.3	18.5
Plowed soil	7	4.6
Water	8.2	8.7
Forest	1	16.9

SAM application results. Thus, two different sets of prototype signatures (direct method and PPI index) lead to slightly different results from that of modeling the spectral variability of the AST image ER using the SAM and MVS method. In addition, the MVS and SAM methods generated a slightly different vegetation cover classification. It is important, in this respect, to note that the accuracy of SAM and MVS methods depends on the selection of the prototype signatures used to model the studied image. Comparison of the results obtained by MVS and SAM shows that SAM allows to classify the dominant forest species with a better precision than the MVS.

References

1. Cyr, L., Bonn, F., & Pesant, A. (1995). Vegetation indices derived from re-mote sensing for an estimation of soil protection against water erosion. *Ecological Modelling*, *79*, 277–285.
2. Biard, F., & Baret, F. (1997). Crop residue estimation using multi-band re-flectance. *Remote Sensing of Environement*, *59*, 530–536.
3. Hill, J., Hostert, P., Tsiurlis, G., Kasapidis, P., Udelhoven, Th, & Diemer, C. (1998). Monitoring 20 years of intense grazing impact on the Greek island of Crete with earth observation satellites. *Journal of Arid Environnent*, *39*, 165–178.
4. Arsenault, E., & Bonn, F. (2001). Evaluation of soil erosion protective cover by crop residues using vegetation indices and spectral mixture analysis of multispectral and hyperspectral data. *Proceedings of the 23rd Canadian Symposium on Remote Sensing. 21–24 août 2001*, Association Québécoise de télédétection.
5. Bannari, A., Teillet, P., Leckie, D., & Fedosejevs, G. (1999). Impact des con-ditions internes et externes au couvert forestier sur les indices spectraux dérivés de simulations spectrales de AVHRR. *Télédétection*, *1*, 157–181.
6. Elmore, A. J., Mustard, J. F., Manning, S. J., & Lobell, D. B. (2000). Quantifying vegetation change in semiarid environments: Precision and accuracy of spectral mixture analysis and the normalized difference vegeta-tion index. *Remote Sensing of Environment*, *73*(1), 87–102.
7. Hostert, P., Roder, A., & Hill, J. (2003). Coupling spectral unmixing and trend analysis for monitoring of long-term vegetation dynamics in Mediterranean rangelands. *Remote Sensing of Environment*, *87*(2–3), 183–197.
8. Clark, R. N., King, T. V. V., Ager, C., & Swayze, G. A. (1995). Initial vegeta-tion species and senescence/stress mapping in the San Luis Valley, Colorado using imaging spectrometer data (pp. 64–69). In: H.H. Posey, J.A. Pendelton & D. Van Zyl (red.), *Proceedings: Summitville Forum 95* (vol. 38, pp. 56–73). Colorado: Colorado Geological Sur-vey Special Publication.
9. Kruse, F. A., Lefkoff, A. B., Boardman, J. W., Heiderbrecht, K. B., Shapiro, P. J., & Goetz, A. F. H. (1993). The spectral image processing system (SIPS) - interactive visualisation and analysis of imaging spectrometer data. *Remote Sensing of Environment*, *44*(2–3), 145–163.
10. Dai, X., & Khorram, S. (1999). Data fusion using artificial neural net-works: a case study on multitemporal change analysis. *Computers, Environment and Urban systems*, *23*, 19–31.
11. Liu, J., Shao, G., Zhu, H., & Liu, S. (2005). A neural network ap-proach for enhancing information extraction from multispectral image data. *Canadian Journal of Remote Sensing*, *31*(6), 432–438.
12. Quarmby, N. A., Petropoulos, G. P., Vadrevu, K. P., Xanthopoulos, G., Karan-tounias, G., & Scholze1, M. (2010). A comparison of spectral angle mapper and artificial neural network classifiers combined with Landsat TM imagery analysis for obtaining burnt area mapping. *Sensors* (Basel) *10*(3), 1967–1985. Published online 2010 Mar 11. https://doi.org/10.3390/s100301967.
13. El_Rahman1, S. A. (2016). Performance of spectral angle mapper and parallelepiped classifiers in agriculture hyperspectral image (IJACSA). *International Journal of Advanced Computer Science and Applications*, *7*(5), 55–63.
14. Margate, D. E., & Shrestha, D. P. (2001). The use of hyperspectral data in identifying desert-like soil surface features in Tabernas area, (red.) *Proceedings of the 22nd Asian Conference on Remote Sensing, 5-9 novembre 2001, Centre for remote Imaging, Sensing and processing (CRISP)* (pp. 736–741). National University of Singapore; Singapore Institute of Surveyors and Valuers (SISV); Asian Association on remote Sensing (AARS), Southeast Spain Singapore.
15. Chikhaoui, M., Bonn, F., Bokoye, L. A., & Merzouk, A. (2005). A spectral index for land degradation mapping using ASTER data: Application to a semi-arid Mediterranean catchment. *International Journal of Applied Earth Observation and Geoinformation*, *7*, 140–153.
16. Abdelrahim, A. M, El-Tyeb, S., Elmahlb, G. (2017). Spectral mixture analysis (SMA) and change vector analysis (CVA) methods for monitoring and mapping land degradation/desertification in arid and semiarid areas (Sudan), using Land-sat imagery, Egypt. *The Egyptian Journal of Remote Sensing and Space Science*, *20*(1), S21–S29. https://doi.org/10.1016/j.ejrs.2016.12.008.
17. Sohn, Y., Moran, E., & Gurri, F. (1999). Deforestation in north-central Yucatan (1985–1995) mapping secondary succession of forest and ag-ricultural land use in sotuta using the cosine of the angle concept. *Photogrammetric Engineering and Remote Sensing*, *68*, 1271–1280.
18. Yang, H., Van Der Meer, F., & Bakker, W. (1999). A back-propagation neural network for mineralogical mapping from

AVIRIS data. *International Journal of Remote Sensing*, *20*(1), 97–110.
19. Sohn, Y., & Rebello, S. (2002). Supervised and unsupervised spectral angle classifiers. *Photogrammetric Engineering and Remote Sensing*, *68*, 1271–1280.
20. Zhang, M., Qin, Z., Liu, X., & Ustin, S. L. (2003). Detection of stress in to-matoes induced by late blight disease in California, USA, using hyperspectral remote sensing. *International Journal of Applied Earth Observation and Geoinformation*, *4*, 295–310.
21. Hasan, E., Fagin, T., Alfy, Z., & Hong, Y. (2016). Spectral Angle Mapper and aeromagnetic data integration for gold-associated alteration zone mapping: A case study for the Central Eastern Desert Egypt (pp. 1762–1776). Received 02 Sep 2015, Accepted 07 Mar 2016, Published online: 13 Apr 2016 Download cita-tion. https://doi.org/10.1080/01431161.2016.1165887.
22. Liu, D., Cao, C., Chen, W., Ni, X., Tian, R., & Xing, X. (2016). Monitoring and predicting the degradation of a semiarid wetland due to climate change and water abstraction in the Ordos Larus relictus National Nature Reserve, China (pp. 367–383). Received 21 Dec 2014, Accepted 30 Jul 2016, Published online: 03 Oct 2016 Download citation. https://doi.org/10.1080/19475705.2016.1220024.
23. Bangira, T., Alfieri, S. M., Menenti, M., van Niekerk, A., & Vekerdy, Z. (2017). A spectral unmixing method with ensemble estimation of endmembers: Application to flood mapping in the Caprivi floodplain. *Remote Sensing*, *9*(10), 1–24. [1013] https://doi.org/10.3390/rs9101013.
24. Tompkins, S., Mustard, J. F., Pieters, C. M., & Forsyth, D. W. (1997). Optimization of endmembers for spectral mixture analysis. *Remote Sensing of Environment*, *59*(3), 472–489.
25. Adams, J. B., Smith, M. O., & Gillespie, A. R. (1993). Imaging spectroscopy: Interpretation based on spectral mixture analysis. In C. M. Pieters & P. Engelet (red.) *Remote geochemical analysis: Element and mineralogical com-position* (pp. 145–166). New York: Cambridge University Press.
26. Bateson, A., & Curtiss, B. A. (1996). Method for manual endmember selection and spectral unmixing. *Remote Sensing of Environment*, *55*(3), 229–243.
27. Baret, F., & Guyot, G. (1991). Potentials and limits of vegetation indices for LAI and PAR assessment. *Remote Sensing of Environment*, *35*, 161–173.
28. Sang-Wook, K., & Chong-Hwa, P. (2004). Linear spectral mixture analysis of landsat imagery for Wetland Land-Cover classification in Pal-dang reservoir and vicinity. *Korean Journal of Remote Sensing*, *20*(3), 197–205.
29. Boardman, J. W. (1993). Automated spectral unmixing of AVIRIS data using convex geometry concepts. *Summaries, Fourth JPL Airborne Geoscience Workshop* (vol. 1, pp. 11–14), JPL Publication 93-26.
30. Nadeau, C. (2000). Analyse des effets atmosphériques dans les données en télédétection du moyen infrarouge sur la classification des minéraux de surface en milieux aride. Mémoire de maîtrise, Département de géographie et télédétec-tion, Faculté des lettres et sciences humaines (116 p.). Québec, Canada: Université de Sherbrooke, Sher-brooke.
31. Boardman, J. W., Kruse, F. A., & Green, R. O. (1995). Mapping target signatures via partial unmixing of AVIRIS data. *Summaries of the 5th JPL Airborne Earth Science Workshop* (pp. 23–26). Pasadena, California: JPL Publication 95-11, Jet Propulsion Laboratory, California Institute of Technology.
32. Foody, G. M., & Cox, D. P. (1994). Sub-pixel land-cover composition estimation using a linear mixture model and fuzzy membership functions. *International Journal of Remote Sensing*, *15*(3), 619–631.
33. Jaafari, M. (1991). *Minéralisations polyphasées associées aux calcaires campaniens du jabel el Ghreffa (district de Jalta)* (p. 120). D.E.A en géologie, Faculté des Sciences de Tunis.
34. Roussev, G., Radivoev, B., & Papov, A. (1976). Gisement de plomb de Jalta. Rapport géologique, compagne de recherche 1974–1975. Société tunisienne d'expansion minière. Convention de renouvellement des réserves des mines en activité du 11 .06 .1974 - Techno-exportstroy - Bulgarproremi - Bulgarie (101 p.).
35. Bonn, F., & Rochon, G. (1992). Précis de télédétection (vol. 1). Principes et méthodes (512 p.). Presses universitaires du Québec.
36. Robin, M. (1993). *La télédétection des satellites aux systèmes d'information géographiques* (318p p.). Coll. Nathan, F.
37. El abed, I. (2002). Apport de la télédétection et des systèmes d'information géographique à l'évaluation de la dégradation des sols par éro-sion hydrique (les abords de Ain Jelloula en Tunisie centrale. D.E.A en géologie (90 p.).
38. Green, A. A., Berman, M., Switzer, P., & Craig, M. D. (1988). A transformation for ordering multispectral data in terms of image quality with implications for noise removal. *IEEE Transactions on Geoscience and Remote Sensing*, *26*(1), 65–74.
39. Scholte, K. H. (2005) Hyperspectral Remote Sensing and Mud Volcanism in Azerbaijan (147 p.). Ph.D. Thesis, Delft University of Technology, Delft.
40. Kruse, F. A., & al. (1997). Techniques developed for geologic analysis of hyperspectral data applied to near-shore hyperspectral ocean data. *Presented at the Fourth International Conference on Remote Sensing for Marine and Coastal Environments*, Orlando, Florida. Retrieved March 17–19, 1997.
41. Boardman, J. W., & Kruse, F. A. (1994). Automated spectral analysis: A geological example using AVIRIS data, northern Grapevine Moun-tains, Nevada. *Proceedings Tenth Thematic Conference, Geologic Remote Sensing*, 9–12 May 1994 (pp. 407–418). Texas: San Antonio.
42. Zhang, B., Wang, X., Liu, J., Zheng, L., & Tong, Q. (2000). Hyperspectral Im-age processing and analysis system (HIPAS) and its applications. *Photogrammetric Engineering and Remote Sensing*, *66*(5), 605–606.
43. Crosta, A. P., Sabine, C., & Taranik, J. V. (1998). Hydrothermal alteration mapping at bodie, California, using AVIRIS hyperspectral data. *Remote Sensing of Environment*, *65*(3), 309–319.

Assessment Approach for the Automatic Lineaments Extraction Results Using Multisource Data and GIS Environment: Case Study in Nefza Region in North-West of Tunisia

Slimene Sedrette and Noamen Rebai

Abstract

Remote sensing technology has long been applied in many fields of studies, especially in structural geology and mineral exploration, where remote sensing is useful for lineaments and structural feature extraction. In this study, Landsat ETM-7 satellite data have been used and band-8 is proved to be the most suitable band in automatic delineation. The automatic lineament extraction process has been carried out with LINE module of "PCI Geomatica" based on automatic detection algorithms. The lineaments analysis is conducted through the study of geospatial information incorporating the length, density, and direction of lineaments in order to acquire knowledge about the characteristics and proprieties of the geological structures. The purpose of this work is to evaluate the result of the automatic lineament extraction method and therefore to investigate its ability to give real results in comparison with the fault map.

Keywords

Lineaments extraction • Assessment • Satellite images • Nefza • Tunisia

1 Introduction

Previous works show that there is a connection between surface indices (faults) and traces of the underlying geology [1]. Consequently, the study of these surface faults over several tens of kilometers can help to understand some deep geological phenomena.

Studies of lineaments of both local and regional significance have been progressing rapidly and several approaches have already been used in numerous studies in the world [1–7].

The importance of multiscale and multisource data correlation for both local and regional knowledge is underscored by some structural geologists including Derouin et al. [8] and Saaidi [9].

O'Leary et al. [10], defined the term lineament as a simple or composite linear feature of a surface whose parts are aligned in a rectilinear or slightly curvilinear relationship and which differs from the pattern of adjacent features and presumably reflects some subsurface phenomena.

Nowadays, there is a range of computer tools aiming at facilitating the interpretation of lineaments. Some even allow the automatic extraction of lineaments, using contrast-detection algorithms within an image. In addition, it is always important to compare the results of satellite images with geo-scientific data using spatially geospatial information.

In the present study, both sets of results from satellite data and published geological map are being compared. The goal is to improve our understanding of the lineament distribution of Nefza region through the use of satellite image and image processing techniques for mapping the linear tectonic features. Indeed, a major emphasis is put on showing that a small scale study (Satellite image) can have the same result as a large scale study (map).

If successful, the use of remote sensing and this automated process may reduce exploration hazard, time, and cost.

2 The Geological Setting

The investigated area (Fig. 1) is located in the North East of Tunisia and enclosed between latitudes 84808E and 91035E and longitudes 365334N and 370530N, covering an area of 513,8 sq. km.

S. Sedrette (✉) · N. Rebai
LR14ES03 Geotechnical Engineering and Georisk Laboratory, National School of Engineering of Tunis, University of Tunis El Manar, BP. 37, le Belvédère 1002, Tunis, Tunisia
e-mail: ssedrette@gmail.com

N. Rebai
e-mail: noamen.rebai@enit.utm.tn

N. Rebai and M. Mastere (eds.), *Mapping and Spatial Analysis of Socio-economic and Environmental Indicators for Sustainable Development*, Advances in Science, Technology & Innovation, https://doi.org/10.1007/978-3-030-21166-0_6

The study area belongs to the eastward extension of the Maghreb orogen, which fossilizes in its dynamic history both the opening and closing of the Tethys, inherited from the movements of the African and Eurasian plates. The sedimentary substrate in this region comprises the Ed Diss thrust sheet (Upper Cretaceous to Eocene) overlain by the Numidian nappe [10]. The latter consists of a thick (1000 m) series of siliciclastic flysch (Numidian flysch), Oligocene to Lower Miocene (Burdigalian).

The geology of the region is characterized by several structural features that are usually associated with the phenomena responsible for the development of polymetallic mineralization, as well as the transfer and dispersion on the surface of various geochemical elements. During Neogene, when the Nefza magmatism emplaced, there was an alternation of oblique compressive regime and extension regime, which is typical of the postcollisional period. Importantly, during the Serravallian and the Tortonian periods, the northern African realm was affected by a very oblique compressive regime [12], and experienced short-lived returns to extensional conditions during the Messinian and Pliocene periods. Since then, the Africa–Europe convergence rate has been very low (0.5 cm.a1) and the plate boundary is still located at the front of the Tellian Zone [11], with some folding just to the north of the coastline.

The complex tectonic setting of this region, determines the increased tectonic activity in relation with the development of polymetallic mineralization, making it an ideal study area upon which the evaluation of the automatic lineament extraction result is built.

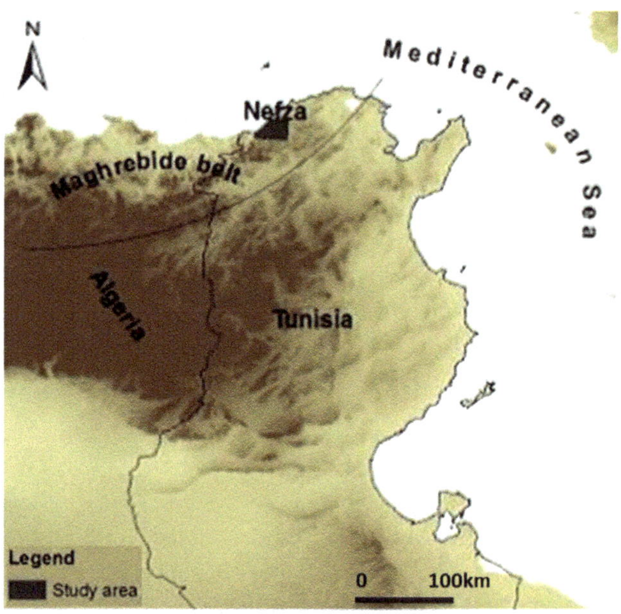

Fig. 1 Location map of the study area

3 Materials and Methods

The first step in our proposed methodology lies in the selection of initial input data for lineament extraction. The Landsat 7 ETM + scene of the study region is uploaded from the USGS Earth Explorer website http://earthexplorer.usgs.gov. These images are chosen in terms of their spectral characteristics and spatial resolution. They were taken in the dry season and without any cloud cover. Their characteristics are as follows: path/row: 192/34; date: 12/05/2009; file type: GEOTIFF; projection: UTM, zone 32 north; ellipsoid: WGS-84.

Panchromatic band has been favored in lineament analysis due to its improved spatial resolution (15 m). Therefore, in our case study, lineaments are automatically extracted from the panchromatic band with the LINE module of "PCI Geomatica" as shown in Fig. 2. The selection of appropriate parameters for automatic lineaments extraction with the PCI Geomatica Line module allows the drawing of the area's lineament map. The lineament extraction algorithm of PCI Geomatica software consists of edge detection, thresholding, and curve extraction steps. In this study, used settings values are selected from several setting tests.

The lineaments map obtained is cleaned automatically by overlaying an anthropic effect present in the area and by removing all the straight edges coinciding with the boundaries of the subset image. This map will be the subject of a spatial characterization using GIS.

The second datasets to be acquired for this study are the digital scanned geological maps at 1:50000 scale. They are imported into ArcGIS to create the fault map of Nefza region by the semi-automatic extraction method.

As a second step, the necessary vector data are obtained in order to calculate each tectonic lineament lengths and directions [13].

For so doing, the fault network is being considered as a vector object with a list of attributes (Fig. 3) that must be manipulated while taking into account their geological sig-

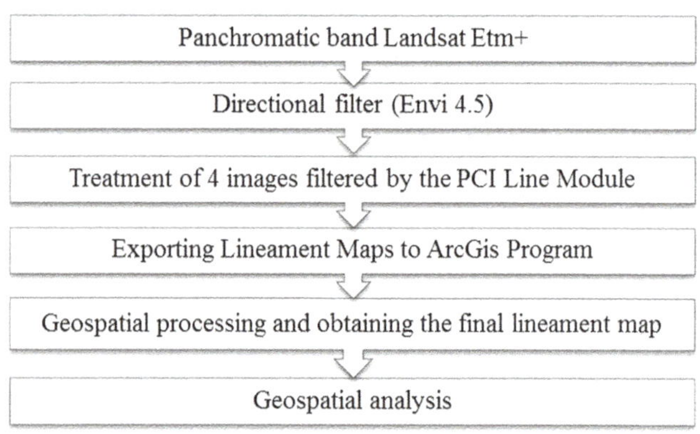

Fig. 2 The flowchart displays the different steps of lineaments extraction and analysis [12]

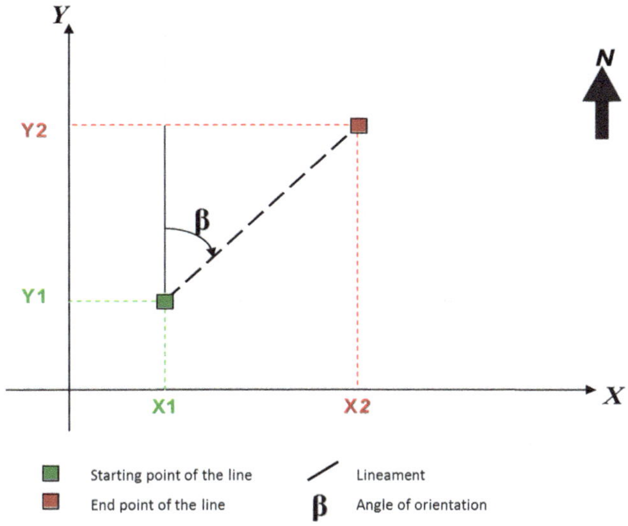

Fig. 3 The calculation principle of the lineament orientation angle [13]

nificance. In our work, we have adopted 3 steps [14], which are respectively:

- Calculation of the faults extremities coordinates;
- Geometric determination of the orientations angles;
- Determination of the lineaments directions.

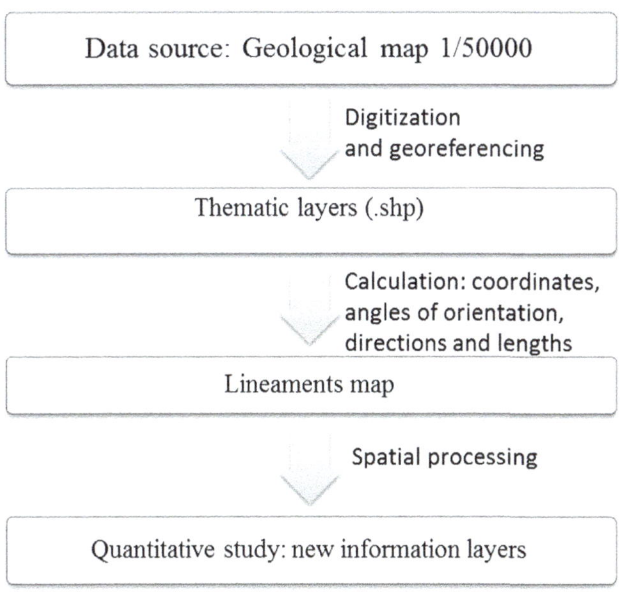

Fig. 4 The flowchart shows the steps of semi-automatic faults extraction from the geological map and the statistical analysis [14]

4 Results and Discussions

The comparison between the automatic lineaments extraction results and those of the map of tectonic lineaments (fault map) issued from the semi-automatic extraction method (Fig. 4) based on Nefza 1/50000 geological map are summarized in Table 1.

The number and the total length of the lineaments using automatic method are higher than the number and the total length of the faults in the fault map. The most important factor for this difference is that the lineaments in the automated one are shorter in length (2.5 km) compared with the length of fault lines (4.15 km).

There is a significant number of lineaments (>50 % of total number) extracted from the satellite image. They are superimposed or slightly offset relatively to the ground structural elements or sometimes in their extension.

Given the accuracy of the satellite image, the gap between satellite and geological lineaments can be explained on the one hand by the geocoding and resampling effect that causes inaccuracy in the processed image and on the other hand by the scale changes (and sometimes projection system) and the thickness line [15].

Also, there were some of lineament lines that could not match any fault line in the fault map.

They may be an expression of other line weakness, geomorphological structure, boundary of vegetation, moisture content, and soil or rock composition. This result is possibly due to the fact that the automatic lineament extraction method does not discriminate lithological, hydrographical, or other features account for most tonal contrast which are used to extract the linear features [15].

In order to show the correlation and the significant geological link between the various lineaments extracted from the satellite image and to improve the analysis and interpretation of the results, we have established a set of necessary steps for the validation of the map obtained.

Through this approach, therefore, we hope to find some interpretations to the different lineaments detected from the panchromatic band.

The first step is based on the comparison between the major families of lineament direction found on the geological map and extracted from the panchromatic band. Table 2 demonstrates that the orientation NE-SW is the major direction in terms of the number of lineaments. This direction is followed by the other three ones as follows: NW-SE, N-S, and E-W in decreasing order.

These four directional families keep the same proportions regardless of the extraction support.

The second step in the validation approach is premised upon density analysis, which allows representing the frequency of lineaments per unit area.

Table 1 Comparative table of lineaments extraction results from the geological map and from the panchromatic band

Lineament	Panchromatic band	Fault map
Number	1617	817
Minimum length (km)	0.26	0.27
Maximum length (km)	2.5	4.154
Average length (km)	0.51	0.538
Total length (km)	804	439

Table 2 Comparison, at the local scale, between the orientation and frequency of lineaments extracted from geological map and satellite images

Orientation	N-S		NE-SW		E-W		NW-SE	
	Number	%	Number	%	Number	%	Number	%
Geological map	100	12	385	47	74	9	258	32
Satellite image	205	13	783	48	158	10	471	29

In our work, we have utilized the ArcGis software's "spatial analyst" tool in order to obtain two lineament density maps.

The distribution of the lineaments high and low density zones can be superimposed on the two density maps, but with different intensities. Indeed, the density of the lineaments is greater on the map of the lineaments resulting from the satellite image than from the geological map (Figs. 5 and 6).

The highest lineaments density is shown in the automatic lineament extraction map especially in areas at high altitudes and ridges (Fig. 5) in the North part and also in areas adjacent to major faults.

Furthermore, the region is marked by great lithological diversity. These contrasts, both at the lithostratigraphic level and at the level of the general orientation of the rock formations, could also partially justify the high density of lineaments extracted from the satellite image.

For verifying this last observation, we have resorted to the functionalities offered by the GIS, which allow the subtraction matrix function to deduce the difference in density between the two raster maps of the distribution of the lineaments according to the following formula:

$$[\mathbf{A}] - [\mathbf{B}] = [\mathbf{C}]$$

[A] = [Density of lineaments extracted from satellite image]
[B] = [Density of fault map]
[C] = [Residual lineaments density map].

The overlay of the lineaments residual map on the map of the lithological outcrop boundaries of the region of Nefza

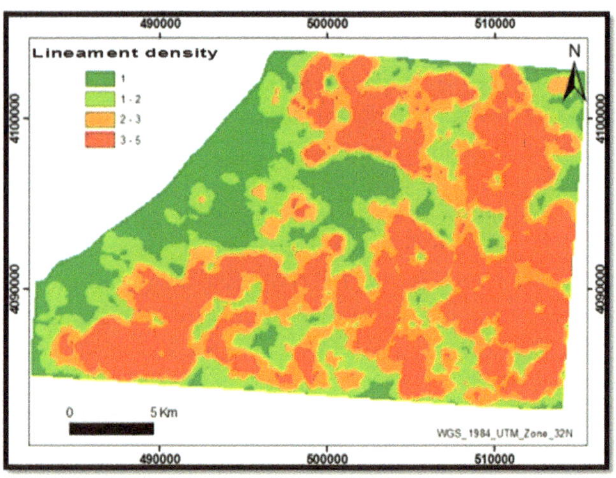

Fig. 5 Density map of automatic lineament extraction [14]

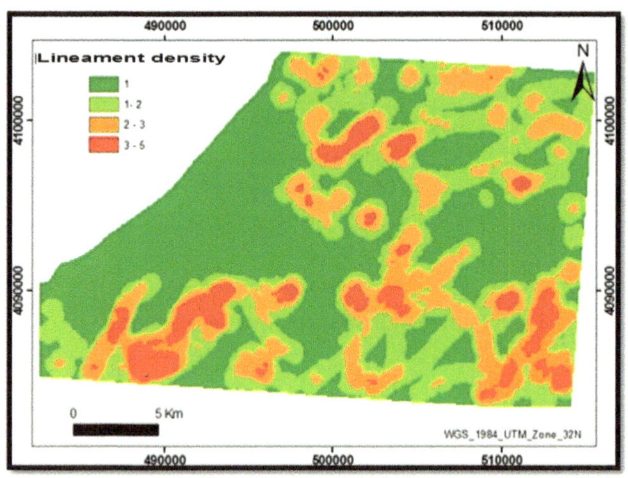

Fig. 6 Density fault map [14]

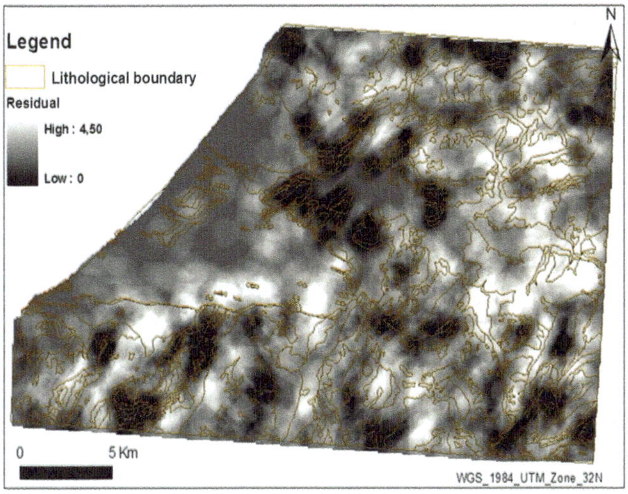

Fig. 7 The lineaments residual map overlaid with the lithological outcrop boundaries

determines the way most of the zones with high residual value (light zone on the residual map) are superimposed with lithological limits (Fig. 7). This fact allows us to validate our findings.

In regions with contrasting reliefs, as in the case of our study area, rivers and reliefs reflect the existence of regional tectonics [16, 17].

The analysis of the topographic surface allows us to distinguish other types of discontinuities. The latter result partly from the inherited or the current tectonic activity. The topographic and morphological lineaments map extracted by the topographic analysis of the summits isolines map (Fig. 8) provides a general representation of all the components of the geomorphological environment: topography, morphological inheritance, morphostructure, and linear elements of the landscape.

An important number of discontinuities are mapped. It is shown that while crossing different areas of Nefza region, they overlap and shift the other preexisting structures.

Other kind of features may be extracted by automatic extraction from high resolution images. Probably the higher order watercourses can escape from the mapping on topographic maps at 1/25000 scale as well as by automatic extraction of the hydrographic network from digital support. Thus, a significant number of this order of drains influences the quantitative result of the extracted lineaments.

In addition, according to the results obtained from the morphometric study of the region and regarding the correlation between tectonic and hydrographic network [18], we can deduce the presence of neotectonics by the rejuvenation of most of the lineaments mapped on the geological map. Following all these steps taken to validate the map of the lineaments resulting from the satellite image, we come to the conclusion that the high density of the lineaments extracted automatically from the panchromatic band can be explained by:

- Their short lengths.
- The limits of lithological outcrops.
- Hydrographic lineaments especially for higher order drains in Strahler's classification.
- The detection of new tectonic lineaments.
- Geomorphological elements and/or various tectonic structures.

However, it is important to note that the results obtained by this approach, concerning major directional families, are consistent with the results of the previous studies undertaken in the region [18–22]. Also, taking into consideration all the published works on the extraction of lineaments from satellite images [2, 3, 18, 23–29], we have observed that geological

Fig. 8 Synthetic map of the main lineaments revealed by the morphostructural approach [18]

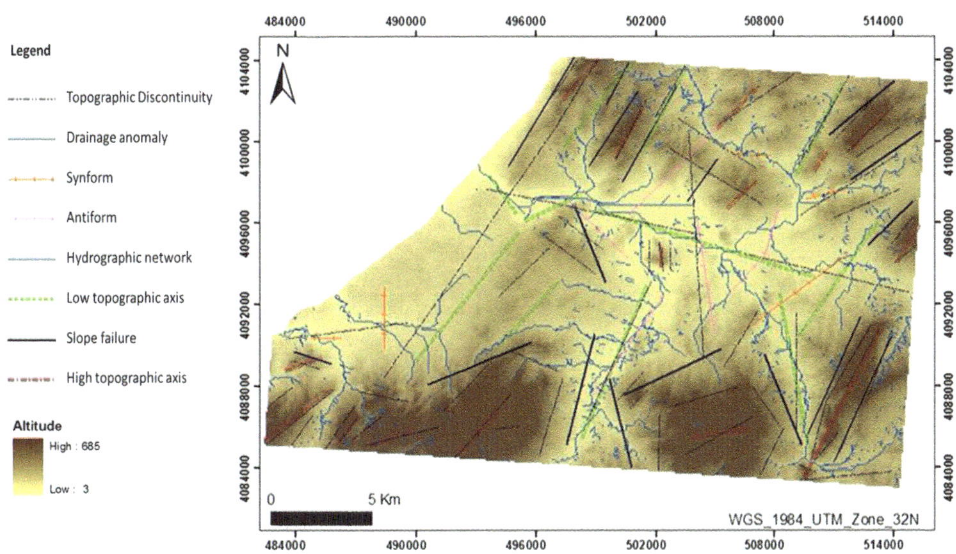

structures control so much the Earth's morphological surface and the lineaments extracted from the satellite image highly characterize these structures [30]. Whereas, nongeological and artificial lineaments may be added to the final lineament map due to the nature of algorithms used. Automatic methods need an advanced assessment approach to improve this technique by recourse to different satellite images, different resolutions, and different geological environments.

5 Conclusion

In this study, the lineaments are extracted from the geological map and LANDSAT images imported into a GIS platform, though we have shown that both approaches have their limits.

In fact, the extraction of lineaments from the geological map remains unsatisfactory in terms of scale. While remote sensing can overcome this limitation, the reliability of extraction in quantitative terms needs to be improved.

For example, the automatic lineament extraction from remote sensing support shows 100% reliable in relation with the major family's directions of lineaments. However, in quantitative terms, the adopted methodology permits the extraction of a larger number (200%) of lineaments.

Several steps still remain to be explored; but this does not deny the fact that we are beginning to see positive results. Lineaments need to be compared with topographic or hydrographic features. Also, an update of the tectonic map is necessary in order to evaluate the presence of the region's active tectonics.

In perspectives, it is necessary to have a good ration between the extractions of the lineaments by using satellite image and geological mapping process. This is to apply morphometric indexes and combine them in an intelligent (Smart) System platform.

References

1. Anderson, F. J. (2008). *Lineament mapping and analysis in the Northeastern Williston Basin of North Dakota*. North Dakota Geological Survey, Geological Investigations No. 70 (21 p.).
2. Yésou, H., Poin, J. C., Besnus, Y., & Saint, J. R. (1993). Improvement of SPOT data for the structural mapping in the tropical environment. Example of the Jbel Pagala gossans region (Togo). Third day. Scient. Res. Remote Sensing. UREF. In: Microcomputer Tools and Remote Sensing of changing environments (pp. 143–164).
3. Kouamé, K. F. (1999). Hydrogéologie des aquifères discontinus de la région semi-montagneuse de Man-Danané Ouest de la Côte d'Ivoire. Apport des données des images satellitales et des méthodes statistique et fractale à l'élaboration d'un système d'information hydrogéologique à référence spatiale. Thèse de 3ème cycle, Université de Cocody-Abidjan, Côte d'Ivoire (194 p.).
4. Vassilas, N., Perantonis, S., Charou, E., & Seretis, K. (1999). The Automatic Lineament Detection from Geophysical Grid Data Using Efficient Clustering and Weighted Hough Transform Algorithms. Workshop on Intelligent Techniques for Spatio-Temporal Data Analysis in Environmental Applications. ACAI'99. Chania, Greece.
5. Anudu, G. K., Essien, B. I., Onuba, L. N., & Ikpokonte, A. E. (2011). Lineament analysis and interpretation for assessment of groundwater potential of wamba and adjoining areas, Nasarawa State, North-central Nigeria. *Journal of Applied Technology in Environmental Sanitation*, *1*, 185–198.
6. Yao, T. K., Fouché-Grobla, O., Yéi Oga, M. S., & Assoma, V. T., (2012). Extraction de linéaments structuraux à partir d'images satellitaires, et estimation des biais induits, en milieu de socle précambrien métamorphisé. *Revue Télédétection*, *10*, 161–178.
7. Bidzang, F. N., Eric, J. O., Martial, S. N., Martial, J. A., & Joseph, M. O. (2015). Apport des linéaments satellitaires pour la recherche des indices aurifères du massif éburnéen de Ngovayang au sud Cameroun (Craton du Congo). *International Journal of Innovation and Applied Studies*, *13*, 368–376.
8. Derouin, J. P., & Chorowics, J. (1989). Méthode d'établissement de cartes géologiques synthétiques à l'aide d'images SPOT: exemple du Bas-Languedoc méditérranéen. *France. Photointerprétation*, 53–59.
9. Saaidi M. (1996). Combinaison de données multi-sources pour l'étude structurale du bassin de Guercif. Liens structuraux avec le Moyen Atlas (Maroc). Photo-interprétation (pp. 89 – 98).
10. O'Leary, D. W., Friedman, J. D., & Phn, H. A. (1976). Lineament, linear, lineation: Some proposed new standard for old terms. *Geological Society of America Bulletin*, *87*, 1463–1469.
11. Benaouali, M., Frizon de Lamotte, D., Roca, E., Bracene, R., Faure, J. L., Sassi, W., et al. (2006). Post-Cretaceous kinematics of the Atlas and Tell systems in central Algeria: Early foreland folding and subduction-related deformation. *Comptes Rendus Geoscience*, *338*, 115–125.
12. Jolivet, L. (2008). *Géodynamique méditerranéenne* (pp. 5–47). Vuibert: Société Géologique de France.
13. Rebai, N., Mabrouk, Y., & Bouaziz, S. (2005). Elaboration d'un SIG intégrant des modèles de représentation relatifs aux données géologiques et structurales du Sahel de Sfax. 9ème Journées de la Géologie tunisienne, Tunis 28–29 septembre 2005.
14. Sedrette, S., & Rebai, N. (2016). Automatic extraction of lineaments from Landsat Etm+ images and their structural interpretation: Case study in Nefza region (North West of Tunisia). *Journal of Research in Environmental and Earth Sciences*, *4*, 139–145.
15. Lacina, C. (1996). *Interprétation structurale des linéaments par traitement d'image satellitaire : cas des sousprovinces d'Abitibi et d'Opatica (Québec)*. Université de Sherbrooke, Département de.
16. Horton, R. (1945). Erosional development of streams and their drainage bassins.
17. Biemi, J. (1992). Contribution à l'étude géologique, hydrogéologique et par télédétection des bassins versants subsahéliens du socle précambrien d'Afrique de l'Ouest: hydrostructurale, hydrodynamique, hydrochimie et isotopie des aquifères discontinus des sillons et aires granitiques de la Haute Marahoué (Côte d'Ivoire). Thèse de Doct. d'Etat, Univ. Nat. Côte d'Ivoire (479 p.).
18. Sedrette, S. (2017). *Extraction semi-automatique des linéaments et des indicateurs morphométriques dans un environnement SIG pour la caractérisation morphostructurale et néotectonique de la région de Nefza dans le Nord-Ouest de la Tunisie* (p. 221p). Thèse de Doctorat, Université de Tunis El Manar.
19. Rouvier, H. (1977). *Géologie de l'extrême nord-tunisien: tectonique et paléogéographie superposées à l'extrémité orientale de la chaîne nord-maghrébine*, (307 p.). Thèse d'Etat, Paris VI.
20. Ben Ayed, N. (1986). Evolution tectonique de l'avant - pays alpine de Tunisie. Stratigraphie, caractérestiques géophysiques et évolution géodynamique. Tunis, Tunis El Manar.

21. Bracène, R., & Frizon de Lamotte, D. (2002). The origin of intraplate deformation in the Atlas system of western and central Algeria: From Jurassic rifting to Cenozoic-Quaternary inversion. *Tectonophysics*, *357*, 207–226.
22. Rekhiss, F. (2007). *Modèle d'Evolution Structurale et Géodynamique à l'Extrémité Orientale de la Chaine Alpine d'Afrique du Nord*, (285 p). Thèse d'Etat en Géologie, Université de Tunis El Manar.
23. Lim, S., Ibrahim, K., & Tjia, D. (2001). Radiometric and Geometric information content of Tiung Sat-l MSEIS data. In: Tiung SAT-1: From Inception to Inauguration (pp. 169–184).
24. Jourda, R. (2005). *Méthodologie d'application des techniques de télédétection et des Systèmes d'Information Géographique à l'étude des aquifères fissurés d'Afrique de l'Ouest, concept de l'hydrotechnique spatiale: cas des zones tests de la Côte d'ivoire*. Abidjan: Thèse Université de Cocody.
25. Jourda, J. P., Djagoua, E. V., Kouamé, K., Saley, M. B., Gronayes, C. C., Achy, J. J., et al. (2006). Identification et cartographie des unités lithologiques et des accidents structuraux majeurs du département de korhogo (nord de la côte d'ivoire): Apport de l'imagerie etm+ de landsat. *Revue Télédétection*, *6*(2), 123–142.
26. Singhal, S., & Gupta, R. P. (2010). *Applied hydrogeology of fractured rocks* (2nd ed., p. 428). Netherlands: Kluwer Academic Publishers (Springer) Dordrecht.
27. Koita M., Jourda H., Ruelland D., Koffi K., Pistre S., & Savane I. (2010). Cartographie des accidents régionaux et identification de leur rôle dans l'hydrodynamique souterraine en zone de socle. Cas de la région de Dimbokro-Bongouanou (Côte d'Ivoire) (pp. 805–820).
28. Anwar, A., Shawki, N., & Abdoh, G. (2013). Landsat ETM-7 for lineament mapping using automatic extraction technique in the SW part of Taiz area, Yemen. Global Journal of human social science geography, geo-science, environmental and disaster management.13, ISS 3, VER 1.0, 2013.
29. Rayan, T. (2013). Automatic Extraction and Geospatial Analysis of Lineaments and their tectonic significance in some areas of Northern Iraq using remote sensing techniques and GIS. *International journal of enhanced research in science technology and engineering*, *2*, 25–40.
30. Sedrette, S., Rebai, N., & Turki, M. M. (2015). Développement d'une approche quantitative d'extraction semi-automatique des accidents tectoniques de la carte géologique de Nefza au 1/50 000 en format shp. *Arabian Journal of Earth Sciences*, *2*, 53–64.

Part II
GEORISK and Environmental Monitoring

Mass Movement Hazard Assessment at a Medium Scale Using Weight of Evidence Model and Neo-predictive Variables Creation

Mohamed Mastere

Abstract

In this paper, mass Movements hazard (MMH) was evaluated by using weights of evidence model (WOE). The Bab region (Central Rif, Morocco) is known as one of the most areas prone to mass movements (MM). High resolution remote sensing data interpretation and a systematic field survey were performed in order to obtain an inventory of MM. A total of 247 MM of the following types: landslides (45%), rock-falls (48%) and debris flows (7%) were identified, which covered an area of 20 km^2 and accounting for 3% of the study area (630 km^2). Then, nine conditioning parameters characterizing topographical, geological, and environmental conditions were mapped, hierarchized and integrated as the main parameters, controlling the occurrence of MM. To eliminate the causal dependency a strategy based on the establishment of three neo-predictive variables (NPV) with a geomorphological significance from the parameters responsible for the conditional independency violation was implemented. Several simulations with several combinations of parameters were performed in order to examine the influence of each variable on the predictive power of MMH. The prediction capability of each model is determined by the area under the curve value and the best combination model was selected to generate the MMH map.

Keywords

Mass movements • Hazard • Neo-predictive variable • WOE, GIS • Morocco

M. Mastere (✉)
Geopac Research Center, Scientific Institute, Mohammed V University of Rabat, Avenue Ibn Battouta, Agdal, Rabat, Morocco
e-mail: mohamed.mastere@gmail.com

1 Introduction

In the three recent decades, researchers in geomorphology and other fields have become increasingly interested in using geographical information systems (GIS) in order to provide hazard maps of mass movements (MM). GIS is a powerful tool and has the facility to manage and assess mass movements hazard (MMH) on a medium scale of (1:25 000–1:50 000), where numerous calculations, data collection, and management are required. Thus, the GIS has become an important tool to predict the probability, location, and frequency of these phenomena in the future.

MMH assessment defines the probability of a potentially damaging instability occurring within a certain period of time in a given area [1]. The "probability" of a MM occurring can refer to: (a) recurrence time; (b) uncertainty of geomechanical parameters or geotechnical models; (c) the frequency, intensity and duration of triggering agents [2]. The assessment methods proposed in literature [3] can generally be divided into two groups: direct mapping based on qualitative methods, in which the degree of hazard is based on the experience of the geomorphologist [4]; indirect mapping based on qualitative methods, in which empirical, statistical or deterministic approaches [5] can be used and based on the information obtained from relationships between instability causative factors and MM distribution.

These different methods and techniques for MMH mapping have been tested by the scientific and technical institutions. A great variety in scale and mapping procedures exists. In fact, the choice of the type and scale of the map depends on many factors. Primarily on the requirements of the last user, the ultimate purpose of the investigation, the degree of precision must the result be presented, and the available resources in the form of money and manpower [3,6]. Despite these efforts, no agreement has yet been reached on the techniques and methods for MMH mapping. This makes it difficult to compare the results that achieved by different authors [7–11]. Overall these authors indicate that research needs

N. Rebai and M. Mastere (eds.), *Mapping and Spatial Analysis of Socio-economic and Environmental Indicators for Sustainable Development*, Advances in Science, Technology & Innovation, https://doi.org/10.1007/978-3-030-21166-0_7

improvements by exploring more explanatory variables and more powerful techniques. MM are one of the natural risks to which Morocco is confronted, especially in the Rif regions. Those seem more frequent in the regions located in faults corridor, where their recent reactivation may potentially contribute to MM genesis. The consequences of this phenomenon can be multiple: habitations destruction, interruption of the roads traffic, and scarce loss in human lives. Nowadays, in the Rif about 53% of the total budget allocated to the Provincial Directorate of Public Works was used for reinforcement work and rehabilitation of roads due to the occurrence of a large number of MM. MMH assessment is thus a topic of major interest for the authorities that manage regional planning of environmental protection and prevention from natural hazards. Consequently, Morocco needs reliable MMH maps to help the decisions-makers for safe management and development. Several funds and foundations were set up to provide a budget about 3G.Dhs (i.e 270 M.) for the prevention and prevision of natural hazards. Although several efforts have been made previously to prevent the MM occurrence in the Rif, but their extent and frequency increase even in the regions qualified with "low" hazard [12–15]. The following reasons are advocated: $-1°$ MM are not considered in their global geodynamic context within the watershed. Namely, taking into consideration seismicity, active tectonic, and irregular rainfall in the study area without forgetting the other environmental factors, $-2°$ The utilization of not suitable methods or classical man-agement [16], $-3°$ A bad choice of scaling-adapted model and of combination of MM causative factors. The Bab Taza region knows every year several MM which have a big economic repercussion. They result in physical and material damage, being able to affect the individuals and their properties; delays and disruptions brought to the traffic (between the villages and with the surrounding zones) in the case of roads interruption; disasters caused to houses and farmlands particularly by the phenomena of badlands, bank undermining, and sliding.

The current study evaluates MMH through GIS techniques using Bayes theorem based on weights of evidence (WOE) to solve most of these problems. It aims to produce the MMH map for Bab Taza (Chefchaouen, Morocco) at the reliable scale of 1/50,000. Finally, in order to validate the MMH map obtained by the WOE method, the degree of model fit will be evaluated by calculating the cumulative percentage of MM area in each hazard class accompanied by expert judgment.

2 Study Area

The Moroccan Rif situated in the northwest of Africa is a part of an Alpine thrust belt that extends from the Betics in southern Spain and curves through the Straits of Gibraltar into North Africa. The regional thrust sheet transport directions swing through an 180° arc. Thrusting is toward the north in Spain, toward the west around the Straits of Gibraltar and the northwest Rif, and toward the south in the eastern Rif and Tell mountains of Algeria and Tunisia. Many workers attribute the arcuate nature to a collision during the Tertiary period between a micro-plate (Alboran plate), the Iberian plate, and the North African plate [17, 18]. The study area is located in the northern edge of central Rif (North West of Morocco) (Fig. 1) and it contains the three main structural domains of Rifain belt which characterize the whole of the Rifain orogen and which are limited by big overlapping accidents with a vengeance toward the NW. From inside to outside and bottom to top, we can find: (1) the Internal Zone, called Alboran Domain; (2) the Flyschs Domain, and (3) the External Zone. Each domain consists of tectonic complexes of stacked nappes.

1. The Internal Domain can be divided on a three units complex, where the Dorsale Calcaire dominated by Triassic-Liassic carbonates and which occupies nearly half of the study area surface.
2. The Flyschs Domain structured on a nappes complex, which stacked from the deformation of an initial subsiding furrow. These nappes are presented in the study area by: the Jebel Tisirène dated from Late Cretaceous—Early Jurassic, the Beni Ider nappe dated from age Paleogen to burdigalian, the Predorsalian units and the Numidian nappe, also dated from Paleogen to Burdigalian.
3. The external domain contrasting with previous domains by the nature of its materials mainly constituted by marly and clayey formations. It was intensely deformed. This domain is represented in the study area by Intra-rifaine zone, especially the Kétama—Tanger externe and Tanger interne unities.

The Soft lithological nature of geological formations associated to the active tectonic system of the Rif belt induce an intense erosive dynamic through the study area. The resulting morphology is characterized by very undulating reliefs and very steep slopes reaching maximum values of 80. This topography provides to this zone a mountainous character marked by a much-boxed valleys and strong unevennesses reaching 2120 m. Finally, this area is characterized by a Mediterranean climate with a hot and dry summer and a sub-humid and cold winter.

3 Data Preparation

3.1 Mass Movements Inventory and Description

Known as one of the most areas prone to MM in the northwest of Morocco, the Bab Taza sector was selected because of the frequency, distribution and diversity of its active MM (Fig. 1).

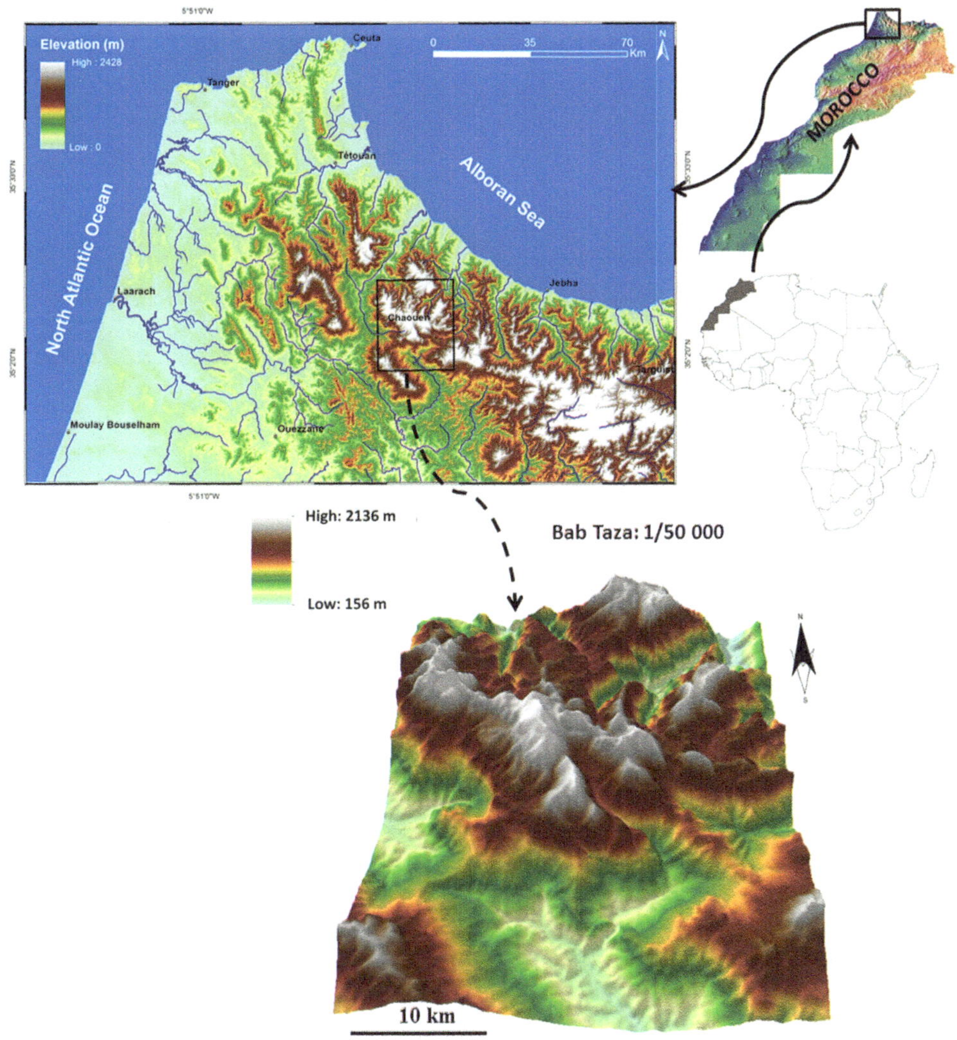

Fig. 1 Geographical situation of the study area with mass movements position

We mapped and inventoried the various MM of the study area. The inventory database was constructed using 1/17,000 and 1/20,000 scales aerial photographs, high-resolution remote sensing data and supported by the available geomorphological maps. Insitu checking of the produced inventory was also performed during extensive field studies. A total of 247 MM were identified, which covered an area of 20 km^2 and accounting for 3 % of the study area (630 km^2). The minimum, mean and the maximum mass movement areas are 0.001, 0.05 and 1 km^2 respectively.

However, typical classification of MM in the Rifean area is not generally easy to establish for the great variability of materials, processes and the morphological complexity. Distinguishing between different types MM requires considerations of several parameters like: velocity and mechanism of movement, material, mode of deformation, geometry of the moving mass and other parameters. For this reason, it is not surprising that there are many classifications in use and conflicts in applications of terms [19]. The earliest widely used classification is that of [20] and most workers since then owe some debt to him for his pioneer effort. More recent classifications are those of [19,21–24].

According to the classification done by [19] which completes and combines between all the previous classifications, we classified the inventoried MM into three categories: i- Rock falls (Fig. 2) that present 48% of the total number of recorded movements. They group a set of fast and rough phenomena which affect stiff and broken rocks such as limestones, clays or crystalline rocks. In the case of sedimentary rocks, the stratification increases the breakage of the rock and thus the predisposition of instability. ii- Landslides (Fig. 2) which occupy 45% of the totality of movements. They correspond to slow movements of a coherent land mass generally along of curved or flat discontinuity surface. According to their types, land-slides may have very different characteristics and achieve very variable dimensions. Most of landslides which we met during the field studies have occurred in sandstone and marly formations or in marly calcareous intercala-

Fig. 2 Pictures of typical mass movements of the study area. **a**: Rock Falls, **b**: Debris flow and **c**: Landslide

tions. iii- Debris flow (Fig. 2) constitute 7% and are mainly encountered in the limestone ridge conglomerate and pelitic formations. They correspond to very fast and sudden movements that start in large-scale masses.

3.2 Mass Movements Influencing Factors

In order to perform MMH mapping it is indispensable to determine the MM causative factors. Therefore, in the first stage, MM related spatial database should be created. Through the knowledge attained from the literature and field investigation the causative factors were defined. Therefore, nine causative factors were selected for the susceptibility mapping and the spatial database of these factors was compiled. Those factors are: lithology, fracturing density, land-use, slope degree, slope aspect, elevation, hydrographical network, rainfall, and earthquakes.

The first dataset was the lithological map. Geological map at 1:50,000 scale covering the study area was simplified and the lithological units were identified and digitized (Fig. 3a). The used classification of lithological formations was proposed by [3] and considered as standard. It is based on the landscape segmentation on geomorphological areas, closely associating lithology and forms.

However, recent and inherited tectonic activity can intervene, either by conditioning or triggering MM in the case of a seismic tremor [25,26], in the fractured areas. The fractures network may support the infiltration of water, involving the increase of interstitial pressure and the reduction of rocks shear strength. When the grounds are saturated the movement is started with formation of major shearing plan. Fracturing map (Fig. 3b) was carried out through high remote sensing data, and orthorectified aerial photographs interpreta-tion, morphostructural analysis of digital elevation model (DEM) and field surveys.

Land use plays an important role in land sliding in a hilly terrain [27]. The land use-landslide relationship can be complex, depending on the nature and type of land use. Land use classes were mapped from the SPOT5 satellite images using supervised and unsupervised classification, visual interpretation and other auxiliary information, such as preexisting maps and field checks. Five land use classes were identified (Fig. 3c):

The three largest datasets are topographical parameters (slope degree, slope aspect, and elevation) (Fig. 3d, e, f) that were obtained from the ASTER remote sensing data (DEM). Before extracting the topographic parameters, we applied some filters allowing correcting artifacts due to the nature of the ASTER sensor. Elevation in each raster cell was the raster cell value of the DEM. For this study, the slope degree was calculated by the inclination computational method using eight neighboring cells. Slope aspect was automatically generated from the DEM and divided into nine classes (Fig. 3f).

The hydrographical density map (Fig. 3g) was prepared by interpreting satellite images and ancillary information. The main drainage pattern of the area is generally dendritic. Up to the 5th order of drainage was found in the area. With this information, a drainage density map was prepared using a density factor computed as the to-tal length of drainage network per a grid cell of 25 × 25 m.

Rainfall is one of the most important causative factors of MM genesis. Over the last few decades, some large and recurrent MM in the surrounding sectors of our study area have affected urbanized areas causing severe damage to properties [28,29]. We used the annual average of rainfall from meteorological stations across Bab Taza region to establish the pattern of annual rainfall variation and generate the rainfall map (Fig. 3h).

In the case of earthquake, there is a release of seismic waves or elastic waves of two types (Longitudinal and transverse) which propagate in the ground. These seismic waves provoke an additional dynamic request of formations [30]. The balance of the strengths is so modified that it can lead to MM. The destabilizations due to earthquakes can manifest in many forms. They can provoke at once landslides, debris flow, rock falls and significant deterioration of infrastructure due to the thixotropic characteristics of soils. In our study area, it does

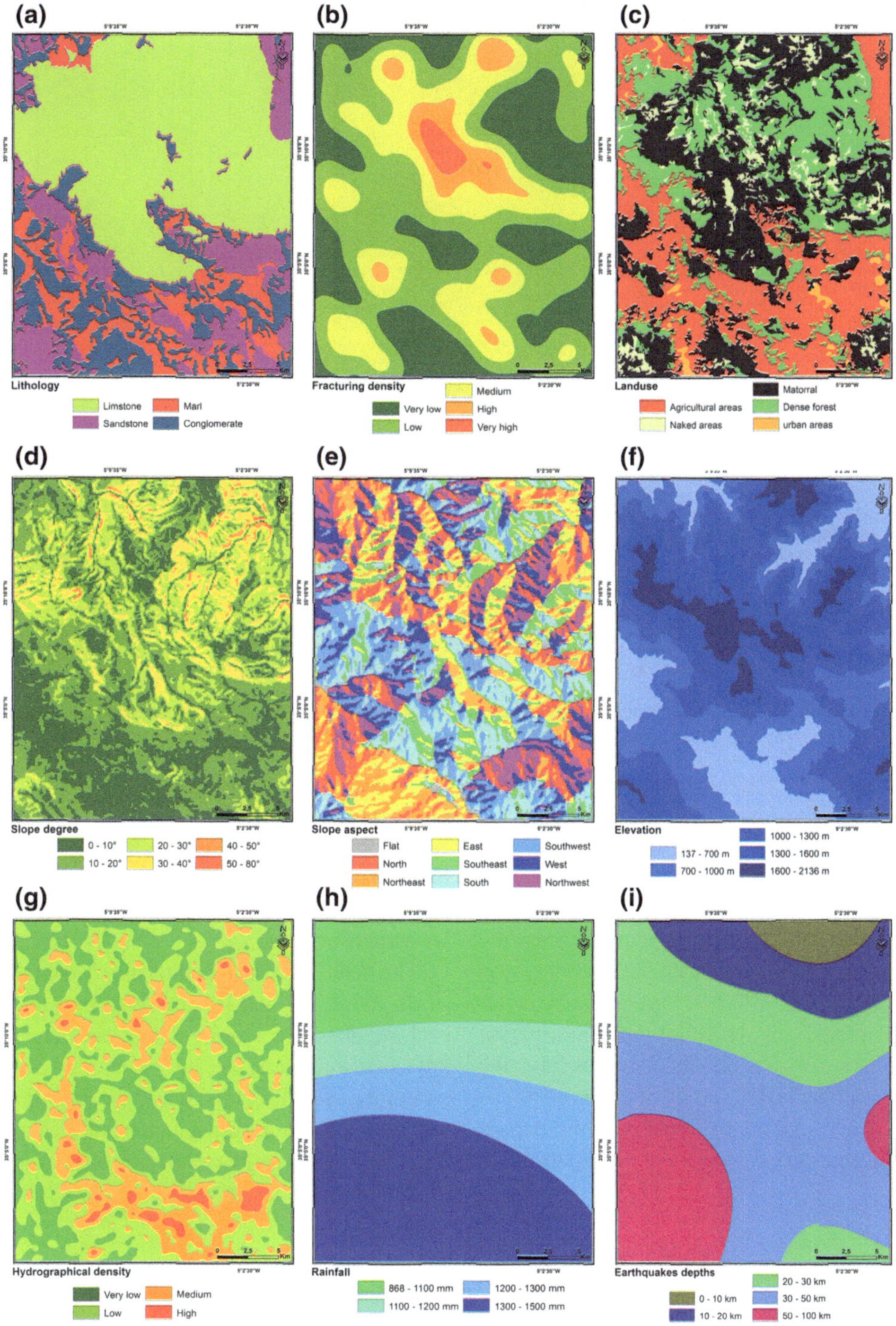

Fig. 3 Mass movements influencing factors used in this study; **a**: lithology; **b**: fracturing density; **c**: landuse; **d**: slope degree; **e**: slope aspect; **f**: elevation; **g**: hydrographical density; **h**: rainfall; and **i**: earthquakes depths

not engender directly MM. The earthquakes do not trigger MM directly. They have effects in the longer term, by increasing the cracking of geological formations and reducing their mechanical resistance. Furthermore, the cracks extended over the longer term tend to increase infiltration of water, promoting the breakdown of materials from the freezethaw cycles and thus result in a MM. However, in seismogenic zones, MMH induced by earthquakes is based on the intensity or magnitude of the earthquake and the nearness of the epicenters or the active faults [31, 32]. A map of earthquakes depths (Fig. 3i) was established and it presents five classes, 0–10, 10–20, 20–30, 30–50, and 50–95 Km.

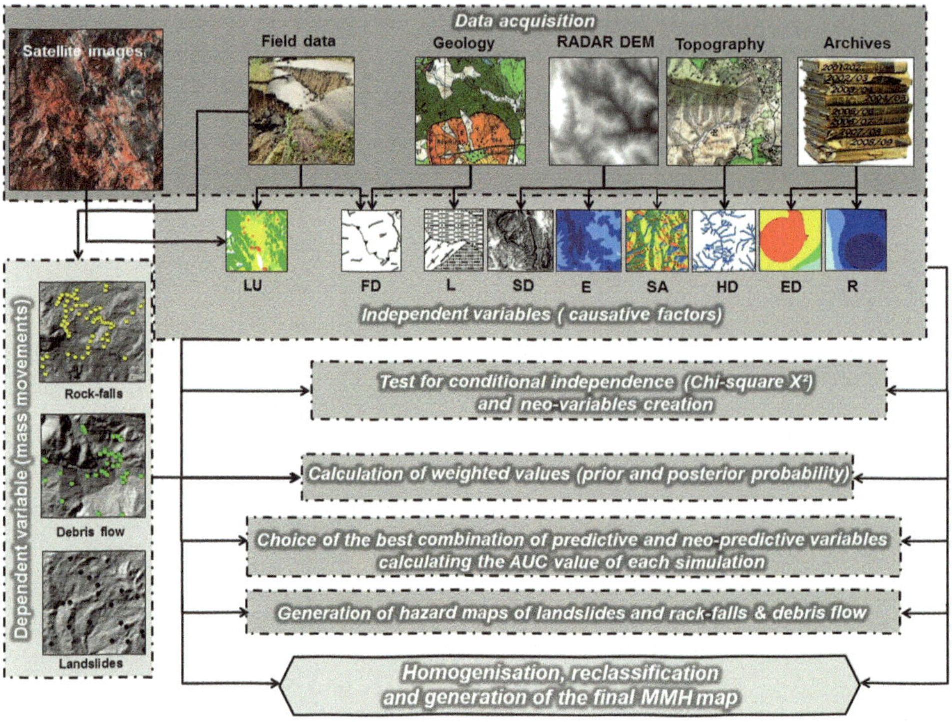

Fig. 4 Flowchart of the modeling process. (LU): landuse; (FD): fracturing density; (L): lithology; (SD): slope degree; (E): elevation; (SA): slope aspect; (HD): hydrographical density; (ED): earthquakes depths and (R): rainfall

4 Methods

The applied methodology is based on the well-known principle of "present and past are keys to the future." The fundamental principle of this approach to MMH mapping is to use the characteristics of existing MM to evaluate the possible areas of future MM. For this reason, MM and MM causative factor databases were constructed (Fig. 4). To store the information of these parameter maps in an uniform thematic database, the size of each pixel for all the layers was 25 × 25 m.

In the weights of evidence model, MM was the dependent variable and the causative factors were the independent variables.

4.1 Weights of Evidence

WOE is a data-driven method which was initially developed for mineral potential assessment. It uses the Bayesian probability model in a log-linear form, and is applicable when sufficient data are available to investigate the statistical relationships between the integrated evidential themes and determine the relative importance each class [33]. It has proven precious for slope failures susceptibility and hazard mapping by several authors [34,35].

In order to apply the WOE approach, historical MM data are indispensable. The MM that occurred in the past is integrated in weighting causative factors. This approach carries the fundamental assumption that future MM will occur under conditions and factors equal or similar to those for comparable MM occurred in the past. It is further assumed that the causative factors for the mapped land-slides remain almost constant over time.

The detailed description of the mathematical formulation of the WOE method is developed by [36,37]. In short, for each variable (X_i) a positive (W^+) and negative (W^-) weight are calculated based on the presence or absence of the MM within the area. Provided that the controlling factors used as variables are conditionally independent, the posterior probability or MMH can be estimated from these weights and the prior probability. Positive (W^+) and negative (W^-) weights can be calculated as:

$$W_i^+ = log_e \left[\frac{\frac{N(LS/X_i)}{N(LS))}}{\frac{N(X_i/\bar{LS})}{N(\bar{LS})}} \right] \quad (1)$$

$$W_i^- = log_e \left[\frac{1 - \frac{N(LS/X_i)}{N(LS))}}{1 - \frac{N(X_i/\bar{LS})}{N(\bar{LS})}} \right] \quad (2)$$

Where:

- $N(LS/X_i)$: is the number of cells in a mapped MM and with the presence of a certain category of a variable (X_i);
- $N(LS)$: is the number of cells in a mapped MM;

Table 1 Results degree of freedom (α) and chi-square (X^2) tests for conditional independence. (A): Contingencies values of theoretical chi-square (X^2) test; (B): Contingencies values of real chi-square (X^2) test; (C): Degree of freedom value which must be superior to 0.05 to reject all conditional independence assumption between each pair of predictor variables

(A)

Variable	SD	ED	R	HD	LU	L	FD	E
SA	40	32	24	24	32	24	32	32
SD		20	15	15	20	15	20	20
ED			12	12	16	12	16	16
R				9	12	9	12	12
HD					12	9	12	12
LU						12	16	16
L							12	12
FD								16

(B)

Variable	SD	ED	R	HD	LU	L	FD	E
SA	13	18	17	5	11	8	3	5
SD		14	18	4	16	17	5	6
ED			75	4	11	18	8	5
R				2	11	17	11	1
HD					4	3	3	6
LU						19	2	3
L							7	11
FD								5

(C)

Variable	SD	ED	R	HD	LU	L	FD	E
SA	1,00	0,97	0,81	1,00	0,99	0,99	1,00	1,00
SD		0,82	0,00	0,99	0,67	0,00	0,99	0,99
ED			0,00	0,97	0,75	0,00	0,94	0,99
R				0,97	0,51	0,00	0,52	0,99
HD					0,97	0,91	0,98	0,90
LU						0,00	0,99	0,99
L							0,82	0,50
FD								0,99

- $N(X_i = \bar{LS})$: is the number of cells with the presence of a certain category of a variable (X_i) and located outside of a mapped MM ;
- $N(\bar{LS})$: is the number of cells outside of a mapped MM.

A positive (W^+) obtained for a certain category of a variable means that the presence of this category in a grid cell will not contribute to MMH. Similarly, a negative (W^-) means that the absence of the category will not contribute to slope stability. For each grid cell, the contrast C indicates how well a category predicts MM occurrence incorporating both weights.

C is calculated as:

$$C = W_i^+ - W_i^- \quad (3)$$

It allows distinguishing important variables from less important. C has a zero value when a category has no statistical relationship with the occurrence of MM.

4.2 Test for Conditional Independence

The MMH assessment using weight of evidence model requires compulsory the evaluation of conditional independence (CI) of the selected predictive variables. It is about a fundamental assumption of this approach [36] and its violation can engender an overestimation or an underestimation of the probability. Among the recommended approaches to estimate the conditional independence, the statistical chi-squared test X^2 is widely used in such studies [38, 39].

In order to include only categories of conditionally independent variables, a pairwise test supported by the X^2 test was performed. The chi-square values were determined (Table 1). It tests also the evidenced relationships of different causative factors.

This test, presented in a contingency table concerns the comparison of the difference of real and theoretical chi-square

Table 2 Spatial correlation between the predictive variables and the calculated weights. (W^+): Positive Weight ($priorprobability$); (W^-): negative weight ($posteriorprobability$); (C): Contrast, difference between positive and negative weights. (HD): Hydrographical density; (FD): Fracturing density; (NPV): neo-predictive variable and (PV): Variable

Predictive/Neo-predictive variables	Area (km)	W^+	W^-	C
NPV: HD – FD				
HD (very low) and FD (very low)	206,962	−0,269	0,108	−0,378
HD (low) and FD (low)	222,170	−0,013	0,007	−0,020
HD (low) and FD (low)	128,463	0,355	−0,114	0,470
HD (medium) and FD (high)	56,830	−0,444	0,034	−0,479
HD (high) and FD (very high)	18,924	0,665	−0,029	0,695
NPV: Lithology—slope degree (SD)				
Conglomerate and 0–10° SD	102,703	0,062	−0,012	0,075
Marl and 10–20° SD	77,669	0,548	−0,107	0,655
Sandstone 20–30° SD	58,470	0,513	−0,070	0,583
Sandstone and 30–40° SD	63,245	0,433	−0,062	0,495
Limestone and 40–50° SD	240,937	−0,913	0,316	−1,230
Limestone and 50–75° SD	88,277	0,096	−0,016	0,112
NPV: Rainfall—Elevation				
137–700 m, 852–987 mm	58,985	−0,478	0,038	−0,516
700–1000 m, 852–987 mm	124,273	−0,241	0,050	−0,292
1000–1300 m, 987–1123 mm	216,387	−0,236	0,103	−0,339
1300–1600 m, 1258–1394 mm	101,164	0,287	−0,064	0,352
1600–2136 m, 1223–1258 mm	134,777	0,376	−0,130	0,506
PV: Slope aspect				
Plat	0,075	–	–	–
North	80,567	−0,276	0,034	−0,310
Northeast	83,877	0,022	−0,003	0,025
East	81,287	−0,101	0,014	−0,115
Southeast	63,791	0,757	−0,133	0,890
South	84,716	−0,143	0,020	−0,163
Southwest	83,607	0,025	−0,003	0,029
West	80,943	−0,280	0,035	−0,315
Northwest	78,112	−0,245	0,030	−0,275
PV: Landuse				
Dense forest	154,763	−0,035	0,011	−0,046
Matorral	235,331	−0,376	0,170	−0,546
Agricultural areas	199,026	0,410	−0,261	0,672
Naked areas	38,848	−0,447	0,023	−0,470
Urban areas	7,107	–	–	–
PV: Earthquakes depths				
0–10 m	27,070	1,012	−0,079	1,092
10–20 m	87,708	−0,586	0,068	−0,655
20–30 m	161,805	−0,505	0,127	−0,632
30–50 m	245,250	0,094	−0,064	0,158
50–100 m	114,787	0,249	−0,064	0,313

(X^2) for each pair of causative factors and for all the predictive variables, according to a degree of freedom superior to 0.05. When the theoretical real chi-square (X^2) is lower than real chi-square (X^2) for a couple of factors, this means that there is a conditional dependence. In this case, the two parameters cannot be integrated into the modeling process. However, it is advisable to combine them in order to create a neo-variable [2,11,33,40].

Thus, the tables A, B and C show the results of theoretical probability contingencies, the real probability and the calculation results of the degree of freedom between each pair of variables. The comparison of theoretical and real (observed) chi-square (X^2) shows that 6 of 36 pairs of variables have a conditional dependence. These pairs are: lithology—slope degree; lithology—earthquakes depths; lithology—rainfall; lithology—landuse; rainfall—earthquakes depths and rainfall—slope degree.

4.3 Neo-predictive Variable Creation

The proven dependency must be eliminated before introducing the independent parameters in the modeling process. To do this, the dependent causative factors can be combined in a new predictor variable (neo-variable) and then introduced into spatial analysis as a single variable [33,40]. It is also possible to aggregate the classes of the same variable, or increase their numbers.

In this study, to eliminate the causal dependency, we proposed a strategy based on the establishment of three neo-predictive variables (NPV) with a geomorphological meaning including the parameters responsible of the CI violation, as proved by repetitive chi-square tests (Table 1). The three NPV which allowed the causal independency elimination are:

i. Hydrographical density—Fracturing density (NPV HD—FD);
ii. Lithology—Slope degree (NPV L—SD);
iii. Rainfall—Elevation (NPV R—E).

The construction of these NPV was strongly conditioned by the deep knowledge of the relationship between each predictor variable and their relationships with the inventoried MM. The combined variables were chosen respecting the geomorphological sense of their association.

4.4 Calculation of Weighted Values

The resulting weights and contrasts as shown in Table 2 reflect the importance of each category of the independent variables using a priori information on MM locations to maximize contrast values, and results of WOE modeling MMH. For W+, positive values contribute to MM while negative values rather indicate stable zones. For $W-$ it is the other way around. The difference between the two weights (W^+ and W^-) is known as the weight contrast, (Eq. 3), and the magnitude of the contrast reflects the overall spatial association between the causative factor and MM. If the weight contrast is positive, the factor is favorable for the MM genesis, and if it is negative, it is unfavorable for the MM. If the weight contrast is close to zero, this indicates that the factor shows little relation to slope stability.

Table 3 AUC results calculation of various simulations of landslides hazard mapping

Combinations of independent variables	AUC
A = NPV HD-FD + NPV L-SD + NPV R-E	0,86
B = PV ED + PV LU + PV SA	0,70
C = NPV HD-FD + NPV L-SD + NPV R-E + PV ED + PV LU + PV SA	0,89
D = NPV L-SD + NPV R-E + PV ED + PV LU + PV SA	0,81
E = NPV HD-FD + NPV R-E + PV ED + PV LU + PV SA	0,82
F = NPV HD-FD + NPV L-SD + PV ED + PV LU + PV SA	0,85
G = NPV HD-FD + PV ED + PV LU + PV SA	0,87
H = NPV HD-FD + NPV L-SD + NPV R-E + PV ED + PV OS	0,78
I = NPV HD-FD + NPV L-SD + NPV R-E + PV ED + PV SA	0,73
J = NPV HD-FD + NPV L-SD + NPV R-E + PV LU + PV SA	0,90
K = NPV HD-FD + NPV L-SD + PV PH + PV ED + PV LU + PV SA	0,83
L = NPV HD-FD + NPV L-SD + PV ED + PV LU + PV SA	0,86

5 Results: Hazard Map Calculation and Selection of the Best Parameters Combination

5.1 Landslides Hazard Map

The relative weights of each independent variable or neo-variable were summed and the posterior probability was calculated for landslides pixel by pixel. This implementation was carried out by performing several simulations with several combinations of parameters in order to examine the influence of each variable on the predictive power of landslide hazard map. The prediction capability of each model is determined by the area under the curve value (Table 3). Only simulations with AUC value higher or equal to 0.70 were analyzed.

According to the obtained AUC results, it appears that all simulations provide generally good results, excepting the 'B' simulation which has the smallest AUC value (0.70). This may be due to the reduced number of variables integrated in this simulation. The combination 'I' presents an underestimated second value, where the land use factor has not been introduced. It demonstrates the importance of this parameter in MMH assessment. Comparing the same simulation (I) with the succeeded one (J), the integration of land use variable increases the predictive power of the model to 0.90 which allowed the generation of landslides hazard map (Fig. 5a).

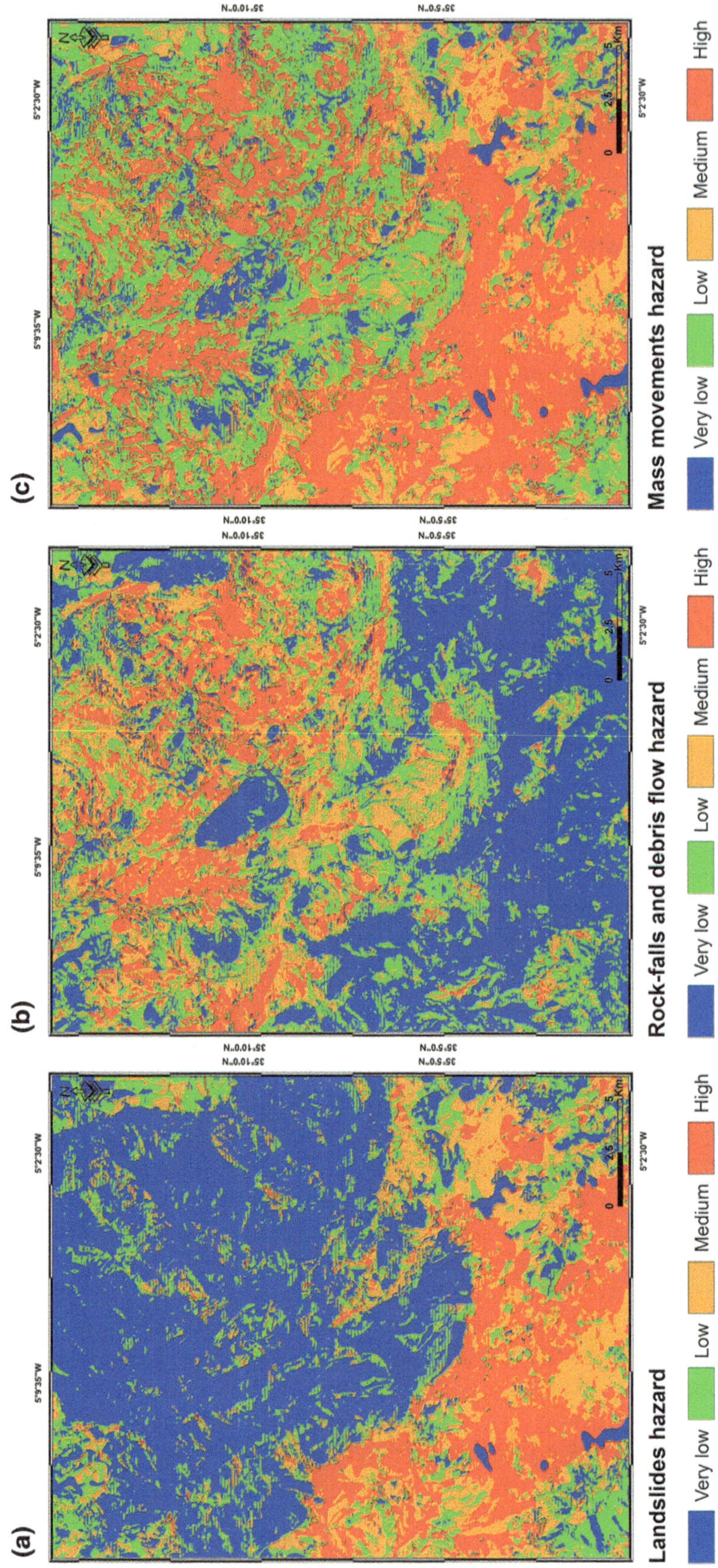

Fig. 5 Hazard maps **a**: landslides hazard map; **b**: rock-falls and debris flow hazard map and **c**: mass movements hazard map

5.2 Rock-Falls and Debris Flow Hazard Map

Rock-falls and debris flows were treated together, because the number of debris flows is much reduced and does not exceed twenty. Also with the exception of a single debris flow case which is in its mature stage and called Akroud debris flow [41], other debris flow phenomena have a rudimentary aspect that can be likened to the behavior of rack-falls, especially as the rock-falls may correspond to one of the evolution stages of a debris flow.

However, the same analysis steps previously adopted for the zoning of the landslide hazard have been followed here. Therefore we carried out several simulations and calculated their AUC values. That allowed us to select the best simulation generate the hazard map of rock-falls and debris flows (Fig. 5b). This is the combination formed by NPV HD-FD + NPV L-SD + NPV R-E + PV LU + PV SA, and which have the largest AUC value (0.81).

5.3 Final MMH Map

To simplify the reading of the obtained results; landslides and rock-falls and debris flow hazard maps have been merged into a single synthetic map defining the levels of MMH in Bab Taza region (Fig. 5c). The low, medium, high and very high hazard classes cover respectively 10, 28, 26, and 36% of the study area. Then, to examine the degree of adjustments of different MMH classes with the inventoried MM and which was not included in the modeling process, we calculated the index of each class fit.

Thus, 47% of MM are classified correctly in the very high hazard class, 30% are classified in the high hazard class, 17% in the moderate hazard level and 6% in the lower hazard level. This is a satisfactory result, considering the misclassified percentage of MM (6%).

6 Conclusions

In the three recent decades a lot of developments have been made in indirect statistically based MMH mapping on medium scale using weights of evidence model. Nevertheless, still many limitations exist. One of them is the elimination of the causal independency without removing the variables responsible for this violation. In particular, when these parameters constitute the main causative factors of MM genesis. This is why we proposed a methodology to assess MMH at medium scale (1: 50,000) using a bivariate approach of WOE model by creating and incorporating NPV. All steps were applied within the framework of a GIS in five steps:

- MM (dependent variable) mapping, where a detailed MM inventory map of the study area was established from high-resolution satellite images, aerial photographs, geomorphological maps, and extensive field studies. According to the classification done by [19], we defined three MM categories: rock falls, landslides and debris flow.
- Influencing factors (independent variables) mapping and classification,
- degree of freedom (α) and Chi-Square (X^2) tests were calculated to check the predictor variables conditional independency and the elimination of its violation by creating and integrating three NPV,
- Model application, choice of the best combination of predictive and neo-predictive variables calculating the AUC value of each simulation,
- Establishment of MMH by merging the separately generated hazard maps of landslides and rock-falls & debris flow.

The predictive capability of the application of WOE model in Bab Taza region for known and unknown MM shows that the three created NPV are sufficient to create an optimum and valid map MMH. That's why it's important to state that a best model is not necessarily which uses the largest number of dependent variables. This means that a careful selection of predictor variables with a good accuracy and precision will considerably reduce gathering efforts without affecting the quality of results.

In addition, the spatial MM distribution is the result of the interaction of many factors; some of them are not mappable or difficult to incorporate in hazard analysis. A reliable and accurate hazard assessment depends on the proper identification of these factors. The integration or not of some of the causative factors may change significantly the capability hazard modeling.

Furthermore, it is important to emphasize the importance of Geo-graphical Information Systems (GIS), because many factors can play a role in the genesis of MM, and the analysis is complex with the requirement of a large number of input variables. The use of GIS has improved the capabilities for MMH mapping over large regions in shortly.

Generation of mass movements hazard (MMH) maps is of great significance for land use planning, engineering works design, civil protection and risk mitigation programs. Ranking the slope stability of the study area in categories that range from stable to unstable, it shows where MM may occur. For this reason, the produced hazard map can be used for planning protective and mitigation measures by the local and regional authorities. Especially, in the orientation and the choice of implementations in development sites, in urban extensions as well as in the setting up of new roads and highways in the frame of the National Development Program of provinces in

Northern Morocco. This study reinforces the use of bivariate statistical models based as an objective approach for MMH susceptibility assessment, assuming the use of specific statistical tests with the creation and incorporation of neo-predictive variables if the conditional independence rule is violated.

The proposed procedure has to be tested in other types of environment in order to verify its spatial robustness. Nevertheless, the WOE model used in this study may also be applicable to other MM prone areas with similar environmental conditions and at various scales, using the same association of predictive variables, to define its reproducibility and examine the associated uncertainty of predictions.

Acknowledgements This research was financially supported by Moroccan government (CNRST: Centre National de la Recherche Scientifique et Technique) and the integrated action project MA/08/192 associated to the partenariat Hubert Curien Volubilis 17174 PK (France). The author would like to thank Yannick Thiery, for his constructive criticism and discussions which contributed to the improvement of this paper.

References

1. Varnes, D. J. (1984). *Landslide Hazard Zonation, a review of principles and practice* (p. 63). UNESCO, Paris: IAEG Commission on Landslides.
2. Thiery, Y. (2007). Landslide susceptibility in the Barcelonnette basin (French South Alps): morphodynamic cartography, spatial analysis and probabilistic modelling. Ph.D. Thesis (p. 445). France: Caen/basse-Normandie university.
3. Van Westen, C. J. (1993). Aplication of Geographic Information Systems to Landslide Hazard Zonation. Ph.D. Dissertation, Technical University Delft. ITC-Publication Number 15, ITC (p. 245). The Netherlands: Enschede.
4. Carrara, A. (1989). Landslide hazard mapping by statistical methods: A "black-box" model approach. In F. Siccardi & R. Bras (Eds.), *International Workshop on Natural Disasters in European-Mediterranean Countries* (pp. 427–445). Perugia: CNR-US NFS.
5. Moon, V., & Blackstock, H. (2004). A methodology for assessing landslide hazard using deterministic stability models. *Natural Hazards*, *32*, 111–134.
6. Van Westen, C.-J., Van Asch, T.-W.-J., & Soeters, R. (2006). Landslide hazard and risk zonation: Why is it still sodifficult? *Bulletin of Engineering Geology and the Environment*, *65*, 167–184p.
7. Rowbotham, D., & Dudycha, D. N. (1998). GIS modelling of slope stability in Phewa Tal watershed, Nepal. *Geomorphology*, *26*, 151–170.
8. Dai, F. C., & Lee, C. F. (2002). Landslide characteristics and slope in-stability modeling using GIS, Lantau Island, Hong Kong. *Geomorphology*, *42*, 213–228.
9. Guzzetti, F., Reichenbach, P., Cardinali, M., Galli, M., & Ardizzone, F. (2005). Landslide hazard assessment in the Staffora ba-sin, northern Italian Apennines. *Geomorphology*, *72*, 272–299.
10. Pradhan, B., Oh, H., & Buchroithner, M. (2010). Weights-of-evidence model applied to landslide susceptibility mapping in a tropical hilly area. *Geomatics, Natural Hazards and Risk*, *1*(3), 199–223p.
11. Poiraud, A. (2014). Landslide susceptibility-certainty mapping by a multi-method approach: A case study in the Tertiary basin of Puy-en-Velay (Massif central, France). *Geomorphology*, *216*, 208–224.
12. Fares A (1994) Essai méthodologique de la cartographie des risques naturels liés aux mouvements de terrain. Applica-tion à l'aménagement de la ville de Taounate (Rif, Maroc). (p. 177) Thèse de Doctorat. Université de Franche Comté.
13. Margaa, Kh. (1994). *Essai de cartographie des risqué naturels: application à l'aménagement de la region d'Al Hoceïma (Rif, Nord-Maroc)* (p. 196p). Thèse Doctorat: Université de Franche-Compté.
14. El Kharim, Y. (2002). *Etude des mouvements de versants dans ta région de Tétouan (Rif occidental): inventaire, analyse et cartographie* (p. 250). Maroc: Thèse de Doctorat, Université Abdelmalek Essaadi, Tétouan.
15. Sossey Alaoui, F. (2005). *Traitement et intégration des données satellitaires optiques et Radar dans un SIG en vue de l'obtention de carte de l'aléa lié aux instabilités de terrain dans la péninsule de Tanger (Rif septentrional, Maroc)* (p. 175p). Doctorat: Université Mo-hamed V, Faculté des Sciences, Rabat.
16. Ardizzone, F., Cardinali, M., Carrara, A., Guzzetti, F., & Reichenbach, P. (2002). Uncertainty and errors in landslide mapping and landslide hazard assessment. *Natural Hazards and Earth System Sciences*, *2*, 1–23.
17. Andrieux, J. (1971). La structure du Rif central. Etude des relations entre la tctonique de compression et les nappes de glisse-ment dans un tronon de la chane alpine. *Notes et Mmoires du Service Gologique du Maroc*, *235*, 1–450.
18. Tapponnier, P. (1977). Evolution tectonique du systme alpin en mditerrane : poinonnement et crasement rigide-plastique. *Bulletin de la Socit Gologique de France*, *7*, 437–60.
19. Selby, M. J. (1993). *Hillslope materials and processes* (2nd ed., p. 451). Oxford: Oxford University Press.
20. Sharpe, C. F. S. (1938). *Landslides and related phenomena: A study of mass movements of soil and rock* (p. 137). New York: Columbia University Press.
21. Varnes, D. J. (1974). The logic of geological maps, with reference to their interpretation and use for engineering purposes. U.S. Geological Survey Professional Paper 837, p. 48.
22. Nemčok A, Pašek J, Rybář J. (1972). Classification of landslides and other mass movements. *Rock Mechanics*, *4*, 71–78.
23. Hutchinson, J. N. (1988). General report: Morphological and geotechnical parameters of landslides in relation to geology and hydrology. In *Proceedings 5th International Symposium on Landslides* (Vol. 1, pp. 3–35). Lausanne.
24. Sassa, K. (1989). Geotechnical classification of landslides. *Landslide News, Japan Landslide Society*, *3*, 21–24.
25. Mastere, M. (2011). Mass movements susceptibility in the Chefchaouen province (central Rif, Morocco): spatial analysis, multi-scale probabilistic modeling and impact on development and planning. Ph.D. thesis (p. 316). University of Western Brittany.
26. Hewitt, K. (2004). Geomorphic hazards in mountain environments. In P. Owens & O. Slaymaker (Eds.), *Mountain Geomorphology* (pp. 187–218). London: Hodder Scientific. Chap. 9.
27. Mastere, M., Van Vliet Lanoe, B., Ait Brahim, L. (2013). Land use mapping and its relation to mass wasting and gullying in North-Western Rif (Morocco). Geomorphology: Relief, process and environment. V3, 335–352.
28. Mastere, M., Van Vliet, Lanoe B., Ait Brahim, L., & El Moulat, M. (2015). A linear indexing approach to mass movements susceptibility mapping: The case of the Chefchaouen province (Morocco). *International Journal of Geomatics and Spatial Analysis*, *2*(25), 245–265.
29. Mastere, M., Van Vliet, Lanoe B., & Ait Brahim, L. (2017). Evaluation de la suceptibilité aux mouvements de terrain par approche pro-babiliste, application à la zone méditerranéenne entre Jebha et Oued Laou. *Press Universitaires de la Méditerranée, Eds Léone F et Vinet F, Géorisques*, *7*, 169–178.
30. Durville J.L, Seve G (1996) Stabilit des pentes: Glissements en terrain meubles. Techniques de l'ingnieur, 254.

31. Papadopoulos, G. A., & Plessa, A. (2000). Magnitude-distance relations for earthquake-induced landslides in Greece. *Engineer Geology, 58*, 377–386.
32. Keefer, D. K. (2002). Investigating landslides caused by earthquakes. A historical review. *Surveys in Geophysics 23*(6), 473–510.
33. Bonham-Carter, G. F. (1994). *Geographic Information Systems for Geoscientists: Modeling with GIS* (p. 416p). Canada: Pergamon Press.
34. Lee, S., Choi, J., & Min, K. (2002). Landslide susceptibility analysis and verification using the Bayesian probability model. *Environmental Geology, 43*, 120–131.
35. Dahal, R. K., Hasegawa, S., Nonomura, A., Yamanaka, M., Dhakal, S., & Paudyal, P. (2008). Predictive modelling of rainfall-induced land-slide hazard in the Lesser Himalaya of Nepal based on weights-of-evidence. *Geomorphology, 102*, 496–510.
36. Bonham-Carter, G. F., Agterberg, F. P., & Wright, D. F. (1989). Weights of evidence modeling: A new approach to mapping mineral potential. In: Agterberg, F. P., Bonham-Carter, G. F. (Eds.), *Statistical Applications in the Earth Science: Geological Survey of Can-ada Paper* (vol. 89-9, pp. 171–183).
37. Agterberg, F. P., Bonham-Carter, G. F., & Wright, D. F. (1990). Statisti-cal pattern integration for mineral exploration. In G. Gaal & D. F. Mer-riam (Eds.), *Computer Applications in Resource Estimation Prediction and Assessment for Metals and Petroleum* (p. 19). Oxford: Pergamon Press.
38. Agnesi, V., Conoscenti, C., Di Maggio, C., Iudicello, C., Rotigliano, E. (2003). Landslide hazard analysis in the Giardo River Basin (Middle-Western Sicily). In *Proceedings of the Workshop on Geo-morphological sensitivity and system response*. (pp. 3–11). Camerino, Italy: Università di Camerino - Università di Modena.
39. Poli, S., & Sterlacchini, S. (2007). Landslide representation strategies in susceptibility studies using weights-of-evidence modeling technique. *Natural Resources Research, 16*, 121–134.
40. Van Westen, C. J., Rengers, N., & Soeters, R. (2003). Use of geomorphological information in indirect landslide susceptibility assessment. *Natural Hazards, 30*, 399–419.
41. Mastere, M., Van Vliet, L. B., Mansour, M., & Ait Brahim, L. (2011). The Jbel Akroud debris flow analysis us-ing digital photogrammetry and DEM. *Tldtection, 10*, 147–156.

Evaluation of Morphometric Indices SL, LP, AD for the Spatial Analysis of Neotectonics and Recent Crustal Deformations Case study: Atlas Central, Tunisia

Noamen Rebai, Ali Chaieb, Abdelkader Moussi and Slimene Sedrette

Abstract

Our study endeavors to evaluate some morphometric indices that are selected according to several statistical studies and applied in Quaternary lithological formations. Our study area is the Tunisian Central Atlas, which is known by a Quaternary deposit that can reach 400 m. Because of certain constraints related to remote sensing limitations for the tectonic lineaments mapping in the quaternary and the lack of terrain index in most cases in these deposits, we are interested in spatially analyzing the response of tectonics to these morphometric indices among others, Stream Length (SL), Longitudinal Profile (LP) and Drainage Anomaly (DA). Our proposed methodology consists of evaluating, on the one hand and independently, the tectonic response to morphometric indices. On the other hand, we aim at spatially and arithmetically combine the density of SL, LP, DA by an equivalent weighting for these three indices in order to locate all the zones, detected by the one of indices or the other, and which are sensitive to recent deformation or neotectonics. The calculation of these three morphometric indices (SL, LP, DA) is their representation in terms of density, which allows us to evaluate each of them. The result obtained is consistent with some field surveys in the Kasserine plain Quaternary deposit. However, the cartographic spatial analysis through the maps of the density distribution of each index, lead us to combine the three indices in order to improve the spatial analysis by an equivalent arithmetic weighting to finally generate the high-density zones to prove the existence of recent deformation or neotectonics (the Kasserine plain) and the low density medium zones according to the synthesis map of the combination of the densities of the three indices.

Keywords

Morphometric index • Density map • Neotectonic • Recent deformation • Quaternary • Epicenter

1 Introduction

Several geological works have been devoted to the structural study of the Central Atlas, mainly for the purpose of oil exploration in this sector [1–5].

The study of morphometry and the relief components of the Tunisian Central Atlas can help us understand the activity and the tectonic instability of this sector. The structural interpretation of this domain according to several authors [3,6,7], shows the existence of alignments NE-SW, NW-SE, NS, EW, related to seismotectonic risks. Indeed, the frequency of even small earthquakes confirms the presence of active tectonics and recent deformation [2,3].

This study has two main objectives. First, it is concerned with evaluating, in an independent way, the tectonic response to the morphometric indices which are: 1-Stream Length (SL), 2-Longitudinal Profile (LP) and 3-Drainage Anomaly (AD). Second, it zooms on spatially and arithmetically combines the density of SL, LP, and AD by an equivalent weighting for these three indices in order to locate all the zones that are sensitive to recent deformation or neotectonics.

N. Rebai (✉) · A. Chaieb · A. Moussi · S. Sedrette
University of Tunis El Manar, National School of Engineering of Tunis, LR14ES03 Geotechnical Engineering and Georisk Laboratory, BP. 37, le Belvédère, 1002 Tunis, Tunisia
e-mail: noamen.rebai@enit.utm.tn

A. Chaieb
e-mail: chaieb.aly@gmail.com

A. Moussi
e-mail: abdelkader.moussi@fst.utm.tn

S. Sedrette
e-mail: ssedrette@gmail.com

N. Rebai and M. Mastere (eds.), *Mapping and Spatial Analysis of Socio-economic and Environmental Indicators for Sustainable Development*, Advances in Science, Technology & Innovation, https://doi.org/10.1007/978-3-030-21166-0_8

370012_1_En_8_Chapter-print ☑ TYPESET ☐ DISK ☐ LE ☑ CP Disp.:22/7/2019 Pages: 94 Layout: T3_2column

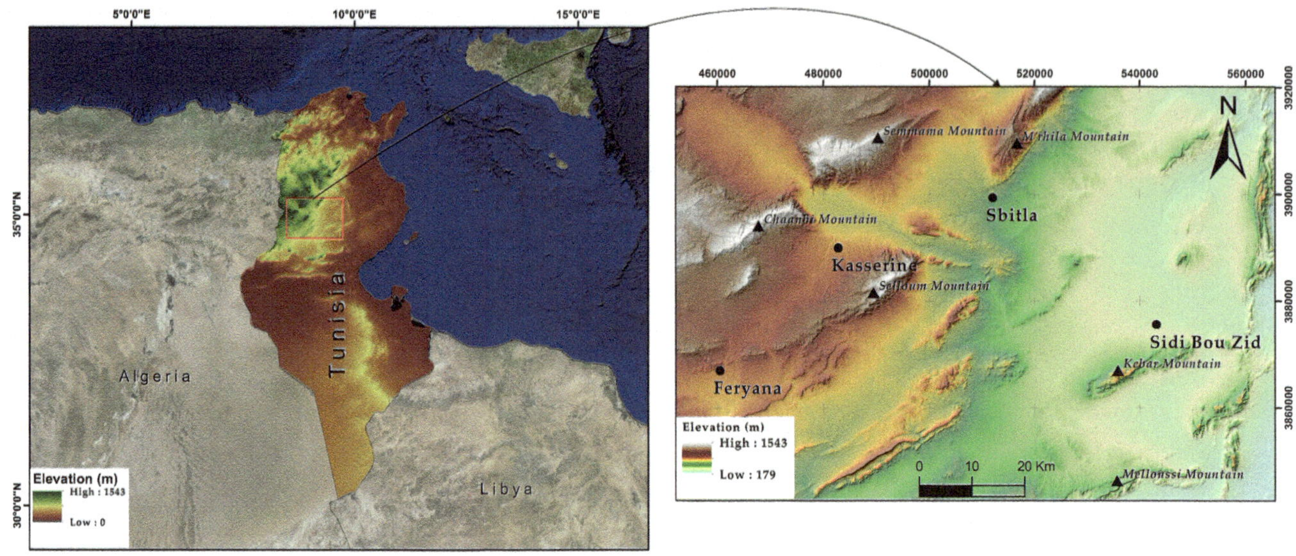

Fig. 1 Location map of the Central Tunisian Atlas area

The problem raised in this study is linked to the lack of a remote sensing approach [8] and structural mapping [9, 10], to detect or illustrate some tectonic accidents that are particularly in the quaternary deposit [11, 12]. However, morphometry can relatively provide some answers by using morphometric indices [13, 14]. In particular, the calculation of these three morphometric indices in question (SL, LP, and DA) is the subject of several bibliographic and statistical studies conducted by some authors [15–17]. Their representation in terms of density, permits their evaluation and validation by field approach. In fact, some results have been obtained from the study of the sector of Kasserine (Fig. 1). A cartographic spatial analysis through the maps of the density distribution density of each index makes us able to combine the three indices by an equivalent arithmetic weighting to finally generate the high-density areas mapping the existence of the recent deformation or the neotectonic (the Kasserine plain) and the low density medium areas.

2 Study Zone

The study area extends from the Tuniso–Algerian border to the NS axis. The latter represents a north-south directional link forming a continuous morphostructural alignment separating the Tunisian Central Atlas from the Sahel platform (Fig. 1). A distinctive raised zone, that is characteristic of this sector, is called the island of Kasserine. It is characterized by important anticlinal structures that are separated by quaternary grabens.

3 Materials and Methodology

From the SRTM DEMs, we have proceeded to an automated approach of the extraction of the hydrographic network, the Versant Basins and the other attributes (slope, orientation, etc.). The development of these attributes is performed according to a morphometric model in a GIS environment. In this context, we are able to compute the morphometric indexes in an automatic way as in the case of the SL [17], executable from the open access library. For the LP case, we have developed an algorithm under Python which facilitates the generation of the knickpoints from the longitudinal profiles of the streams. Concerning the semi-automatic approach, it is applied for the case of the Drainage Anomaly. It allows the characterization of watersheds and the analysis of the behavior of the hydrographic network [18] throughout our study, topographic and seismic data are mapped in the system (WGS84). We have used and represented along the study area, 107 epicenters extracted from the MNI catalog of which 14 are represented in Table 1 and which cover Foussana–Kasserine area.

4 Results and Discussion

4.1 Mapping of the Drainage Anomaly Density

Drainage anomalies express abrupt changes in the hydrographic network architecture and generally emphasize the

370012_1_En_8_Chapter-print ☑ TYPESET ☐ DISK ☐ LE ☑ CP Disp.:22/7/2019 Pages: 94 Layout: T3_2column

Table 1 Coordinates and magnitude of fourteen recent recorded earthquakes in the plain of Kasserine (National Meteorology Institute Catalog "INM"

NUM	Magnitude	X_UTM (m)	Y_UTM (m)
1	5	477251.080234209	3898578.30467933
2	3.2	511829.429036371	3898557.41625562
3	3.8	513671.080220347	3884143.06588395
4	3	498177.858019264	3887459.90806468
5	4.1	468163.208036654	3901932.83128073
6	3.9	469088.038643561	3906365.74884635
7	3.3	472724.751860403	3906354.01171868
8	3.2	509097.326047814	3900772.26861818
9	3.4	486347.308644437	3896341.97430908
10	3.6	500000	3906312.74874531
11	3.4	494544.958391684	3906314.3992533
12	3.9	513654.365155487	3895232.97346845
13	2.5	473637.173542816	3907460.32936123
14	2.6	503646.065191824	3883024.52817507

existence of often hidden structural features of drainage anomalies [19].

The map results obtained demonstrate the concentration of anomalies mainly in the Quaternary plains of Foussana, Kasserine and the North of Jebel Birino and Jebel Douleb (Fig. 2). This index groups the anomalies related to the Longitudinal Profiles of the drains and the hydrographic network architecture, considered by the cross-sectional profiles. DA differs from the LP in relation to the two types of longitudinal and cross-sectional profiles.

The density of DA is relatively correlated with the epicenters located in the West of Jebel Kebar and along Kasserine fault. On the other hand, there is almost no density of DA in the South-Western part below Oueddaga Mountain while DA exists at the Birino Mountain level that is marked by the absence of the quaternary deposits and epicenters.

4.2 Longitudinal Profile Index Mapping (LP)

The longitudinal profiles are crucial to understand the evolution of the relief. Also, the analysis of the graphical plot of the profile allows us to have morphostructural interpretations.

In this work, we have taken the term "K-point" [20]. This point can be an indicator for locating tectonic discontinuities generally related to a fault and structural interpretations [21].

Knickpoint density mapping is limited to plains and is almost absent in mountainous areas (Fig. 3). This can be explained only by the existence of low-order streams of 1 to 2, subject to the steep slope that also characterizes the relief summits.

The LP density is relatively high. It correlates with the epicenters located in the west of Jebel Kebar and along Kasserine fault (Les Grabens de Foussana and Kasserine), (INM catalog). However, it is quite weak in the South-West part below Oueddaga Mountain. The LP density is relatively average at the Melloussi Mountain level where there are Quaternary and epicenters, thus marking the probable presence of the recent deformation.

4.3 Mapping of the "Stream Length Gradient" Index (SL)

The calculation of the length-slope index (Stream Length Gradient Index) is the ratio between the length of the drain and the variation of slope with respect to the half-length of the section of the drain. It is defined by $SL = (\Delta H/\Delta L)L$ where ΔH represents the variation of the watercourse elevation for a distance ΔL, $\Delta H.\Delta L$ expresses the gradient of slope for a section and L represents the distance between the point upstream of the drain and the midpoint of the drain segment [16, 17, 22, 23].

SL Index Knickpoint density mapping is limited to plains and almost absent in mountainous areas (Fig. 4). This is explained in the same way as the LP Index. SL density is relatively high and correlates with epicenters located in the west of Jebel Kebar and along Kasserine Fault (Table 1) [24]. Moreover, contrary to the DA and LP Indices, the density of the SL is relatively present in the South-West part below Oueddaga Mountain, and with a low density at the Melloussi Mountain This presence is marked in the Quaternary deposit and the epicenters.

370012_1_En_8_Chapter-print ☑ TYPESET ☐ DISK ☐ LE ☑ CP Disp.:22/7/2019 Pages: 94 Layout: T3_2column

Fig. 2 Map of drainage anomalies density of the Tunisian Atlas

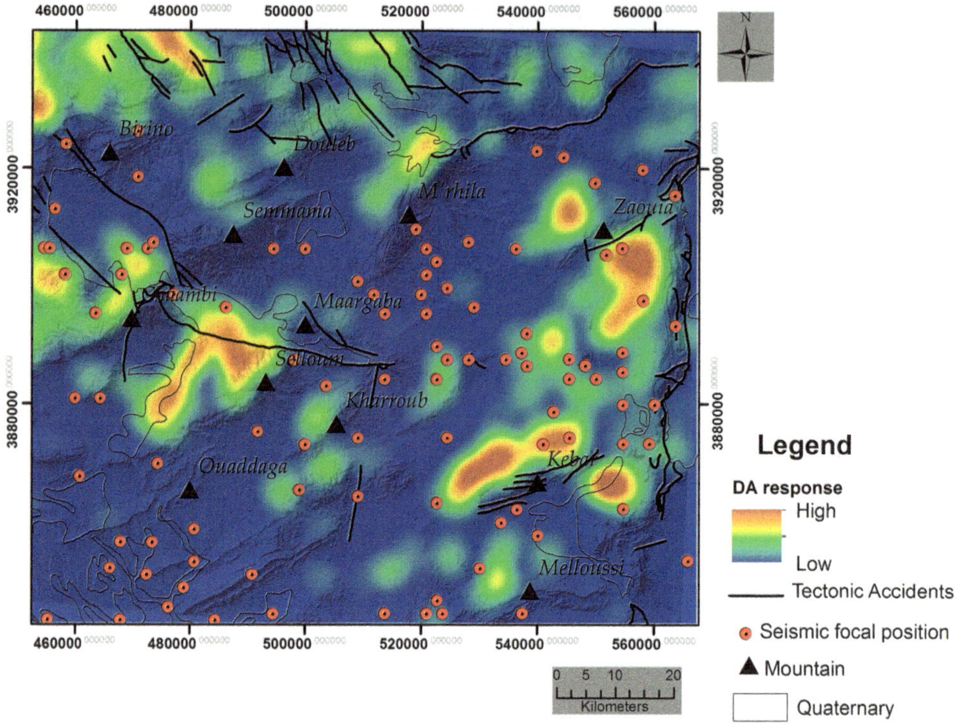

4.4 Mapping of the Combination of the Three Indices

For the better representation of the morphometric anomaly related to the hydrographic network and the topography, we have combined the results of the three parameters with the hypothesis of assigning the same weighting factor for each of them.

The synthesis map obtained shows that the application of these three indices allows to better locate the anomalous zones especially in the Quaternary plains (Fig. 5). These areas are known by recent deformation, confirmed by the presence of epicenters of earthquakes and by the network of active faults of different directions as cited in the work of [2–4,25].

The density map of the combined morphometric indices AD, LP, and SL denotes that there is a high density in Kasserine–Foussena region, which is marked by active tectonics in "Quaternary" deposits [11,12]. This result is well reported by each index independently.

Similarly, there is relatively a high density of the combined indices at the Mghila Mountain, Douleb Mountain level, and between these mountains in the plain area with the presence of Quaternary. This is due to the arithmetic correlation between the three indices where each gives an answer to the tectonic activities with an average or a high density. This answer, which is not affirmative in marking the presence of neotectonics, is marked by the absence of epicenters among others. However, this high density can be the result of the behavior of the drain passing from one lithology to another. This passage will cause a relatively abnormal branching, thus giving an anomalous answer.

Once again, the three indices SL, LP, AD have a relatively average density with the presence of epicenters at the level of the quaternary plain between Melloussi Mountain and Kebar Mountain.

As has been described, the density of SL and LP, contrary to the DA density, is relatively average and present in the southwestern part below Oueddaga Mountain. In this area, there is also a fairly high density of epicenters marking a correlation that expresses a high probability of the presence of recent deformation. This result shows that the indexes SL and LP have demonstrated in distinguishing the identification of the recent deformation, particularly in the quaternary, that the index DA has not detected it.

The result of the combination of the three morphometric indices shows a high density of knickpoints reflecting instability in the plain of Foussana–Kasserine, the anticline of Mghilla Mountain and Douleb Mountain, the SE part of Kebar Mountain, the SE part of Zaouia Mountain and particularly in the quaternary deposit of Oued Htab–Kasserine. The previous research on this area does confirm the presence of recent tectonic activity throughout Oued Htab confirmed by [24,26].

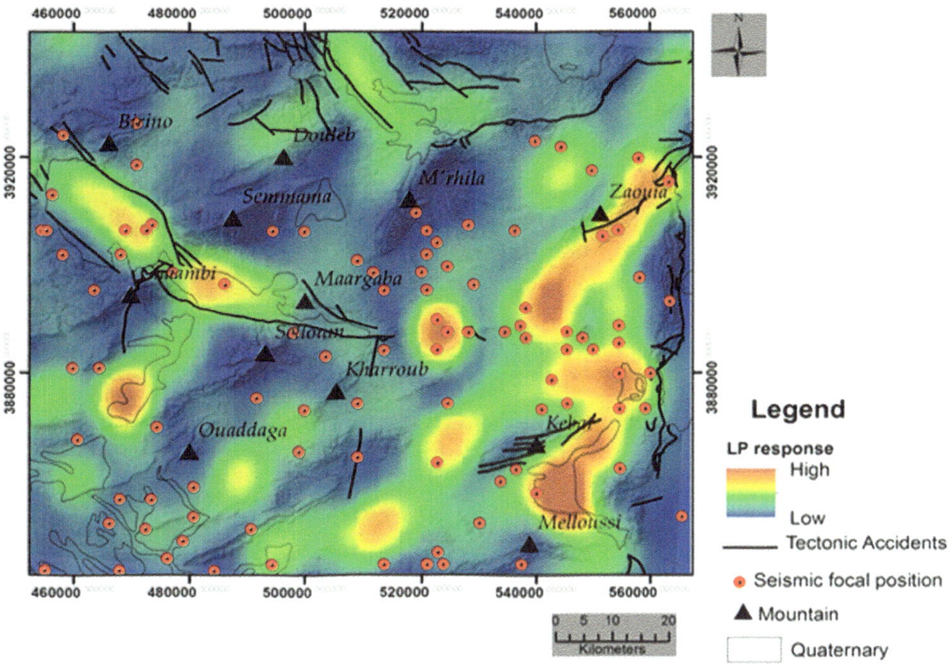

Fig. 3 Mapping of the density of knickpoint longitudinal profiles of drains

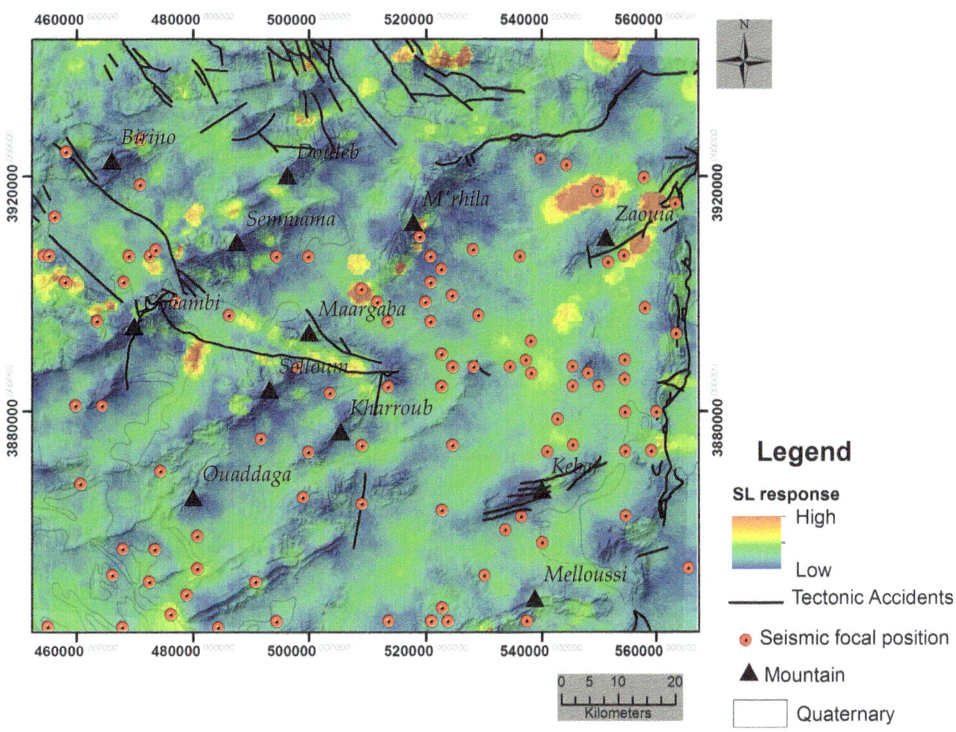

Fig. 4 Mapping of knickpoints calculated by the SL

370012_1_En_8_Chapter-print ☑ TYPESET ☐ DISK ☐ LE ☑ CP Disp.:22/7/2019 Pages: 94 Layout: T3_2column

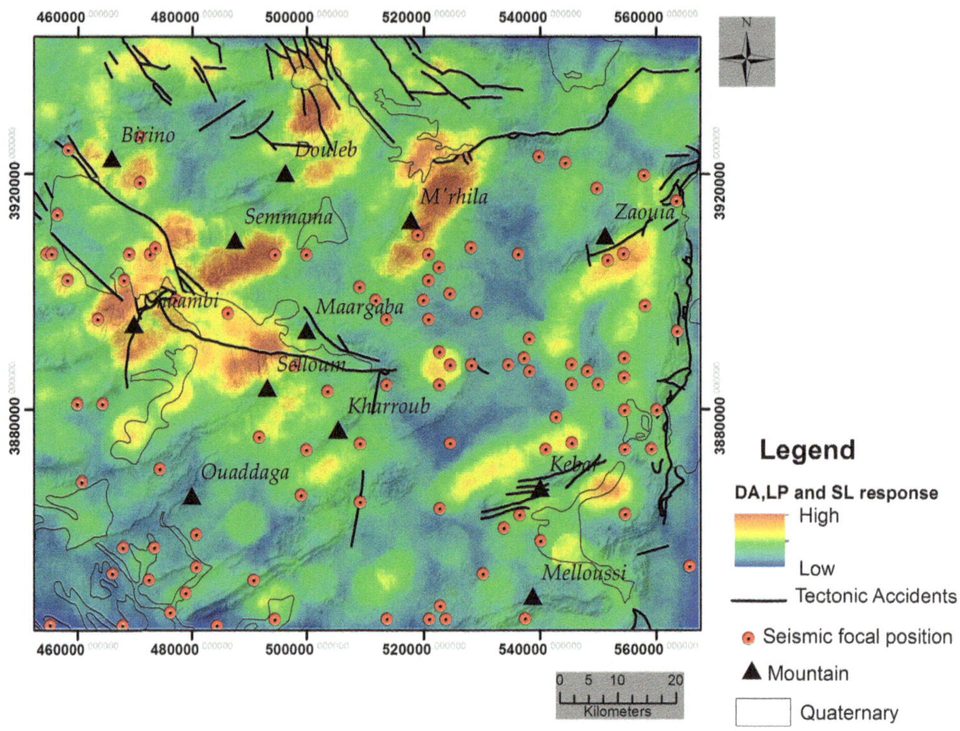

Fig. 5 Mapping the response of the three morphometric indices

5 Conclusion

The investment of several authors in the use of morphometric indices remains a topic of current interest in order to improve the understanding of the tectonic response in particular. Several indices have been exploited, but the statistical approach of some studies on the reliability of these indices to detect the recent deformation or the neotectonics, clearly shows that there are essentially three indices of interests. These indices (SL, LP, and AD) are the subject of our study, throughout which we have focused on evaluating their contribution both independently and when combined.

We tested these indices on an area known by the presence of the Quaternary where the detection of accidents remain rather difficult because of the lack of arguments and the fact that the mapping of these accidents by survey is not so obvious. Similarly, by remote sensing, the detection of lineaments of tectonic accidents is still imprecise for a large scale. On the other hand, morphometry is a relatively effective way to signal these lineaments and to record the recent deformation. It is important to note, in this respect, that morphometry is a means that allows the validation of these lineaments or the mapping of areas potentially sensitive to recent deformation. The presence of alignments or density of epicenters of earthquakes in an area coinciding with morphometric indices (Knickpoints) demonstrates the tectonic response to these morphometric indices.

The density of each morphometric index in the quaternary is substantially identical in the test area of central Tunisia. It strongly coincides with the presence of epicenters at a gradient of low to medium to high density. The confirmation of this result is obtained through field validation by the authors [11,24,26]. However, the index AD remains almost absent in the South-West zone below J. Oueddaga, where there is a fairly high density of epicenters. This correlation expresses a high probability of the presence of recent deformation. This result clarifies the way the SL index is distinguished by a high density compared to the LP index and the index AD which did not give a tectonic answer. This distinction strengthens the hypothesis of the combination of the indicators' densities to best detect the areas sensitive to neotectonics and recent deformation. These conclusions highlight five areas of interest in our sector. This induces us to carry out field observations in details that will be mandatory for the validation of the investment of the combination of the three morphometric indices. In the same vein, we report that the result obtained is consistent with field observations at the level of Foussana–Kasserine plain.

References

1. Burollet, P. F. (1956). Contribution à l'étude stratigraphique de la Tunisie centrale. *Annales de Mines de la Géologie, Tunisie*, *18*, 350 p.
2. Chihi, L. (1984). *Etude tectonique et microtectonique du graben de Kasserine (Tunisie centrale) et des structures voisines Jebel Selloum et Jebel Maargaba*. Thèse Troisième Cycle, Université Paris Sud, Centre d'Orsay, France, 116 p.

3. Ben Ayed, N. (1986). *Evolution tectonique de l'avant pays de la chaîne alpine de Tunisie du début du Mésozoïque à l'actuel*. Thèse ès Sciences, Université du Paris-sud, Orsay, France, 347 p.
4. Abbes, C. (2004). *Structurations et évolutions tectono-sédimentaires Mésozoïques et Cénozoïques, associées aux accidents reghmatiques, à la jonction des marges Téthysienne et Nord- Africaine (Chaîne Nord-Sud, Tunisie centrale)*. Thèse ès Sciences, Université Tunis El Manar, Tunisie, 440 p.
5. Ouali, J. (2007). *Importance du réseau reghmatique dans la tectonogenèse de la Tunisie atlasique à travers l'étude de l'axe Nord-Sud*. Thèse ès Sciences, Université Tunis El Manar, Tunisie, 399 p.
6. Dercourt, J., Zonenshain, L. P., Ricou, L. E., Kazmin, V. G., Le Pichon, X., Knipper, A. L., et al. (1985). Présentation de neuf cartes paléogéographiques au 1 / 20.000.000 s'étendant de l'Atlantique au Pamir pour la période du Lias à l'Actuel. *Bulletin de la Société Géologique de France*, *8*(5), 637–652.
7. Delteil, J., Zouari, H., Chikhaoui, M., Creuzot, G., Ouali, J., Turki, M. M. (1991). Relation entre ouvertures téthysienne et mésogéenne en Tunisie. *Bulletin de la Société Géologique de France*, *162*(6), 1173–1181.
8. Sedrette, S., & Rebai, N. (2016). Automatic extraction of lineaments from Landsat Etm+ images and their structural interpretation: Case study in Nefza region (North West of Tunisia). *Journal of Research in Environmental and Earth Sciences*, *04*(2016), 139–145.
9. Jacques, P. D., Salvador, E. D., Machado, R., Grohmann, C. H., & Nummer, A. R. (2014). Application of morphometry in neotectonic studies at the eastern edge of the Paraná Basin, Santa Catarina State, Brazil. *Geomorphology*, *213*, 13–23.
10. Ambili, V., & Narayana, A. C. (2014). Tectonic effects on the longitudinal profiles of the Chaliyar River and its tributaries, Southwest India. *Geomorphology*, *217*, 37–47.
11. Dlala. (1995). *Evolution géodynamique et tectoniques superposées en Tunisie : Implication sur la tectonique récente et la séismicité*. Thèse d'Etat, Tunis II, 389 p.
12. Chihi, L. (1995). *Les fossés néogènes à quaternaires de la Tunisie et de la mer pélagienne : étude structurale et leur signification dans le cadre géodynamique de la Méditerranée centrale*. Thèse Doct. Ès Science. Géol., Tunis, 325 p.
13. Larue, J.-P. (2011). Longitudinal profiles and knickzones: The example of the rivers of the Cher basin in the Northern French Massif Central. In *Proceedings of the Geologists' Association* (Vol. 122, pp. 125–142).
14. Phillips, J. D., & Lutz, J. D. (2008). Profile convexities in bedrock and alluvial streams. *Geomorphology*, *102*, 554–566.
15. Troiani, F., & Della Seta, M. (2008). The use of the Stream Length-Gradient index in morphotectonic analysis of small catchments: A case study from Central Italy. *Geomorphology*, *102*, 159–168.
16. Pérez-peña, J. V., Azañón, J. M., & Azor, A. (2009). An ArcGIS extension to calculate hypsometric curves and their statistical moments. Applications to drainage basin analysis in SE Spain. *Computers and Geosciences*, *35*(6), 1214–1223.
17. Queiroz, G. L., Salamuni, E., & Nascimento, E. R. (2015). Knickpoint finder: A software tool that improves neotectonic analysis. *Computers and Geosciences*, *76*, 80–87.
18. Hassen, M. B. (2012). *Analyse de la déformation récente dans l'atlas méridional de la Tunisie par géomorphométrie et interférométrie RADAR (DINSAR)*. Thèse de doctorat, Université de Tunis El Manar.
19. Le Pape, S. (1998). Analyse et quantification du Réseau Hydrographique. Le réseau hydrographique comme objet vectoriel. Mémoire de dipl ôme d'ingénieur E.S.G.T.
20. Lin, Z., & Oguchi, T. (2006). DEM analysis on longitudinal and transverse profiles of steep mountainous watersheds. *Geomorphology*, *78*, 77–89.
21. Delcaillau, B., Carozza, J. M., & Laville, E. (2006). Recent fold growth and drainage development: The Janauri and Chandigarh anticlines in the Siwalik foothills, Northwest India. *Geomorphology*, *76*, 241–256.
22. Keller, E. A., & Pinter, N. (2002). *Active tectonics* (2nd ed.). New Jersey, Upper Saddle River: Prentice Hall.
23. Hack, J. T. (1973). Studies of longitudinal stream profiles in Virginia and Maryland. USGS Professional Paper, Washington (pp. 45–97).
24. Chaieb, A., Rebai, N., Ghanmi, M. A., Moussi, A., & Bouaziz, S. (2017). Spatial analysis of river longitudinal profiles to cartography tectonic activity in Kasserine Plain Tunisia. *Geographia Technica*, *12*(2), 30–40.
25. Philip, H., Andrieux, J., Dlala, M., Chihi, L., & Ben Ayed, N. (1986). Evolution tectonique mio-plioquatrenaire du fossé de Kasserine (Tunisie centrale) : Implications sur l'évolution géodynamique récente de la Tunisie. *Bulletin de la Société géologique de France, Paris*, *II*(4), 559–568.
26. Khemiri, S. (2014). *Modélisation morphostructurale et intérêt hydrogéologique du fossé de Foussana (Tunisie Centrale) : contribution du SIG 3D, de la télédétection et des données géophysiques*. Thèse de Doctorat en Géologie (p. 345).

Hydrological Response to Snow Cover Changes Using Remote Sensing over the Oum Er Rbia Upstream Basin, Morocco

Abdelghani Boudhar, Hamza Ouatiki, Hafsa Bouamri, Youssef Lebrini, Ismail Karaoui, Mohammed Hssaisoune, Abdelkrim Arioua and Tarik Benabdelouahab

Abstract

Water supply for the arid irrigated plains in Morocco depends largely on the upper mountainous basins where significant amounts of precipitation fall as snow. In the Oum Er-Rbia River Basin (OER), snow covers the highest elevations from November to April. Despite the importance of this component in the hydrological cycle, snowmelt contribution to streamflow is still poorly understood and no monitoring stations exist in this zone. Therefore, studying the spatiotemporal change of snow cover through satellite observations to investigate its influence on the hydrological response of this scarce region is thus required to better manage water resources. This chapter explores basic characteristics of snow cover area (SCA) in the upstream area of the OER River (Tillouguite sub-basin) using MODIS daily snow cover products (MOD10A1). Correspondence between streamflow, accumulated air temperature and SCA changes during the winter and spring periods was examined from 2001 to 2009 at a weekly time step. The result shows an inverse linear relation between the maximum SCA and the mean normalized stream flow values, and a significant relation between the relative streamflow and cumulated temperature, especially during spring melt season depending on the length of the melt period. These primary results could be used to develop simplified predictable models for spring discharge in ungauged watershed using remote sensing and accumulated air temperature.

Keywords

Moroccan Atlas • Snow cover • MODIS10A1 • Streamflow • Scarce data

1 Introduction

Mountainous regions contain about 70% of the earth's continental fresh water [1] and considered as the world's natural "water tower" [2]. There is thus a strong interaction between mountains and lowlands; about 40% of world's population lives in the catchments influenced by mountainous ranges [3]. Therefore, change in the total amount of water supplied by these areas directly affects human society and has implications for all living species. Vulnerability of Water resources to any change becomes more important, especially in countries with arid conditions, where water is scarce [4]. In this context, seasonal snow cover constitutes an important component of hydrological cycle in several mountainous area around the world [5–10]. However, in Morocco, the Atlas Mountain ranges supply significant amount of water to support irrigation for the large downstream lower plains; (e.g. Tadla, Tensift, Souss, Saiss, Tafilalt, and Draa). These irrigated plains

A. Boudhar (✉) · H. Ouatiki · H. Bouamri · Y. Lebrini · I. Karaoui · M. Hssaisoune · A. Arioua
Water and Remote Sensing Team "GEVARET", Faculty of Sciences and Techniques, Sultan Moulay Slimane University, Beni-Mellal, Morocco
e-mail: ab.boudhar@usms.ma

H. Ouatiki
e-mail: hamza.ouatiki@gmail.com

H. Bouamri
e-mail: hafsa.bouamri@gmail.com

Y. Lebrini
e-mail: y.lebrini@gmail.com

I. Karaoui
e-mail: ismailkaraoui@gmail.com

M. Hssaisoune
e-mail: hsaissoune.med@gmail.com

A. Arioua
e-mail: A.ARIOUA@usms.ma

A. Boudhar
Center for Remote Sensing Applications (CRSA), Mohammed VI Polytechnic University, Ben Guerir, Morocco

T. Benabdelouahab
National Institute of Agronomic Research of Morocco, 25 Avenue Ennasr BP 415 RP, Rabat, Morocco
e-mail: tarik.benabdelouahab@gmail.com

N. Rebai and M. Mastere (eds.), *Mapping and Spatial Analysis of Socio-economic and Environmental Indicators for Sustainable Development*, Advances in Science, Technology & Innovation, https://doi.org/10.1007/978-3-030-21166-0_9

produce more than one-third of all foodstuffs in Morocco and therefore contribute significantly to food security. In winter, both liquid and solid precipitations fall at high altitudes and snowfall represents between 20 and 80% of total precipitation measured on the southern and the northern slopes of the High Atlas range [11, 12] respectively.

According to the IPCC report, the global average temperature had a rise of 0.3–0.7 °C over the past 100 years [13], several studies highlighted a significant increase in temperatures with a general tendency toward drier conditions [14–17]. Whereas, rainfall trends are generally less significant with heterogeneous change depending to the geographical location [18–20]. Similarly to rainfall, snow cover area in the Atlas mountain is characterized by a high inter-annual signal [21] with an increase of snow cover duration in February–March and a decrease in April–May between 2000 and 2013 using MODIS snow product [22]. However, these changes will likely have a strong effect on water availability at the basin scale. In Northern Morocco, future projection using climate change signal under scenarios 4.5 and 8.5 indicates a decrease of respectively 30–57% in surface runoff for the mid-term (2041–2062) [23]. Studies in Morocco have also relived that populations are very vulnerable to extreme hydrological events [24, 25] and the change of the hydrological response will directly affect strategies and policies of water resource management and socio-economic development. Water resources in the high altitude of the Oum Er Rbia (OER) basin, one of the major water resources basin in Morocco, play an important role in crop irrigation, drinking and industrial water supplies and hydropower production. However, OER catchment meets the region needs and demands of drinking water to surrounding urban centres such as Casablanca, Marrakech, Béni Mellal, Settat, and Berrechid. The total annual volume allocated to drinking water is about 386 million cubic meters per year. Agriculture sector, however, consumes more than 85% of available water for agriculture use [26] In the climatic context of the OER basin, water availability is one of the main limiting factors in achieving good crop yields. The basin supplies water to about 369.600 irrigated hectares, within and surrounding the basin, and requires 3.400 million cubic meters per year and generate about 2/3 of the total hydropower in the country.

Notably absent are studies concerning snow hydrology in the headwater part of the OER catchment despite the importance of this component to the water balance. Thus, monitoring snow cover dynamics is a crucial step to anticipate streamflow response. Measuring snow parameters through snow surveys and meteorological stations is the efficient way to quantify and obtain accurate information about seasonal snowpack. Due to difficult topography and inaccessibility conditions, ground measurements are lacking for most of the high terrain mountains. This is the case for the upper part of the OER basin; a spatial distribution of network stations is inadequate and marked by the absence of snow measurements. Hence, remote sensing data rest a unique way to derive some snow parameters in such rough terrain. In hydrological application, this spatial information of snow cover serve as an input for snowmelt models [27–30].

The purpose of this study is to investigate the potential of using remote sensing data to evaluate the hydrological response to snow covered area (SCA) and temperature change in scarce data basin. The spatial and temporal signal of SCA, obtained from daily snow product MODIS10A1, is analyzed to understand the snow dynamics in the Tillouguite sub-basin located in the upper part of the OER River. To quantify the potential snowmelt contribution to streamflow, the relationship between SCA and hydro-meteorological data is assessed. Defining the link between streamflow and SCA in these areas will provide information regarding the influence of snowmelt rate on discharge over the entire mountainous basins in Morocco.

2 Study Area

The Tillouguite sub-basin is located in the upstream of Bin El Ouidane Dam in Tadla Azilal region between the High Atlas mountainous range and the Tadla plain, at about 60 km in the south of Beni Mellal city in the center of Morocco (Fig. 1). The spatial domain of the sub-basin covers an area of 2400 km^2 in which the elevation ranges from 1000 to about 3300 m above sea level (a.s.l). The Tillouguite sub-basin is one of the major contributors of water to downstream area, especially for the Bin El Ouidane dam storage. This reservoir, considered as the largest in the region, is used for domestic water supply, irrigation, industry and hydropower generation.

Topographic characteristics, elevation, slope and aspect, were derived from SRTM digital elevation model (DEM) with a spatial resolution of 3 arc-second. The SRTM (Shuttle Radar Topography Mission) acquired elevation data in February 2000 on a near-global scale to constitute the most complete high-resolution digital topographic database of Earth landmass [31]. In order to overlay MODIS snow cover data, the original DEM was resampled to 500 m using a bilinear resampling method that gives better results for continuous data such as elevation and slope [32].

In the mountainous areas, hydrological and meteorological conditions are related to elevation; therefore, the Tilloguite sub-basin was divided into 6 elevation zones with an elevation difference of 400 m. Figure 2 shows the distribution curve of the elevation zones for the studied sub-basin. The distribution of basin area along with altitudinal variation indicated that the Tilloguite sub-basin lies within the elevation range 1000–3400 m a.s.l. (zones Z1–Z6), and that more than 70% of the basin area lies between zones Z4 and Z5. Further, it also showed that 87% area falls within middle elevation zones (Z3–Z5), whereas only 2 and 10% in the higher and lower elevation zones respectively (Z6 and Z1–Z2).

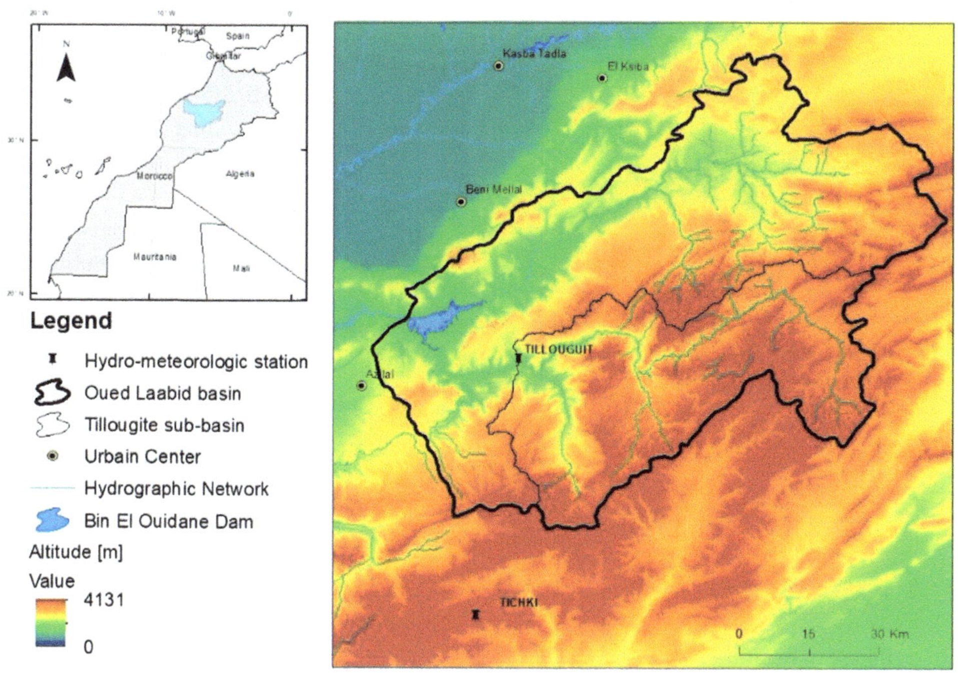

Fig. 1 Geographic situation of the studied basin

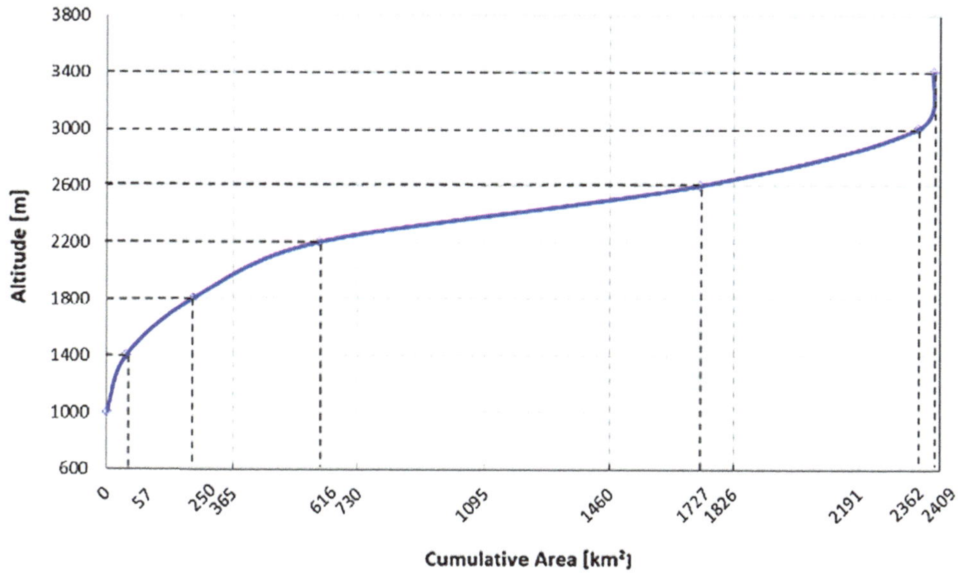

Fig. 2 Hypsometric curve for the Tillouguite sub-basin

The situation of the hydro-meteorological stations used in this work and the topographic variation of the basin are shown in Fig. 1. In general, the climate of the study area is semi-arid with a great variability of precipitations in time and space [33]. Due to the topographic effect, the basin receives snow-fall during winter season [22].

Average temperature, in the Tillouguite sub-basin outlet, is about 28 °C in summer (July) and 7 °C in winter (January), while in the higher elevation it is below freezing point throughout winter season. The basin receives annual average precipitation of 700 mm in the lower part of the basin, which more than 80% occurs between October and May. According to the altitudinal effect, this amount decreases with elevation with an important percentage fall as snow. The variation of 10 years (2000–2010) average monthly rainfall and discharge is given in the Fig. 3. Globally, the distribution of rainfall is irregular over the year with two distinct periods; a dry period (summer) from May to October and a rainy period from November to April. Seasonal variability of runoff, illustrated with standard deviation error bar, reflects the heterogeneity form of rainfall. It can be seen also that the maximum runoff is observed in February, April and March, whereas minimum values are measured in Jun to October.

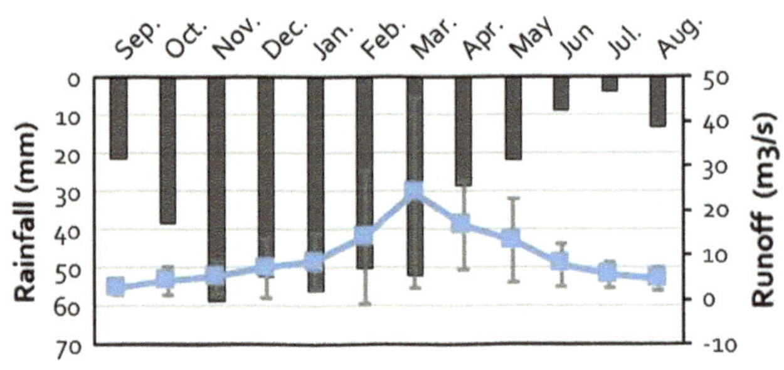

Fig. 3 Mean monthly rainfall (black bars) and runoff (blue line) between 2000 and 2010 over the Tillouguite sub-basin, error bar is a standard deviation of runoff records

3 Data Sets and Methods

3.1 MODIS Snow Product

Given a lack of in situ snow measurements in the high mountainous area in the studied basin, we used remote sensing data to monitor snow cover extent and estimate there spatial and temporal variation. For this, we selected "The Moderate Resolution Imaging Spectroradiometer (MODIS) snow product" that freely available. We used specifically in this study, the MODIS Terra snow cover daily L3 global gridded products (MOD10A1) version 5 at a spatial resolution of 500 m (http://nsidc.org/data/modis/data_summaries). Under cloudiness conditions, the quality of the snow cover products is strongly affected using optical remote sensing. In order to reduce cloud coverage effect, we adopted a methodology based on a spatio-temporal filtering algorithm [22]. To obtain a continuous SCA at a daily time scale, we applied a linear interpolation method over days without information. The spatial and temporal variation of SCA over the basin was analyzed from 13 hydrological seasons and derived for six elevation zones in the basin.

3.2 Hydro Meteorological Measurements

Hydro meteorological measurements; rainfall, discharge; used in this study were obtained from the Tillouguite station managed by the Hydraulic Basin Agency of OER and located within the basin at the elevation of 1100 m a.s.l (Fig. 1). However, temperature data from the years 2000–2010, which were recorded at the Tichki station installed outside the basin at 3200 m a.s.l, were supplied by IMPETUS program [12]. To understand the interrelationship between SCA obtained from remote sensing and hydroclimatic factors, all datasets at different time step and over the same periods as MOD10A1, from 1 December to 30th April every season between 2000 and 2009 are analyzed. The daily streamflow data are averaged into weekly values and precipitations are accumulated for the same time step. Due to strong variability of streamflow peak over the year, we used a relative value by dividing the daily discharge by the maximum value observed in the studied period (from December to April) for each season. For temperatures measurements we used the cumulated degree-days (CDD) for each week calculated from the daily mean temperature as a cumulatively summed degree-day over every 7 days when temperature exceeds 0 °C. When daily temperature is below 0 °C, the degree-day is set equal to 0.

4 Results and Discussion

Rainfall and snowfall are the main factors affecting streamflow response in the mountainous arid basins. To investigate this relationship, especially the dependence between surface water availability and snowmelt, we analysis the stream-flow variability with snow cover extent, rainfall and temperature at different time step.

4.1 Spatial and Temporal Variation of Snow Cover

Figure 4 shows the spatial distribution of SCA occurrence over Tillouguite sub basin, calculated from daily SCA MODIS product from 1st December to 30th April during 13 seasons (from 2000 to 2013). Except around the outlet and riverbeds, almost all the area of Tillouguite sub-basin experienced snowfall during the hydrological seasons with an inter-annual heterogeneity. Although, the occurrence of snow events is not consistently even over space and time. Areas close to the banks of the basin's main rivers rarely register snowfalls with an occurrence below 20% throughout the seasons. However, in the high elevation pixels the snow begins to be more frequent. It ranges from 30% for a drier season as 2009/2010 to reach more than half of the 2012/2013 season.

To describe SCA dynamics at spatial and temporal scales, we calculate the percentage of snow cover extent every 400 m elevation band from 1000 to upper 3400 m. Figure 5 shows the total snow-covered days of the 13 hydrological seasons in six elevation zones. There is an obvious annual pattern for snow cover duration where the topographic control of snowfall is clearly illustrated. Although, a large discrimination between low and high altitudes is showed. Generally, snow is present above 1400 m, except 2006/07 and 2010/11 seasons where SCA appear above 1800 m. The duration of snowpack shows a marked increase with altitude, ranging from an average of 47 days in 2009/2010 to 144 days in 2012/2013. In the high elevation zone (upper 3400 m), snow-pack can rest on ground between 110 and 230 days.

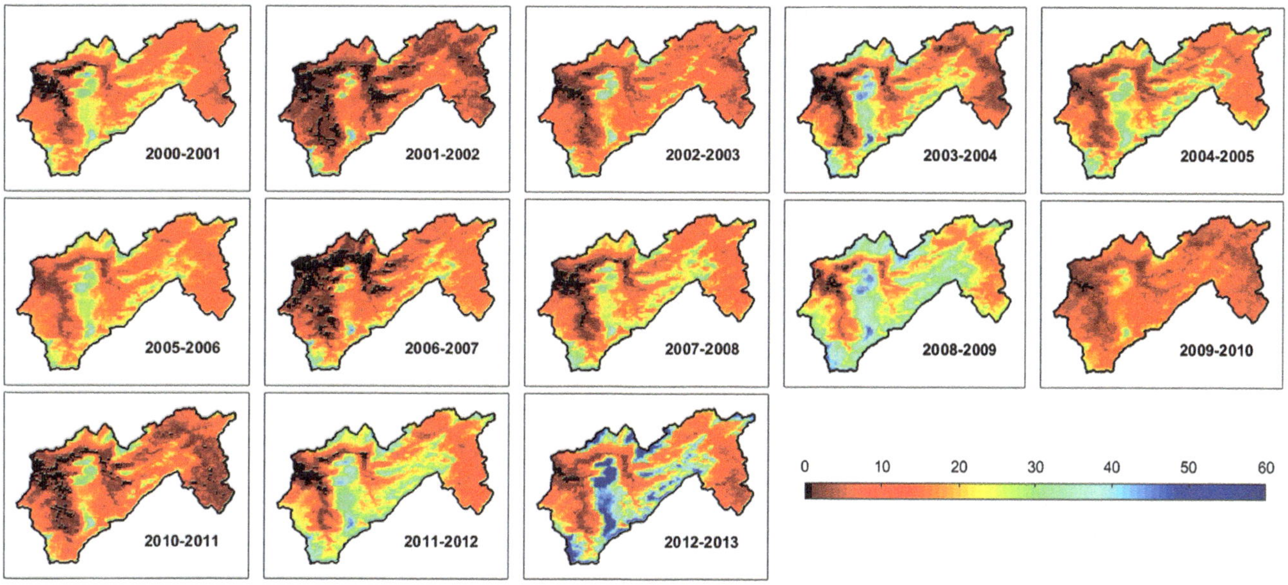

Fig. 4 Annual SCA occurrence frequency over Tillouguite sub basin

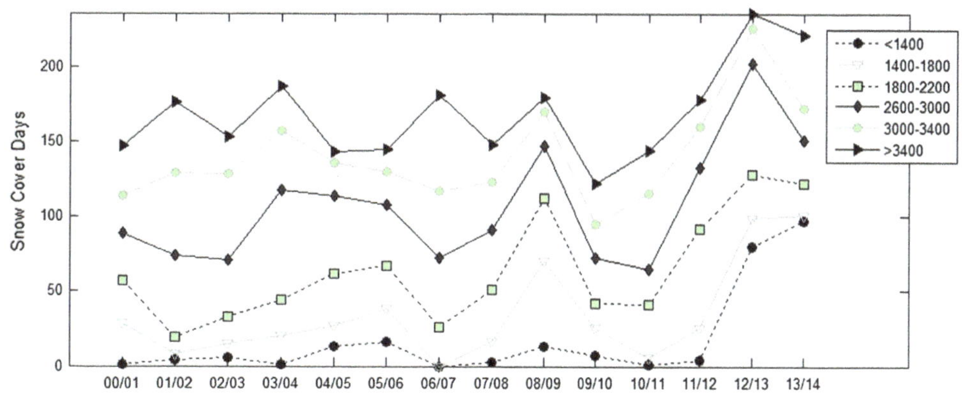

Fig. 5 Snow cover duration between 2000 and 2013 period for six elevation zones in the Tillouguite sub basin

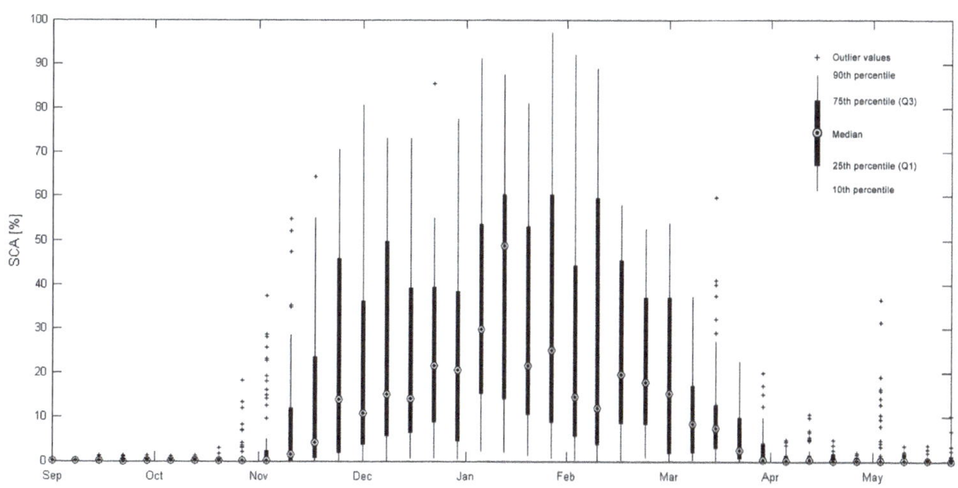

Fig. 6 Behavior of SCA proportion at weekly inter-annual variation between 2000 and 2013 hydrological years

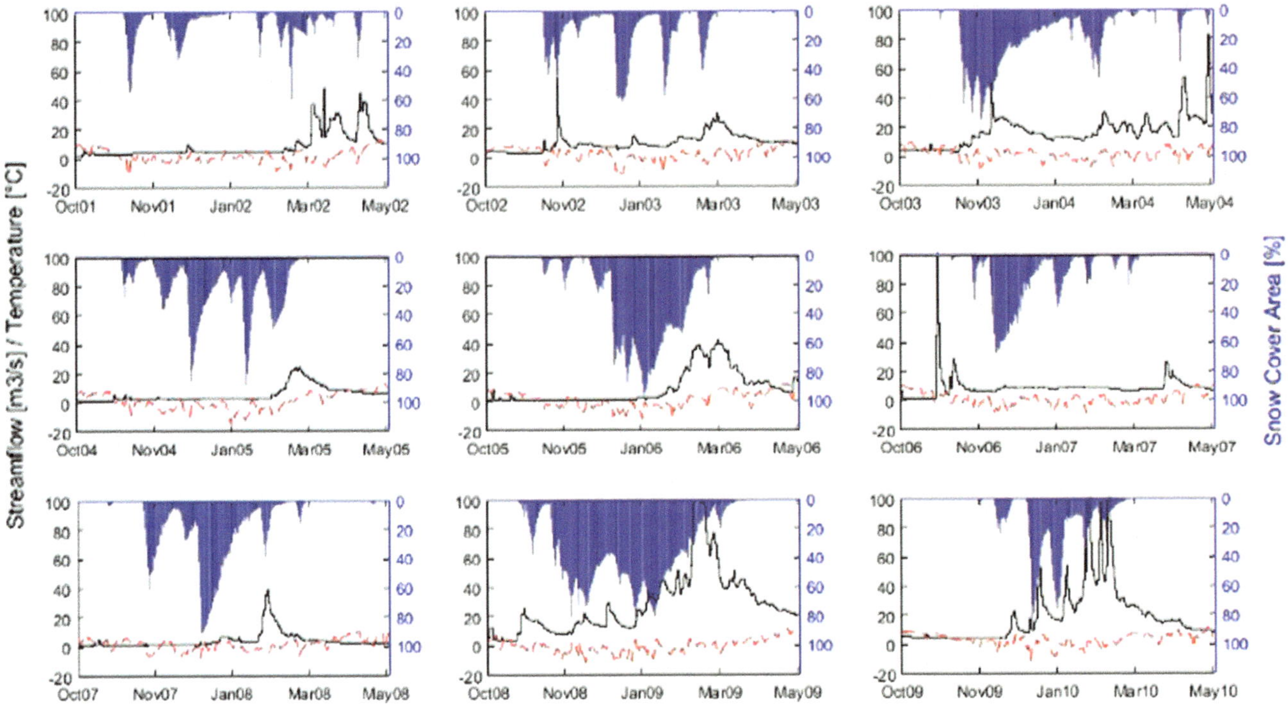

Fig. 7 Daily streamflow variation versus SCA percentage and temperature of nine-season

The box plot graph in Fig. 6 allows us to analysis the behavior of SCA along the studied period in weekly time step. It is relieved that, SCA varies strongly from season to other. Using the median values of SCA proportions, snow season starts effectively in the second week of November and finishes in early April. The 75th percentile indicates that the maximum of snow cover occurs during winter months (December, January and February), when about 36 to 60% of the basin is snow covered.

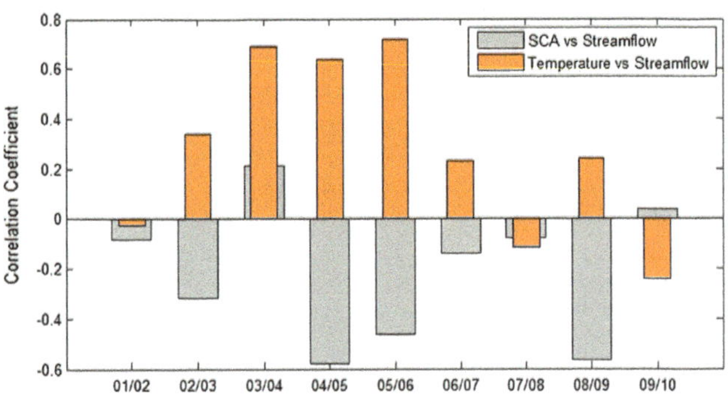

Fig. 8 Annual correlation coefficient between maximum SCA, CDD and normalized streamflow

4.2 Hydrological Responses to Snow Cover Change

Figure 7 compares daily snow cover variation, streamflow and mean temperature from October to April. It is observed that the streamflow variation is strongly linked to temperature and snow cover change. Although, each discharge peak is mostly associated with a snow surface regression period, except 2006/07 season. This relationship, clearly identified for the month from January to March, indicates that snowmelt contributes to runoff for season with a large snow cover duration (e.g. 2008/2009 and 2005/2006).

In order, to understand properly the relationship between temperature, SCA and streamflow, we used a concept of cumulative degree-day (CDD) and normalized streamflow. For each season (from December to April), the linear correlation coefficient between maximum SCA, mean normalized streamflow and CDD was calculated. This analysis was also assessed for different lag time prior to the observing date of streamflow peak (not presented in this paper). As shown in Fig. 8, normalized streamflow is inversely correlated with maximum SCA, meaning that a regression of SCA in a week derives an increase of streamflow. This results is agrees using CDD variable, where the correlation coefficient reach a value of 0.78 in 2005/2006. We conclude that, if air temperature increased, then SCA decreased accordingly excepting 2001/2002 and 2009/2010 seasons. In addition, it is clearly marked that SCA fluctuations, controlled by air temperature, affect the hydrologic response. This relationship varies with the length of the snowy season and the melt period (Fig. 9). For example, in 2004/2005 season, a regression of the weekly maximum SCA from 50 to 7 % generate a normalized streamflow peak from 0.3 to 0.9 in about two weeks.

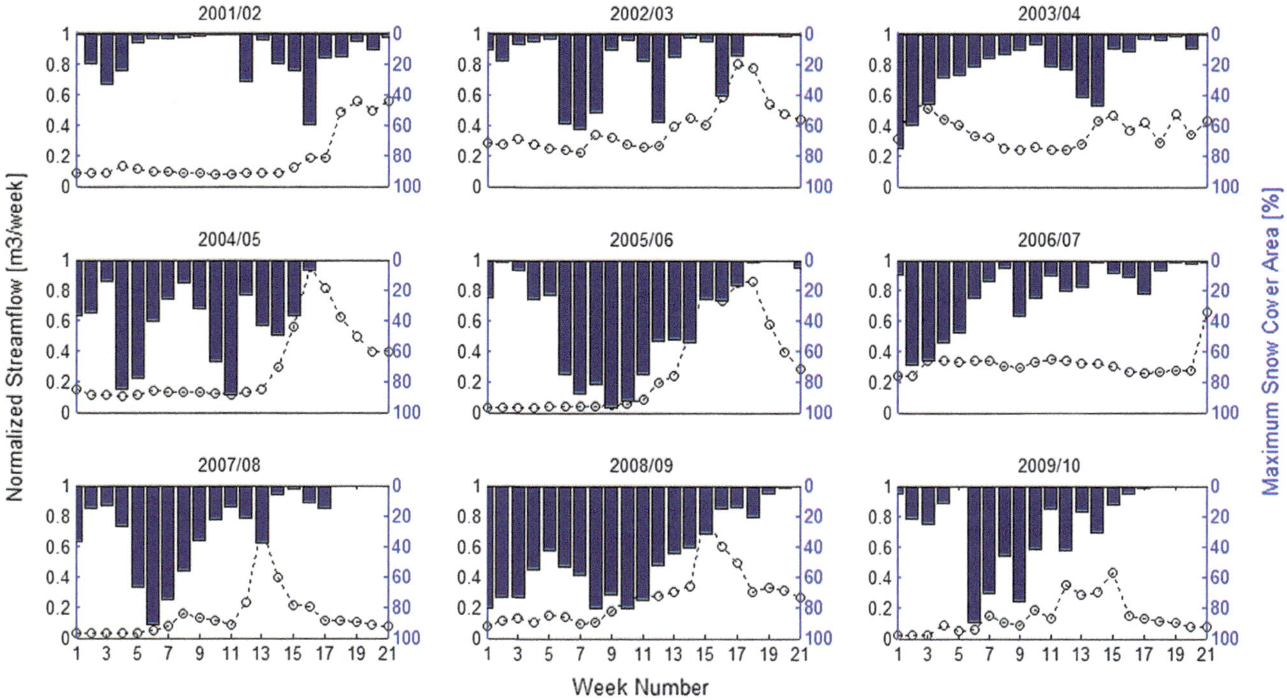

Fig. 9 Relationship between normalized streamflow and maximum SCA at weekly time step

5 Conclusion

Investigating SCA dynamics is the first step to evaluate snowmelt contribution on river streamflow and water cycle in the scarce data basin such as mountainous area in Morocco. In this study, SCA variation at spatial and temporal scales was analyzed over the Tillouguite sub-basin using daily MODIS snow product (MOD10A1) between 2001 and 2009. Generally, there are large differences in snow cover percentage and snow cover duration from year to year, with respect to the topographic effect. The interrelation between streamflow variation together with temperature and SCA change was examined in detail at different time step. It is found that the weekly maximum SCA, for a most of seasons, showed a negative correlation with accumulated degree-days, but the relationship varied with the length of the melt period. SCA is also negatively related to relative streamflow excepting some season (e.g. 2001/2002 and 2007/2008) where rainfall contribution is largely marked.

Despite this importance of snowpack in mountainous basin of Morocco, there are still many gaps in the existing networks of snow measurements at both national and regional scale. Therefore, it may be noted that this is for the first time a snow study has been undertaken for the OER basin. These primary results will be used to develop a simplified model, applicable to individual watersheds, to predict early spring discharge using remotely sensed snow cover information and accumulated air temperature. This approach may be useful for a practical utility to forecast potential snowmelt in ungauged basin with lack of in situ observations. The approach's simplicity should also make it relatively easy and operational for water resources managers, especially for the Bin El Ouidane reservoir located downstream of the basin and used principally for irrigation and hydropower production. In addition, remotely sensed SCA at the basin scale through long-term period is useful as an indicator of climate change impact to the snow storage and is essential for managing water resources.

References

1. Casassa, G., Haeberli, W., Jones, G., Kaser, G., Ribstein, P., Rivera, A., et al. (2007). *Current status of Andean glaciers*. Amsterdam: Elsevier.
2. Viviroli, D., Dürr, H. H., Messerli, B., Meybeck, M., & Weingartner, R. (2007). Mountains of the world, water towers for humanity: Typology, mapping, and global significance Water resources research *43*.
3. Beniston, M. (2005). Mountain climates and climatic change: An overview of processes focusing on the European Alps. *Pure and Applied Geophysics*, *162*, 1587–1606. https://doi.org/10.1007/s00024-005-2684-9.
4. Viviroli, D., & Weingartner, R. (2004). The hydrological significance of mountains: From regional to global scale. *Hydrology and Earth System Sciences Discussions*, *8*, 1017–1030.
5. Boudhar, H. L., Boulet, G., Duchemin, B., Berjamy, B., & Chehbouni, A. (2009). Evaluation of the Snowmelt Runoff Model

in the Moroccan high Atlas mountains using two snow-cover estimates. *Hydrological Sciences Journal*, *54*, 1094–1113.

6. De Jong, C., Lawler, D., & Essery, R. (2009). Mountain hydroclimatology and snow seasonality - Perspectives on climate impacts, snow seasonality and hydrological change in mountain environments. *Hydrological Processes*, *23*, 955–961. https://doi.org/10.1002/hyp.7193.
7. López-Moreno, J., & García-Ruiz, J. M. (2004). Influence of snow accumulation and snowmelt on streamflow in the central Spanish Pyrenees/Influence de l'accumulation et de la fonte de la neige sur les écoulements dans les Pyrénées centrales espagnoles. *Hydrological Sciences Journal*, *49*.
8. Tahir, A. A., Chevallier, P., Arnaud, Y., Ashraf, M., & Bhatti, M.T. (2015). Snow cover trend and hydrological characteristics of the Astore River basin (Western Himalayas) and its comparison to the Hunza basin (Karakoram region). *Science of The Total Environment*, *505*, 748–761. https://doi.org/10.1016/j.scitotenv.2014.10.065.
9. Emre Tekeli, A., Akyürek, Z., Arda Şorman, A., Şensoy, A., & Ünal Şorman, A. (2005). Using MODIS snow cover maps in modeling Snowmelt Runoff process in the Eastern part of Turkey. *Remote Sensing of Environment*, *97*, 216–230. https://doi.org/10.1016/j.rse.2005.03.013.
10. Zhang, G., Xie, H., Yao, T., Li, H., & Duan, S. (2014) Quantitative water resources assessment of Qinghai Lake basin using Snowmelt Runoff Model (SRM). *Journal of Hydrology*, *519*, 976–987.
11. Boudhar, A., Boulet, G., Hanich, L., Sicart, J. E., & Chehbouni, A. (2016). Energy fluxes and melt rate of a seasonal snow cover in the Moroccan High Atlas. *Hydrological Sciences Journal*, *61*, 931–943. https://doi.org/10.1080/02626667.2014.965173.
12. Schulz, O., & de Jong, C. (2004). Snowmelt and sublimation: Filed experiments and modelling in the Hight Atlas Mounatains of Morocco. *Hydrology and Earth System Sciences*, *8*, 1076–1086.
13. Pachauri, R. K., Meyer, L., Plattner, G.-K., & Stocker, T. (2015). IPCC, 2014: Climate change 2014: Synthesis report. In *Contribution of Working Groups I, II and III to the Fifth Assessment Report of the Intergovernmental Panel on Climate Change (IPCC)*.
14. Abahous, H., Ronchail, J., Sifeddine, A., Kenny, L., & Bouchaou, L. (2018). Impacts of the water resources variability on cereal yields in the region of Souss-Massa Southern Morocco. *Groundwater and global change in the Western Mediterranean area* (pp. 1–8). Berlin: Springer.
15. Esper, J., Frank, D., B üntgen, U., Verstege, A., Luterbacher, J., & Xoplaki, E. (2007). Long-term drought severity variations. *Morocco Geophysical Research Letters*, *34*.
16. Filahi, S., Tanarhte, M., Mouhir, L., El Morhit, M., & Tramblay, Y. (2016). Trends in indices of daily temperature and precipitations extremes in Morocco. *Theoretical and Applied Climatology*, *124*, 959–972.
17. Khomsi, K., Mahe, G., Tramblay, Y., Sinan, M., & Snoussi, M. (2015). Trends in rainfall and temperature extremes in Morocco. *Natural Hazards & Earth System Sciences Discussions*, *3*.
18. Donat, M., et al. (2014). Changes in extreme temperature and precipitation in the Arab region: Long-term trends and variability related to ENSO and NAO. *International Journal of Climatology*, *34*, 581–592.
19. Driouech, F., Déqué, M., & Sánchez-Gómez, E. (2010). Weather regimes—Moroccan precipitation link in a regional climate change simulation. *Global and Planetary Change*, *72*, 1–10.
20. Driouech, F., Rached, S. B., & El Hairech, T. (2013). Climate variability and change in North African countries. In *Climate change and food security in West Asia and North Africa* (pp. 161–172). Berlin: Springer.
21. Boudhar, A., et al. (2010). Long-term analysis of snow-covered area in the Moroccan High-Atlas through remote sensing. *International Journal of Applied Earth Observation and Geoinformation*, *12*, S109–S115. https://doi.org/10.1016/j.jag.2009.09.008.
22. Marchane, A., et al. (2015). Assessment of daily MODIS snow cover products to monitor snow cover dynamics over the Moroccan Atlas mountain range. *Remote Sensing of Environment*, *160*, 72–86. https://doi.org/10.1016/j.rse.2015.01.002.
23. Tramblay, Y., Ruelland, D., Bouaicha, R., & Servat, E. (2014). Projected climate change impacts on water resources in northern Morocco with an ensemble of regional climate models.
24. Di Baldassarre, G., Montanari, A., Lins, H., Koutsoyiannis, D., Brandimarte, L., & Blöschl, G. (2010) Flood fatalities in Africa: From diagnosis to mitigation. *Geophysical Research Letters*, *37*.
25. Douglas, I., Alam, K., Maghenda, M., Mcdonnell, Y., McLean, L., & Campbell, J. (2008). Unjust waters: Climate change, flooding and the urban poor in Africa. *Environment and Urbanization*, *20*, 187–205.
26. Schyns, J. F., & Hoekstra, A. Y. (2014). The added value of water footprint assessment for national water policy: A case study for Morocco. *PLoS One*, *9*, e99705.
27. Gómez-Landesa, E., & Rango, A. (2002). Operational Snowmelt Runoff forecasting in the Spanish Pyrenees using the Snowmelt Runoff Model. *Hydrological Processes*, *16*, 1583–1591. https://doi.org/10.1002/hyp.1022.
28. Kult, J., Choi, W., & Choi, J. (2014). Sensitivity of the Snowmelt Runoff Model to snow covered area and temperature inputs. *Applied Geography*, *55*, 30–38. https://doi.org/10.1016/j.apgeog.2014.08.011.
29. Li, X., & Williams, M. W. (2008). Snowmelt Runoff modelling in an arid mountain watershed. *Tarim Basin, China Hydrological Processes*, *22*, 3931–3940. https://doi.org/10.1002/hyp.7098.
30. Senzeba, K. T., Bhadra, A., & Bandyopadhyay, A. (2015). Snowmelt Runoff modelling in data scarce Nuranang catchment of Eastern Himalayan region. *Remote Sensing Applications: Society and Environment*, *1*, 20–35.
31. Farr, T.G., et al. (2007) The shuttle radar topography mission. *Reviews of Geophysics*, *45*.
32. Baboo, S.S., & Devi, M.R. (2010). An analysis of different resampling methods in Coimbatore, District. *Global Journal of Computer Science and Technology*.
33. Ouatiki, H., et al. (2017). Evaluation of TRMM 3B42 V7 Rainfall Product over the Oum Er Rbia Watershed in Morocco. *Climate*, *5*, 1.

A Topo-Bathymetric Survey of the Morphological Evolution of a Microtidal Barred Beach. Case Study: The Coastal Prism of Korba (Mediterranean Coast; Northeast of Tunisia)

Zouhour Yahyaoui, François Sabatier, Noamen Rebai and Saâdi Abdeljaouad

Abstract

For mapping the three-dimensional shape of the submarine bars, studying their dynamics and assessing the spatio-temporal evolution of sedimentary balances, the coastal prism of Korba (emerged and submerged beach) has benefitted from topo-bathymetric monitoring. This monitoring was carried out on an annual basis between July 2006 and July 2009. Numerical Terrain Models (N.T.M.) and transverse profiles were then analyzed. The main results show, firstly, that the shoreface of Korba is characterized by a homogeneous evolution across all profiles of its two-bar system. Moreover, the extent of changes in the sand volume of the beach suggests that the evolution takes place without significant sedimentary loss and that the beach is in a dynamic equilibrium. Accretion of the bars and migration towards the shore are found in periods of small waves (between July 2006 and July 2007). However, following a period of minor agitations, a filling of the outer trough is observed. It is proceeded by the migration of the outer bar towards the coast (between July 2007 and July 2009). These results confirm the traditional model of self-organization of a barred beach. These data and results, then, allow us to establish conceptual models and simulations of the evolution of the microtidal, barred beach of Korba.

Keywords

Microtidal beach • Submerged beach of Korba • Nearshore bars • Topo-bathymetric monitoring • Transverse dynamic

1 Introduction

In coastal systems, the nearshore area presents an integral part of coastal accumulations in the continuity of beaches and dunes. The morphodynamic study of each of these coastal compartments is therefore crucial to understand the evolution of the sedimentary system as a whole. The interactions between the nearshore zone, the seabed and the wide beach from one part and sedimentary exchanges between various compartments from another part, determine therefore, coastal morphological dynamics and evolutions resulting from the shore [1]. Indeed, this type of study seems to be fundamental to improve coastal management strategies and their equilibrium conditions. This depends on a deep knowledge of coastal sedimentary stocks and their dynamics [2–6].

Unlike beach/dune systems, those are the subject of many studies and spatio-temporal surveys [7–10], the nearshore zone or submerged beach is still relatively little studied. This is a complex area, home to many hydro-sedimentary phenomena related to the transformation of tidal coast, gravitational waves, induced currents and winds. Its morphology is controlled by many factors such as waves, sedimentary supplies, storms and sea level changes [11,12]. These factors, operating on very different scales of time and space, complicate any research on this environment.

Z. Yahyaoui (✉) · S. Abdeljaouad
Laboratoire des Ressources Minérales et Environnement (LRME), Département de Géologie, Université Tunis El Manar/Faculté des Sciences de Tunis, 2092 El Manar 2, Tunis, Tunisie
e-mail: zouhour80@yahoo.fr

S. Abdeljaouad
e-mail: saadi_abdeljaouad@yahoo.fr

F. Sabatier
Aix-Marseille Université, Centre Européen de Recherche et d'Enseignement des Géosciences et de l'Environnement, UMR CNRS 6635, Europole de l'Arbois, B.P. 80, 13545 Aix-en-Provence Cedex 04, France
e-mail: sabatier@cerege.fr

N. Rebai
LR14ES03 Geotechnical Engineering and Georisk Laboratory, University of Tunis El Manar, National School of Engineering of Tunis, le Belvédère 1002, B.P. 37, Tunis, Tunisia
e-mail: noamen.rebai@enit.utm.tn

N. Rebai and M. Mastere (eds.), *Mapping and Spatial Analysis of Socio-economic and Environmental Indicators for Sustainable Development*, Advances in Science, Technology & Innovation, https://doi.org/10.1007/978-3-030-21166-0_10

The most characteristic sedimentary forms of underwater beaches are sand bars. Many authors have described these bars that can be single, double [13], multiple, rectilinear [14], scalloped [15] and transverse [16]. Their operation is critical to the coastal area as a whole because they are at the heart of the sedimentary exchanges of the beach at large. They form mobile sand deposits and may help protect a role in protecting the coastline during storms [17].

On the Tunisian coast, the submerged beaches and the pre-littoral bars have been little studied. We have just the works of: [18] at the North coast of Mahdia (eastern Tunisia), [19], in the Gulf of Tunis and [20], at the coast of Tabarka (northern Tunisia). However, no morphodynamic study has been done in relation to the Cap Bon more specifically the eastern façade (Gulf of Hammamet) in its submerged area since 1885. Hence, the beach of Korba (eastern facade of Cap Bon) was selected to conduct this study.

This article delineates the results of the first topo-bathymetric missions on the prism coastline of Korba which is marked with a barred microtidal beach with a wave medium energy and submitted free of any maritime planning. The objectives of our study are the following:

- to characterize the morphology of the coastal prism of Korba and to analyse the bathymetric evolution of the nearshore zone on an annual scale,
- to clarify the residual sedimentation of the underwater beach of Korba by calculating sedimentary balances on the entire coastline prism,
- and finally to calculate the volumes mobilized by creating a 'blank' in Surfer for each morphological unit (inner bar, outer bar, upper beach) and then to quantify sedimentary exchanges between the prism units.

2 Study Area

The coastline of Korba is a low, rectilinear and sandy coast. It extends over 2 km from Korba Lagoon (Sabkhet Ech Charquia) in the Northeast to Korba Wadi (Oued de Korba) in the Southwest (Fig. 1). The coastline of Korba is identified by a fairly wide beach (between 50 and 100 m) bordered by foredunes which altitude rarely exceeds 4 m. The upper beach is in dynamic equilibrium and the line of dunes is accreting [9].

The prevailing wind in the area comes from the northwest sector with a speed of 20 m/s. In fact, the northwest winds dampen wave propagation by reducing their height and slowing down their progression. However, there is a tendency for winds from the sea in the summer of a North-eastern direction, contrary to the winter during which the wind from the land in dominant.

The beach of Korba is characterized by a microtidal regime in which tidal range during high water seldom exceeds 45 cm. On top of this periodic and regular paced tide, there are variations in the water level that are of a meteorological origin. These variations are called increases and decreases and their study depends on statistical methods. The average maximum tidal range is 0.30 m for a return period of 5 years and 0.40 m for a return period of more than 10 years. The average maximum decreases estimated at this point do not exceed −0.30 m [22,23].

3 Methodology

The coastal prism of Korba benefitted from three campaigns of topo-bathymetric surveys by profiles on an annual basis between July 2006 and July 2009 (Fig. 2a). The bathymetric measurements are taken by a boat that has a shallow Tritech sonar ST500 and a DGPS connected to a laptop computer, which controls the acquisition and storage of the data. The vertical accuracy is of the order of decimetres. The accuracy in the horizontal plane is about one meter (DGPS). The acquisition rate allows for theoretical measuring steps of 1.25 m. at the operational speed of five knots. Thus, on a length of 3 km, 65 radials spaced 20 m are performed transversely to the sand bars, arranged perpendicular to the coastline from Korba Lagoon (Sabket Ech Charquia) to Korba Wadi.

Topographical measurements are provided using the total station. The different measurements are given with the following precision:

- the azimuth angles are measured with an error margin of 5.104°,
- distances are measured at +/−1 cm close in tracking, up to about 1000 m,
- the differences in height are measured at +/−1 cm.

Thus, for each date (2006, 2007 and 2009), a Digital Terrain Models (DTM) is calculated by triangulation. These numerical models are used to extract interpolated beach profiles (Fig. 2b) and develop differential evolution maps, reflecting the 3D morphological changes undergone by the sedimentary prism (underwater beach and upper beach), in a general way, and the submerged beach (inner bar and outer bar) of Korba in a particular way. A common limit at −12 m of the coast is taken to delimit the differentials. This limit is supposed to integrate all the movements of the submerged beach, since the closure depth limit beyond which there is no further significant movement of the bottom is located between 6 and 10 m according to [24,25].

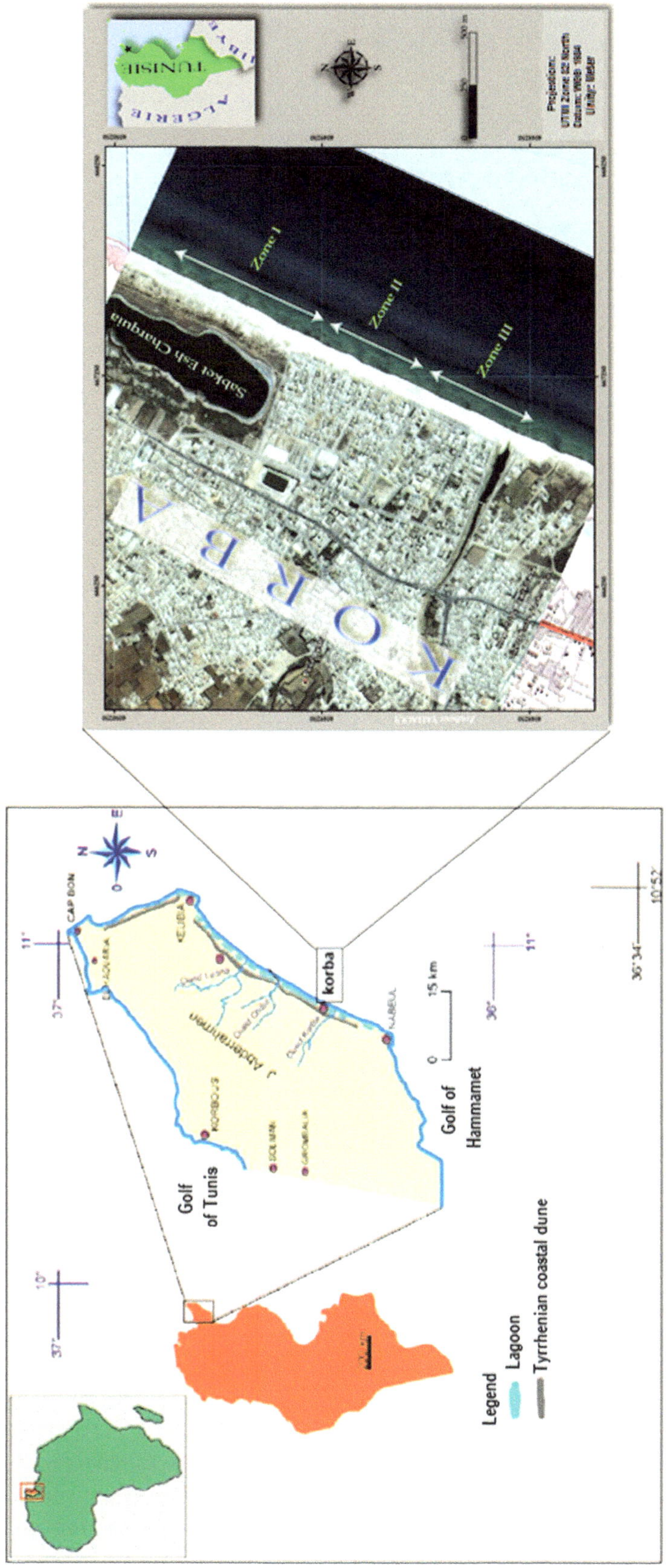

Fig. 1 Geographical location of the study area and morphological description of the coastal prism of Korba ([21], modified)

Fig. 2 Positioning plan of topo-bathymetric surveys and location of transverse profiles from Digital Terrain Models (DTM)

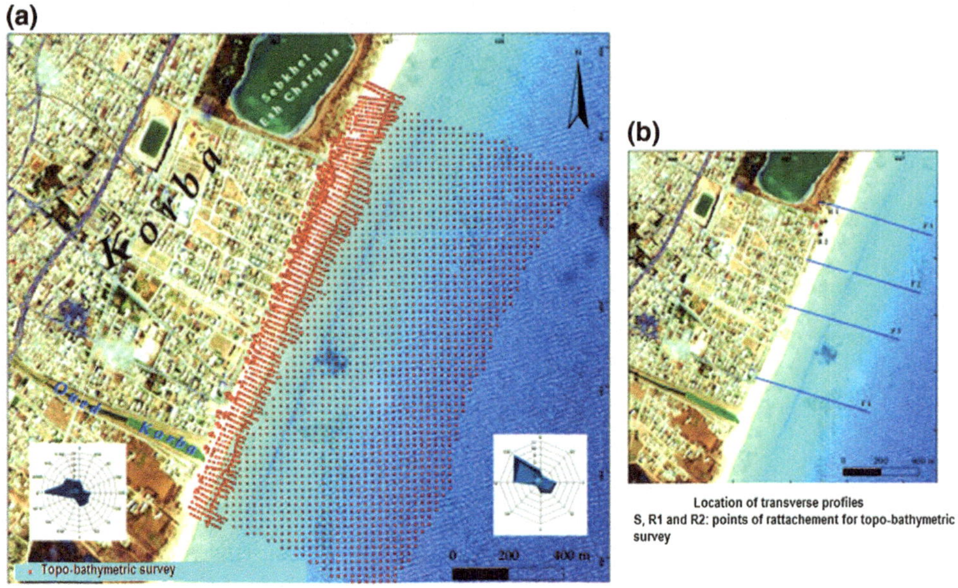

Digital Terrain Models (DTM) in 3D and the differential evolution maps enable us to assess the sedimentary morphodynamics of our study site at an annual scale and to deduce the changes in volumes and transverse movements.

4 Results and Interpretations

4.1 Morphology of the Submerged Beach of the Coastal Prism of Korba

This description is based on the first topo-bathymetric survey conducted in July 2006 (Fig. 3). Indeed, the establishment of the bathymetric map of July 2006 reveals a regular structure of the front-side of Korba. The front-side (or submerged beach) of Korba, shows the existence of a two-bar system: a straight inner bar that is sometimes scalloped and a straight external one. The two bars are separated by a very marked trough, so the studied beach was classified in dissipative beach (=6). At a depth of 6 m., the profile is stable and from that point, called 'depth of closure', the submerged beach does not undergo any changes [26].

4.2 Transverse Evolution

The submerged beach of Korba has a relatively gentle slope of the order of 0.8 on average with two remote sedimentary bars at a distance of 50–200 m for the inner bar and 350–400 m for the outer one. Both bars are always straight but the latter the inner bar may take a slightly scalloped or disjointed character.

The analysis of the general evolution of the coastal prism is based on the exploitation of transverse profiles extracted from DTM (Fig. 4b).

The results show that the front side of Korba has the same evolution on all cross-shore profiles of the bar system and troughs along the study site. To recapitulate, in what follows, we will describe the annual morphological changes from July 2006 to July 2009.

4.2.1 Morphological Evolution Between July 2006 and July 2007

During this period, the outer bar is well-marked with a slight migration to the coast; same for the inner bar. Topographically speaking, the entire study site outlines the same state of evolution. Thus, the individualization of the inner and outer bars forms the major morphological evolution. The latter is coupled with an upward movement of the crests of not only the inner and outer bars but also the inner and outer troughs (Figs. 4, 5, 6 and 7).

The individualization is the result of vertical accretion of the crests that reach up to 50 cm for the outer bar and 40 cm for the inner one. It is important to note that in July 2007, this accretion was followed in some cases by slight migration towards the coast. This accretion underlines the effect of a normal operating mode of the submerged beach after a period of good weather. The inner bar migrates to the coast to append to the shore. This apposition will result in both the broadening of the beach and the forming of a berm along the shore underlining the summer profile.

From the perspective of volumetric variations, positive budget was recorded along this period. The surplus is the

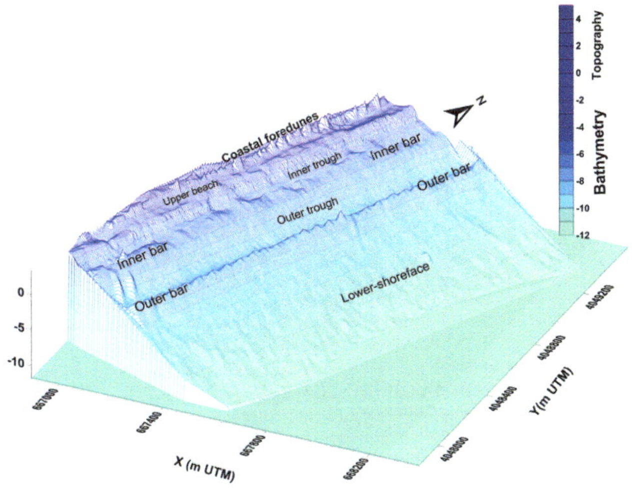

Fig. 3 Morphology of the coastal prism of Korba according to the campaign of July 2006

order of 290 m^3.m. l in the Northern sector (Zone II) of the study site (Fig. 4), 100–200 m^3.m.l in the central area (Figs. 5 and 6) and about 235 m^3.m.l in the Southern area (Zone III) of the study site (Fig. 7). Accordingly, the Northern sector (Zone I) has the biggest sedimentary surplus.

During this period we have recorded a sediment supply to the coast at the level of the upper beach, that is, an accretion of the visible beach.

4.2.2 Morphological Evolution Between July 2007 and July 2009

From July 2007 to July 2009, the profile has had a configuration different from that observed between July 2006 and July 2007. There was a diminishment of the outer bar. During this morphological change, there was vertical erosion which could reach 1 m followed by a vertical accretion of about 30 cm in the outer trough filled with sediment. As a result, the inner bar sometimes tended to migrate to the coast (Figs. 5, 6 and 7) and, at the other times, remains in place (Fig. 4).

At the upper beach we also observe a fattening of the foredune in the period from July 2007 to July 2009. This causes a vertical accretion of up to 50 cm on the Northern sector of the study site (Fig. 4). During this period, the morphological evolution observed at the submerged beach is expressed particularly through three major elements: the lowering of the outer bar, the filling of the outer trough and the segmentation of the inner bar. It reflects, according to the literature [27–29], the impact of an exceptional phenomenon (energetic storm).

Regarding volumetric changes, the registered sediment balance along the study site is in deficit. The latter has an order of −80 m^3.m.l at the Northern sector (Fig. 4), between −180 and −140 m^3.m.l in the central sector (Figs. 5 and 6) and about −200 m^3.ml in the Southern area of the study site (Fig. 7).

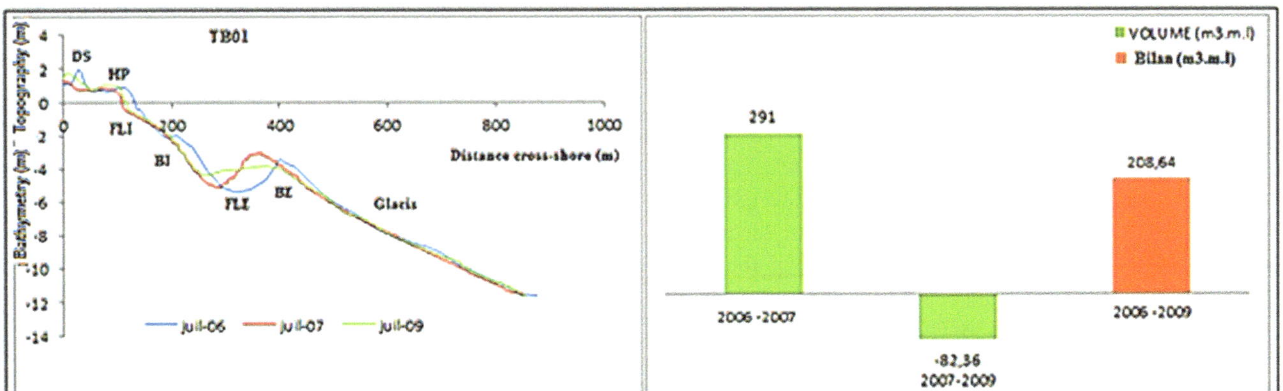

Période	localisation	FLI	BI	FLE	BE	DS
juil.06	Localistion (m)	121	193	306	380	30 26
	Altitudes (m)	-1 33	-1 66	-5 38	-3 32	2 73
juil.07	Localistion (m)	132	182	294	374	39
	Altitudes (m)	-0 8	-0 97	-5	-3 16	2 31
juil.09	Localistion (m)	121	143	286	380	22 7
	Altitudes (m)	-0 85	-0 57	-4	-3 86	2 78

Fig. 4 Topographic and volumetric evolutions of topo-bathymetric profiles in the North sector (Zone I) between July 2006 and July 2009. The table associated with each profile represents the location and altitude of the morphological unit that makes up the emerged beach (DS: Dune Summit, HP: Upper Beach) and submerged beach (BE: Outer Bar; FLE: Outer trough; BI: Inner Bar, FLI: Inner trough)

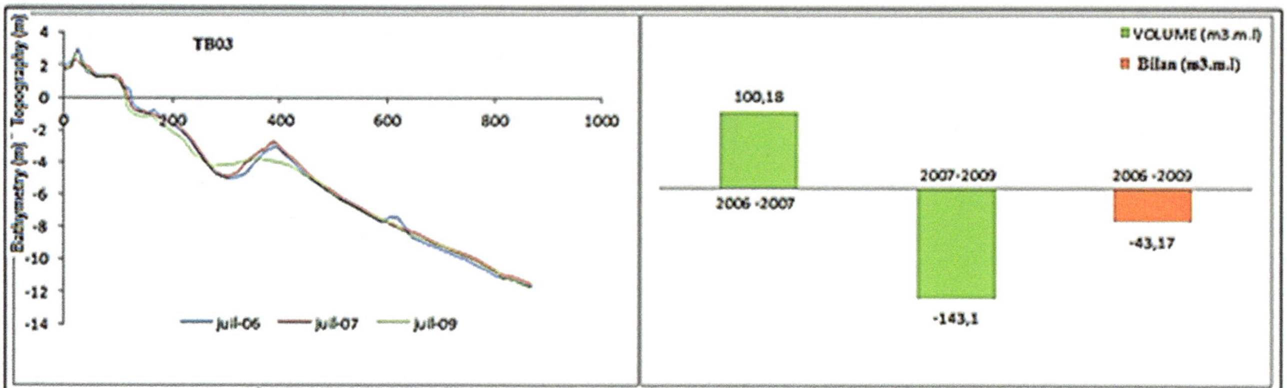

Période/localisation		FLI	BI	FLE	BE	DS
Juil-06	Localistion (m)	150	185	330	410	54
	Altitudes (m)	-0.5	-0.9	-5	-3	2.88
Juil-07	Localistion (m)	145	200	315	408	54
	Altitudes (m)	-0.6	-1.3	-5	-2.8	2.98
Juil-09	Localistion (m)	146	186	315	423	50
	Altitudes (m)	-0.9	-1.4	-4.2	-4	2.88

Fig. 5 Topographic and volumetric evolutions of topo-bathymetric profiles located in the central area (Zone II) of the study site between July 2006 and July 2009

4.2.3 Morphological Evolution Between July 2006 and July 2009

Volumetric changes prove that the recorded sediment balances along the period of 2006–2009 were in most cases in excess, thus favouring the accretion of the front side (Figs. 4, 6 and 7) (at the TB03 profile, this balance is in deficit of around -40 m^3.m.l; Fig. 5). The outer bar tended to recover its equilibrium position by approaching the coast with the filling of the outer trough. The latter's evolution, either by emptying or filling, is a good indicator of the morphodynamic evolution of a microtidal beach with bars [29].

4.3 Residual's Evolution and Sediment Budget

The realization of Residual Digital Terrain Models (RDTM) by subtracting two Digital Terrain Models from successive surveys permits the calculation of the volume of the sedimentary stock of the submerged beach including internal and external bars and quantification of their altitudinal changes (Figs. 8, 9 and 10). Thus, in this section we will calculate the total volumes of all the study area on an annual scale in order to identify not only areas in a state of erosion but also zones in a state of both accretion and stability. In addition, we will determine the volumes localized at the submerged beach and at the upper beach in order to quantify their interaction. Finally, we will try to specify the volumes mobilized at the inner and outer bars to define and measure their morphological evolution on an annual basis.

4.3.1 Volumes Moved Across the Coastal Prism

The morphological changes occurring between two successive stages involve sediment movements of tens of thousands m^3.

The calculation of volumes along the littoral prism shows that during the period from July 2006 to July 2007 (Fig. 11), the coast inherited large amounts of sediment, resulting in a positive sediment balance. However, the balance was in deficit from 2007 to 2009. As a result, the coastal prism of Korba received significant inflows of sediment for the entire survey period 2006–2009 (Fig. 11). Moreover, an oscillatory nature of volume variations was underlined during the monitoring period reflecting a state of dynamic equilibrium [31].

4.3.2 The Volumes Moved to the Submerged Beach and Emerged Beaches

The acquisition of the topo-bathymetric profiles in the coastal prism of Korba, in the context of an annual monitoring, enables us to compare the annual evolution of the upper beach with the underwater beach for similar periods.

The calculation of volumes, by creating 'breaking' in Surfer software, of the emerged and submerged beaches, indicates that during the period from July 2006 to July 2007, the submerged beach (nearshore, troughs, and lower-shoreface) where characterized by a significant accretion of approximately 345,840 m^3. At the same time, at the emerged beach (foredune/upper beach), this accretion is much lower: it is of the order 44640 m^3, which results in a positive sediment budget (Fig. 12). This outcome highlights the strength and predominance of extracting currents capable of draining large amounts of sandy material towards the open sea. However,

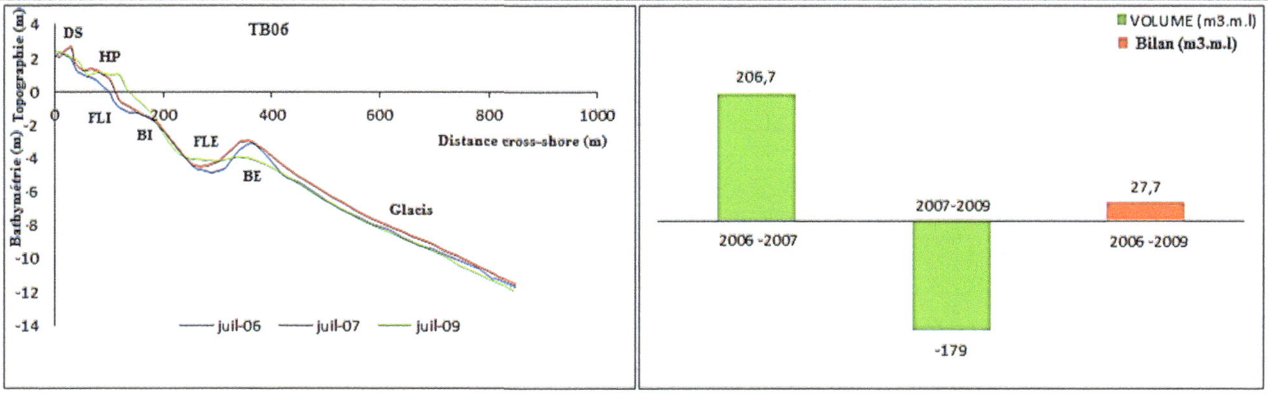

Période/localisation		FLI	BI	FLE	BE	DS
juil-06	Localistion (m)	185	210	340	415	52
	Altitudes (m)	-1	-1.4	-5	-3.3	3.33
juil-07	Localistion (m)	155	184	320	400	54
	Altitudes (m)	-0.4	-0.85	-4.8	-3	3.22
juil-09	Localistion (m)	150	184	315	390	48
	Altitudes (m)	-1	-0.85	-4	-3.9	3.41

Fig. 6 Topographic and volumetric evolutions of topo-bathymetric profiles located in the central area (Zone II) of the study site between July 2006 and July 2009

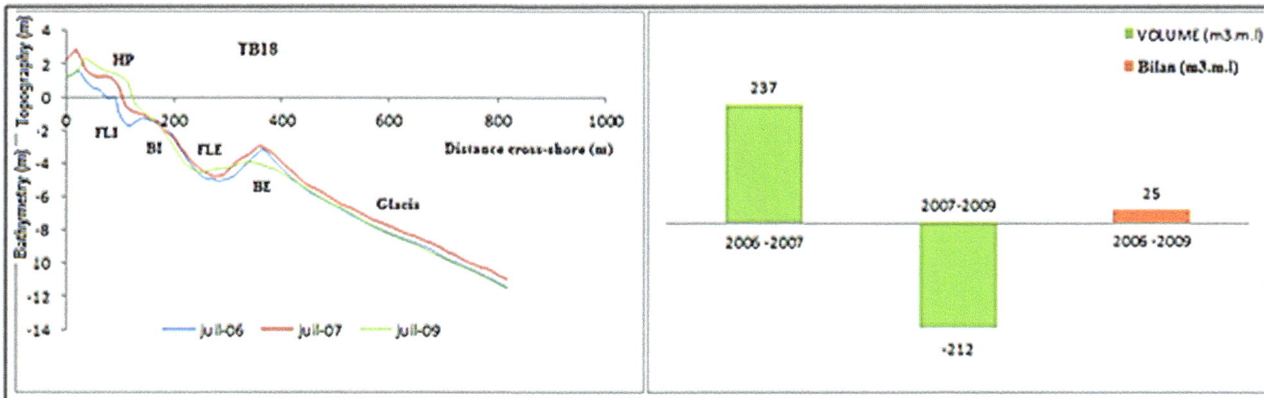

Période/localisation		FLI	BI	FLE	BE	DS
juil-06	Localistion (m)	132	172	287	370	35
	Altitudes (m)	-1.6	-1.45	-4.95	-3.04	2.33
juil-07	Localistion (m)	120	178	280	350	35
	Altitudes (m)	-0.57	-1.17	-4.65	-3.06	2.76
juil-09	Localistion (m)		166	285	365	33.5
	Altitudes (m)		-1.05	-4.28	-4	2.1

Fig. 7 Topographic and volumetric evolutions of topo-bathymetric profiles in the southern sector (Zone III) between July 2006 and July 2009

this result shows that the largest share of the sedimentary stock is located at the submerged beach.

During the period from July 2007 to July 2009, the calculated volumes indicate a clearly negative sediment budget at the submerged beach of about −25,560 m^3, because of sediment loss at the outer bar and therefore its lowering. These data prove (i) that the pre-coastal sand bar changes significantly on an annual basis; and (ii) it responds with high sensitivity to weather events; this sudden change is certainly due to the action of a very active storm [27,28]. However, at the emerged beach, there has been an excess of sediment highlighting the relative sediment balance between the quantities of eroded (23690 m^3) and accreted (33120 m^3) sediments. The positive budget recorded at the emerged beach reflects a sedimentary contribution from the open sea to the coast. This evolutionary scheme reflects a summer profile with sedimentary supplies from the underwater beach to the upper one.

4.3.3 Volumetric Variations at the Submerged Beach and Emerged Beaches

During the monitoring period (July 2006–2009) (Fig. 12), a clearly positive result (61,680 m^3) was registered at the sub-

Table 1 Typology of the morphology of the submerged beach by characteristic phases from July 2006 to July 2009

Characteristic phases	Typology of the submerged beach		
	Inner bar	Outer trough	Outer bar
July 2006–2007	Slightly scalloped and close to the shore	The trenches well-marked with depths exceeding 4 m	Rectilinear and parallel to the shore
July 2007–2009	Segmented and very close to the shore, with orientation tending towards NNE/SSW	Strong filling of the outer trough	Rectilinear and parallel to the shore and moving towards the shore

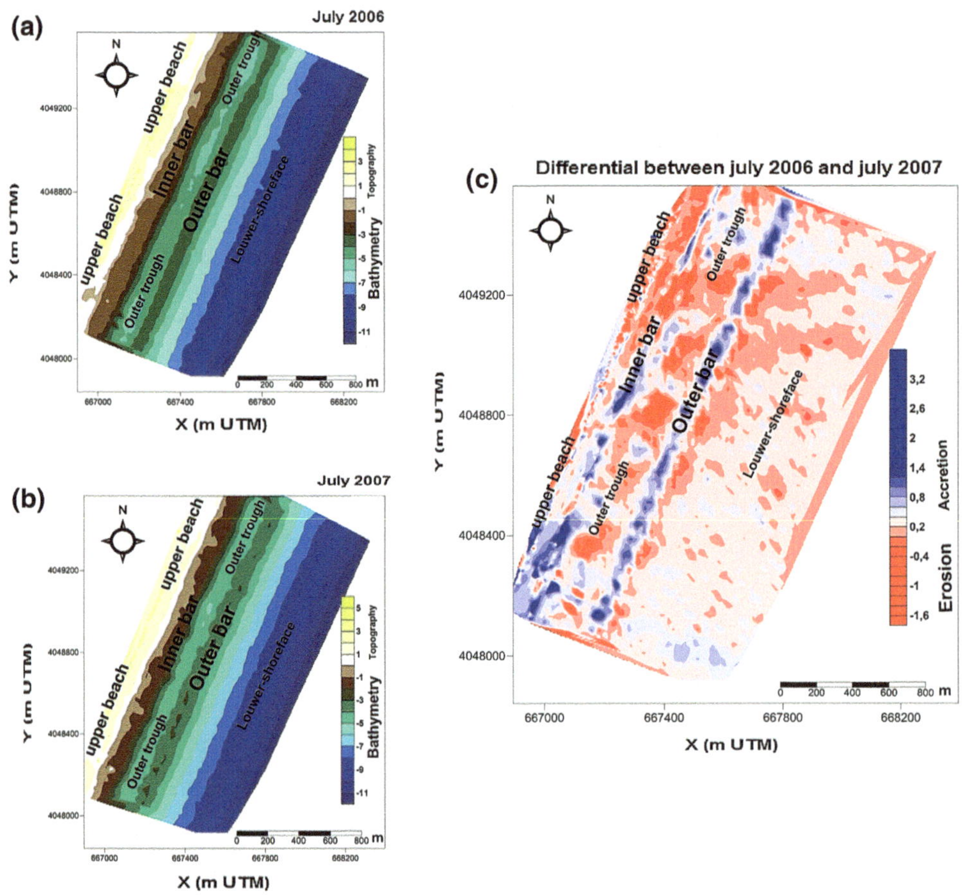

Fig. 8 Residual evolutions recorded during the period from July 2006 to July 2007 at the beach in Korba [30]

merged beach with a relative balance between the subtracted volume (−186 580 m^3) and the added volume (124,900 m^3). However, at the emerged beach, a negative balance, of the order of −45,030 m^3, was recorded (Fig. 12). The volume eroded at the upper beach was added to the submerged beach. Clearly, this evolutionary scheme reflects a winter profile. The submerged beach is fed with sediments taken from the upper beach and deposited in favour of the submerged beach.

The research finding unveils the fact that during the monitoring period of 2006–2009, low volume variations were recorded between the amount of sediment accumulated at the submerged beach and the amount eroded at the emerged beach. Thus the coastal prism of Korba reflects the behaviour of a complete system: sand removal in certain morphological units is balanced by deposits in other units, which explains the low volume variations. The sedimentary stock is, in this way, preserved in the sedimentary prism. This implies mainly the interdependence between the two compartments (submerged and emerged beaches) in sedimentary exchanges and then the dynamic equilibrium characterizing the coastal system of Korba.

4.3.4 Volumetric Variations at the Inner Bar and the Outer Bars

During the first phase of the evolution that extended from July 2006 to July 2007 (Fig. 8), both nearshore bars (inner and outer bars) recorded a significant gain in sediment estimated at 10300 m^3 for the inner bar and 18570 m^3 for the outer one. However, during the second evolutionary phase from July

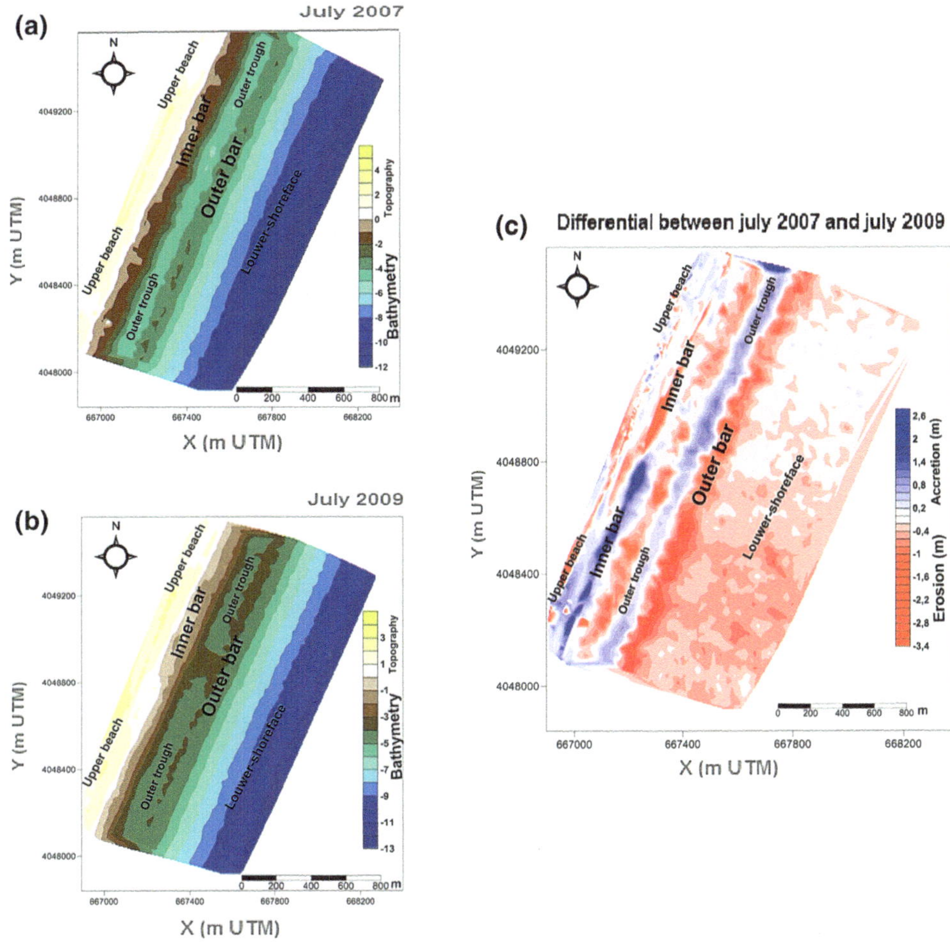

Fig. 9 Residual evolutions recorded during the period from July 2007 to July 2009 at the beach of Korba [30]

2007 to July 2009 (Figs. 13 and 14), the two bars were marked with a negative budget of around 6900 m^3 for the inner bar and about 21570 m^3 for the outer bar. This was certainly related to a stormy phenomenon that led to the segmentation of the inner bar in 2009 and the lowering of the outer bar and thus its movement of mass towards the coast (Fig. 9). Furthermore, it emphasized the filling of the outer trough.

The budget from 2006 to 2009 was in excess at the inner bar (Fig. 13), in deficit at the outer bar and excess in the outer trough (Fig. 14).

5 Discussion

5.1 The Evolutionary Phases of the Beach of Korba

Referring to other works on beaches with sand bars in a microtidal environment [18,29], the one can speculate about the evolutionary phases of the site of Korba. Indeed, two phases of morphological evolution were distinguished in the study period indicated by the characteristics of the submerged beach in response to the agitation conditions in the study area:

- the first phase between July 2006 and July 2007,
- and the second phase of evolution between July 2007 and July 2009 (Table 1).

The morphological evolution of the bars of the microtidal beach of Korba can be connected to the weather-marine system of the monitoring period. Their position and shape vary significantly. Their shape and orientation relative to the shoreline certainly describe some aspects of adjustments [32]. At the level of the bars, which are in most cases rectilinear, signs of feasting can sometimes be recorded. On the one, this state of the underwater beach of Korba (Figs. 8, 9 and 10) is equivalent to that described in Sète in France [29] and in Mahdia beach in Tunisia [18]. On the other hand, the outer trough between the two bars is also a good indicator of morphodynamic evolution. So during its filling in the second phase of evolution (July 2007–2009), a lowering of the outer bar and its movement towards the coast were perceived. Hence, the lowering of the outer bar, resulting in a gradual loss of sed-

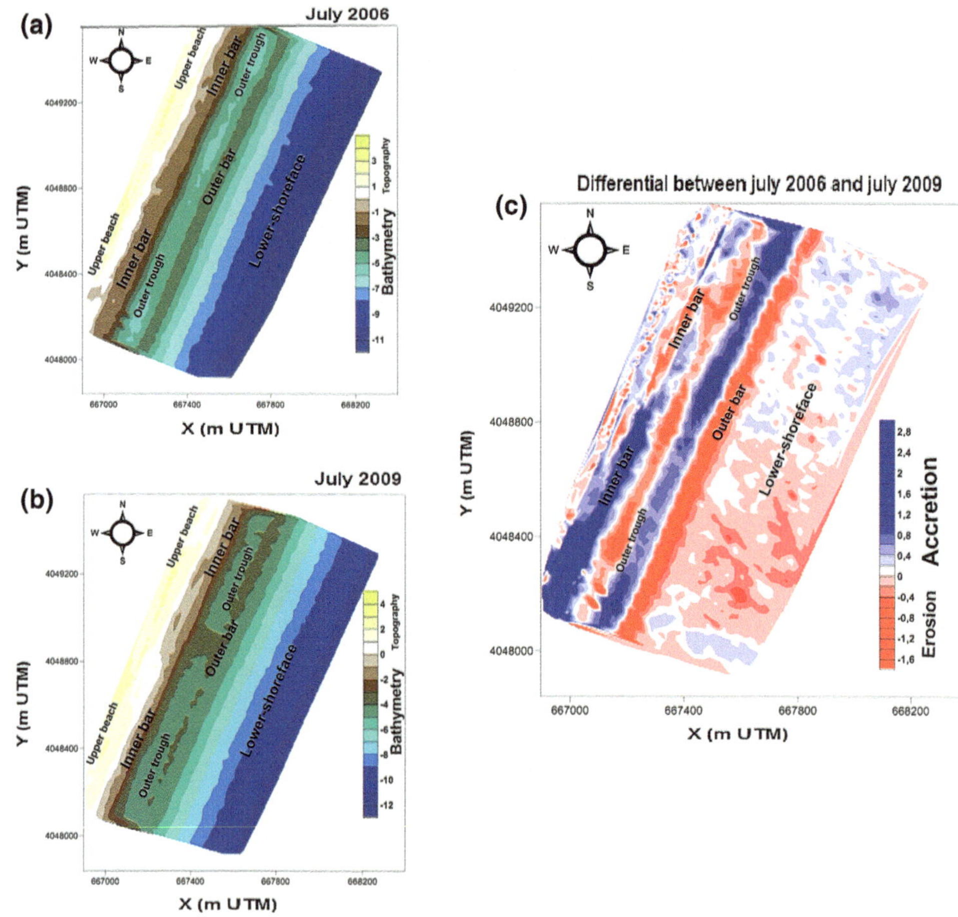

Fig. 10 Residual evolutions recorded during the period from July 2006 to July 2009 at the beach in Korba [30]

iments under the influence of a movement of the centre of the mass of the bar towards the shore and nonlinear waves (non-breaking waves). This state of evolution of the outer bar is observed between sea storms and during periods of low agitation [27,33].

5.2 Dynamics of Nearshore Bars

Hydrodynamic processes that are likely to play an important role in the formation of the bars (brisance, infragravity waves...) are also involved in the development and movement of the bars.

5.2.1 Mode of Compensation

In relation to the differentials between July 2006–2007 and July 2007–2009, there are marked changes in the external bar that was initially growing to evolve into deficit thereafter (July 2007–2009). This deficit is relayed by an onshore migration of this bar. Indeed, many examples studied on microtidal beaches at two bars [29,34] prove that the outer bar is more difficult to mobilize than the inner bar; only the most extreme heavy seas are capable of inducing morphological changes there.

5.2.2 Segmentation of the Inner Bar at the Submerged Beach of Coastal Prism of Korba

The segmentation of the inner bar in July 2009 appears to be related to the wave action according to Certain (2002) [29]. Thus, on the coast of Korba, the bars are generally positioned parallel to the shore. However, the inner bar can sometimes adopt an arrangement in echelons formed by successive seg-

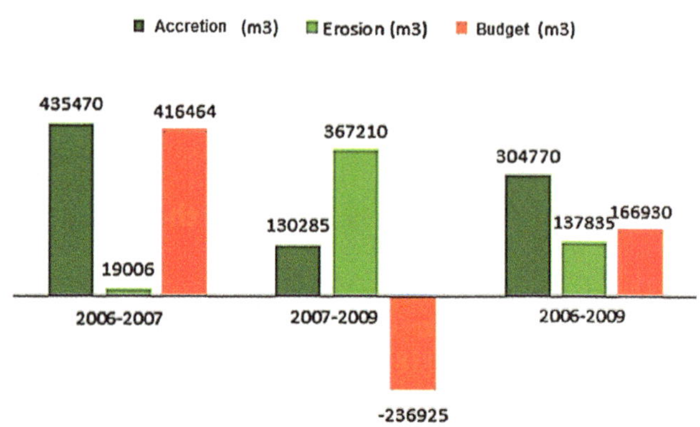

Fig. 11 Volumes moved in passing from one evolutionary phase to another throughout the prism

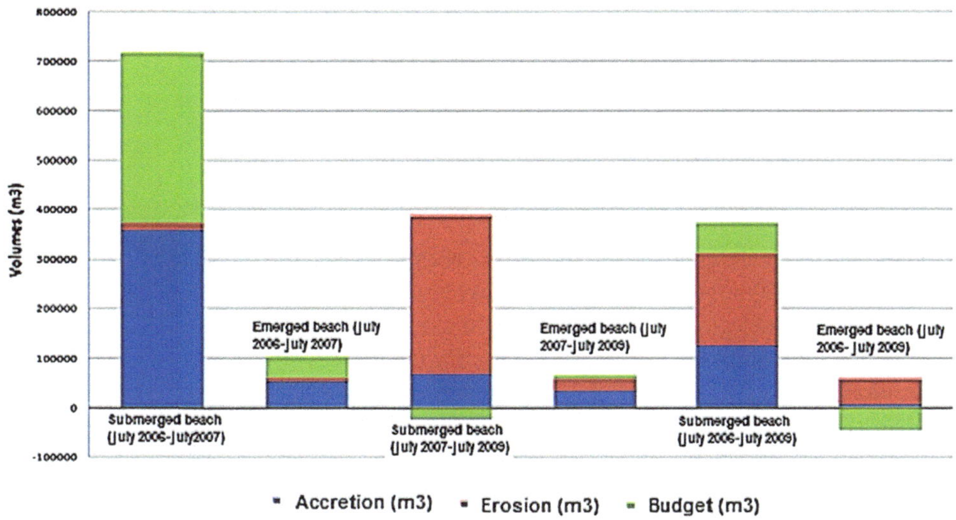

Fig. 12 Localized volumetric changes recorded during the monitoring period at the emerged beach and submerged beach

ments oblique to the coastline. The segmentation of the inner bar into two segments appears to be a response to the incident wave of the North-eastern direction. It is therefore apparent that the inner bar undergoes a rapid reorientation in response to the impact of the wave. This readjustment can be at the origin of this segmentation of the bar (Fig. 15). The reactivity of bottom sediments to wave energy changes, reflected in the profiles of Korba between 2006 and 2009, resulted in active dynamics of the bars. To understand these dynamics, we have recourse to two complementary conceptual models [18,29] of which we will try to show their applicability in the case of the nearshore of Korba:

- Applicability of 'Net Offshore Migration' (N.O.M.) model

This model corresponds to a mode of evolution described by Certain (2002). It is about the diminishment of outer bars and their migration towards the open sea. This mode is provoked by exceptional events. Under the influence of strong storm waves, the bars retreat strongly. The outer bar loses its material in the direction of the coast for the benefit of the inner bar under the effect of the asymmetry of the waves and decreases gradually. During the gradual disappearance of the outer bar, the inner bar, that has fed from the recovered material, while being more exposed to the effects of heavy seas, recedes and gradually occupies the position of the outer bar. This arrangement tends to be restored around an equilibrium position. A second episode, in which the bars are less mobile oscillate around another equilibrium positions, begins. The application of this model in the coastal prism of Korba is delicate insofar as we do not have a sufficient number of surveys and aerial photographs made in appropriate conditions. In fact, surveys on the before and after storm episodes are supposed to be done in order to evaluate their impact on the variation of the bars behaviours. In the available records, we can only see that the outer bar is always present and may retreat back to sea as in July 2009. None of our data permits us to report the disappearance of the outer bar or its replacement by a bar in the middle position between the inner bar and outer bars, suggesting the post-storm replacement described by the model. Under these conditions, without denying the existence of this mode, we cannot confirm it with the current state of the available results.

- Applicability of 'oscillation around a position of equilibrium' (O.P.E.) model

This second model [18,29] takes into account the fact that during an annual cycle, the dynamics of the bars are based on seasonal weather and sea conditions. During periods of low energy, sediment transport towards the shore predominates. The sediment from the underwater beach can feed the supralittoral zone and the upper beach. The bars approach the shore. Waves of small amplitude are more active on the inner bar; low return currents are not able to move substances out to sea. However, the wave currents induce a flow towards the shore that brings sediments back to the beach. The inner trough, filling the coastline, advances towards the sea while making the beach grow.

In the most dynamic wave conditions, return currents are stronger. The offshore movement of material induces the building of a well-developed inner bar further seaward. This establishes the winter equilibrium. The return to summer conditions creates the opposite pattern, initiated by occasional accretion of the inner bar. This event is observed in the submerged beach of the study area. Accordingly, we understand (between 2007 and 2009) that the inner bar grows on the Southern part of the shore and is diminishing in its Northern part (Fig. 9). However, this simple pattern of oscillations, punctuated by a succession of sea states, can be disturbed by movement of the outer bar under long-lasting conditions of low agitation (for example, between 2007 and 2009). The outer bar moves slowly to the beach by filling the outer trough. This case is typically observed in the submerged beach of Korba, but this feature does not cause any change in the overall operation of submerged beach: the outer bar, moved up towards a more exposed level, returns at the end of the exceptionally calm episode to its equilibrium position [18, 27–29, 35].

Thus, during the monitoring period, we have been that the outer bar is present throughout the various stages of the survey, even when it is subjected to deformation by diminishment and migration to the coast. It is the same case for the inner bar, even if it undergoes changes of orientation and possible segmentations. The 'O.P.E.', scheme is therefore confirmed in the submerged beach of Korba.

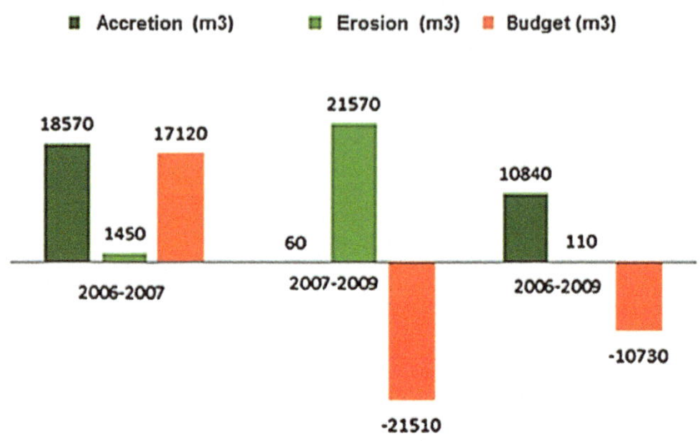

Fig. 14 Volumetric variations recorded during the monitoring period at the outer bar of the submerged beach of Korba

5.3 Relationship Between the Emerged Beach and Submerged Beaches

The width of the beach appears to be closely correlated to the position of the inner bar and in particular to its lateral dynamics (Fig. 9). This is a regularly observed process that causes the variation in the width of the beach.

The evaluation of the sediment balance at the emerged and submerged beaches brings into light a close relationship between these two systems regarding sediment exchanges. Indeed, erosion at the upper beach is often followed by accretion at the submerged beach. This situation is observed during periods of bad weather and the opposite is seen during periods of good weather, hence the conservation of the sedimentary stock and consequently the preservation of the equilibrium of the beach (Fig. 12).

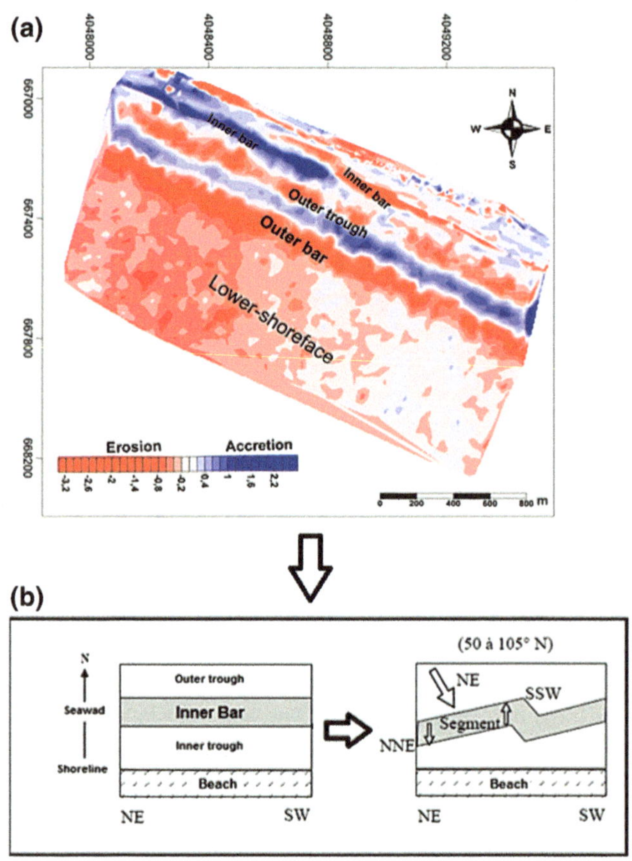

Fig. 15 Adjustment by segmentation of the inner bar in terms of the impact of the incident wave (according to conceptual model developed by Certain (2002) [29]. **a** Korba (Northeast of Tunisia) **b** (Gulf of Lyon, France)

6 Conclusion and Perspectives

In conclusion, the first bathymetric surveys in the nearshore area of coastal prism of Korba have helped us to clarify

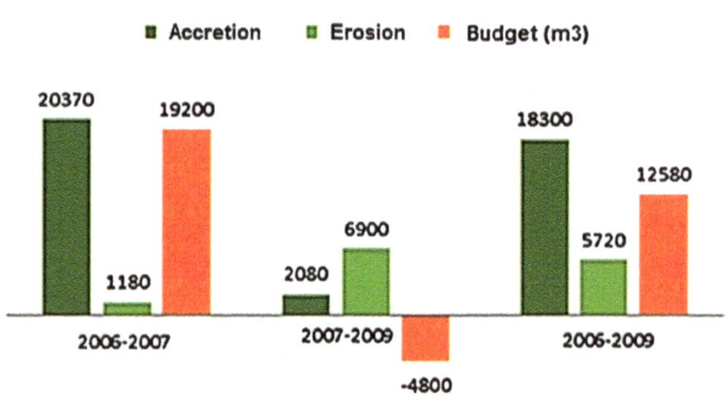

Fig. 13 Volumetric variations recorded during the monitoring period (2006–2009) at the inner bar of the submerged beach of Korba

the morphology of the area and the features of the two nearshores. The coastline of Korba is identified by a sandy prism, the upper part of which has an underwater bar system. The shape and number of these bars depend on several factors including the slope and orientation of the incident waves [36].

The morphology of the sandy prism of Korba is distinguished by the presence of two bars: the inner and outer bars which distributions are homogeneous throughout the study site. In the most common configurations, the two bars are straight and parallel to the coast.

During the survey period of 2006–2009, low volume variations were recorded between the accumulated sedimentary amount at the submerged beach and the amount eroded from the upper beach. Consequently, the coastal prism of Korba behaves as a complete system in a way that the loss of sand in certain morphological units is compensated by deposits in other units, which explains the low volume variations. So, the sedimentary stock is preserved in the sedimentary prism. This brings us to conceive, firstly, the interdependence between the two coastal compartments (the emerged and submerged ones) regarding sedimentary exchanges and, secondly, the dynamic equilibrium characterizing the coastal system of Korba.

The mobility of the bars is part of a dynamic equilibrium (negative loop). Subject to storm conditions, these bars migrate offshore and then return to the coast. In times of low agitations the external bar can migrate to the shore with a loss of sediment and a filling of the external trench. This feature has no effect on the overall functioning of the coast and the outer bar can recover its position in a second storm. This method seems to mark a phase of oscillation around an equilibrium position (O.P.E.).

During periods of low agitations, the inner bar can become segmented while reorienting itself to deal with the incident waves. According to the literature on microtidal coasts, it becomes obvious that the mechanisms associated with self-organization of the beach are responsible for the segmentation of the inner bar.

The beach width appears to be close with the position of the inner bar through its seasonal lateral dynamics.

It is worth concluding, in this respect, that this work has enabled real progress in relation to the modelling and analysis of the morphology of the sandbars (foredunes, berm and nearshore bars) of the coastal prism of Korba. Every kind of improvement has been conducted with the goal of moving towards a method of optimal study that is both precise and three-dimensional; a method that covers the entire coastal area and allows for observations at a high temporal frequency. In the same vein, it appears essential to better monitor events affecting the morphology of the coastal system (foredunes, beach and nearshore bars). A higher frequency of measurements is expected to identify specific and quantified evolutionary sequences. Also, the installation of hydrodynamic devices seems also necessary. To sum it up, collecting this information while linking morphological and hydrodynamic evolutions can help the establishment of a morphodynamic classification of the beach of Korba. This classification has to be a summary presenting the main states of the beach and the environmental conditions that facilitate the changes. Accordingly, an improvement in the accuracy of the budget quantification is a crucial element that occurs through:

- the use of advanced altimetry observations by Lidar [37],
- the simulation of the evolution of the morphological and sedimentological system of the beach by the mathematical formulation of the modelling parameters. This must be a reference tool that simplifies the operation of the environment and a framework for simulations of mathematical models.

References

1. Short, A. D. (1999). *Handbook of beach and shoreface morphodynamics* (p. 379). England: Wiley-Blackwell.
2. Pinot, J. P. (1998). La gestion du littoral. Tome 1: Littoraux tempérés: côtes rocheuses et sableuses. Paris, Institut Océanographique (Vol. 2/1, 399 p.).
3. Costa, S., Gourmelon, F., Augris, C., Clabaut, P., & Latteux, B. (2005). Apport de l'approche systémique et pluridisciplinaire dans l'étude du domaine littoral et marin de la Seine-Maritime (France). *Norois, 196*(3), 91–108.
4. Aernouts, D., & Hequette, A. (2006). Évolution du rivage et des petits fonds en baie de Wis-sant. *Géomorphologie, 1*, 49–63.
5. Horillo-Caraballo, J. M., & Reeve, D. E. (2008). Morphodynamic behaviour of a nearshore sandbank system: The Great Yarmouth Sandbanks, UK. *Marine Geology, 254*(1), 91–106.
6. Furgerot, L., Mouaze, D., Tessier, B., Perez, L., Haquin, S., Weill, P., et al. (2016). Sediment transport induced by tidal bores. An estimation from suspended matter measurements in the See River (Mont-Saint-Michel Bay, northwestern France). *C. R. Geoscience, 348* (this issue).
7. Amrouni, O., & Abdeljaouad, S. (2009). Étude de la dynamique sédimentaire spatiotempo-relle à court terme du système dune/plage de la côte nord de Mahdia (Sahel Tunisien): rôle des agents de transport et de granulométrie. *Revue Méditerranéenne de l'Environnement, 3*, 1–15.
8. Costa, S., & Suanez, S. (2013). Géomorphologie des littoraux français. In D. Mer-cier (dir.), Géomorphologie de la France, Paris, Dunod, coll. Sciences Sup. 65–77.
9. Yahyaoui, Z., & Rbai, N. (2014). Étude de l'évolution morphosédimentaire à court et moyen terme du système dune/plage de Korba (façade orientale du Cap Bon, Tunisie). *Revue In-ternationale de Géomatique, 4*, 471–500.
10. Yahyaoui, Z., Rbai, N., & Abdeljaouad, S. (2014). Etude descriptive de l'évolution saison-nière des dunes bordières de la plage de Korba (façade orientale, Cap Bon) par la méthode des Modèles Numériques de Terrain Résiduel. *Bull. Inst. Natn. Scien. Tech. Mer de Sa-lammbô*, **41**, 143–158 (2014).
11. Desmazes, F., Michel, D., Howa, H., & Pedreros, R. (2002). Etude morphodynamique du domaine pré-littoral nord-aquitain, site atelier du Truc Vert. In *Proceedings of 7ème Journées Natio-nales GCGC, Anglet* (pp. 155–162).
12. Amrouni, O., Medina, R., & Abdeljaouad, S. (2018). Morphodynamics of a microtidal sandbar beach under storm condition: case of the Mahdia nearshore. *Tunisia*, 1–19.

13. Holman, R. A., & Sallenger, A. H. (1993). Sand bar generation: A discussion of the Duck experiment senses. *Journal of Coastal Research, 1*(5), 76–92.
14. Davis, R. A, Jr., & Fox, W. T. (1972). Coastal processes and nearshore sand bars. *Journal of Sedimentary Petrology, 42*(2), 403–412.
15. Goldsmith, V., Bowman, D., & Kiley, K. (1982). Sequential stage development of crescentic bars: HaHoterim beach, southern Mediterranean. *Journal of Sedimentary Petrology, 52*, 233–249.
16. Komar, P. D. (1998). The modelling of processes and morphology in the coastal zone relfexions on the maturity of our science. *Shore and Beach, 66*, 10–22.
17. Lippman, T. C., & Holman, R. A. (1990). The spatial and temporal variability of sand bar morphology. *Journal of Geophysical Research, 95*(C7), 11575–11590.
18. Amrouni, O. (2008). Morphodynamique d'une côte sableuse à barres: Mahdia (Côte Orientale de la Tunisie), Thse, Faculté des Sciences de Tunis (318 p.)
19. Louati, M., & et Zargouni, F. (2009). Modélisation topo-bathymétrique et transit sédimen-taire. Exemple des plages sableuses de la baie de Tunis, Nord-Est de la Tunisie. Géomorphologie: relief, processus et environnement (pp. 211–222).
20. Halouani, N. (2012). *Etude de la dynamique sédimentaire de la frange littorale de Ta-barka-Berkoukech: Modélisation de l'évolution de son trait de côte*. Thèse de Doctorat, Faculté des Sciences de Tunis (189 p.).
21. Bouden, S., Chaabani, F., & Abdeljaouad, S. (2004). Caractérisation géochimique des sédi-ments superficiels de la lagune de Korba (Cap Bon, Nord-Est de la Tunisie). *Geo-Eco-Trop, 28*, 15–26.
22. Baccar, L., Moussa, M., & Ben, H. C. (2001). *Conservation des zones humides littorales et des écosystèmes côtiers du Cap Bon, Partie relative à: l'hydraulique des zones humides de Maâmoura*. Rapport de diagnostic des sites: Tazarka et Korba, (106 p.).
23. Zâara, Ch. (1996). *Étude de la stabilité du littoral du golfe de Hammamet*: Impact de la dynamique sédimentaire sur la morphologie côtière DEA (112 p.).
24. Sabatier, F., Stive, M., & Pons, F. (2004). Longshore variation of depth of closure on a micro-tidal wave-dominated coast. In *International Conference of Coastal Engineering 2004, American Society of Civil Engineering, Lisboa* (pp. 2 329–2 339).
25. Raynal, O., Brunel, C., Certain, R., Aleman, N., Robin, N., & Sabatier, F. (2012). Le bilan sé-dimentaire, un outil pour la conception d'un plan de gestion des sédiments du littoral du Languedoc-Roussillon. *XIIème Journées Nationales Génie Côtier - Génie Civil* (pp. 371–380).
26. Sabatier, F., & Provansal, M. (2000). Sandbar morphology on the Espiguette Spit, Meditterranean Sea, France. In *Marine Sandwave Dynamics* (pp. 179–187).
27. Larson, M., & Kraus, N. C. (1992a). Dynamics of longshore bars. *Coastal Engineering, Chapitre, 10*, 2219–2232.
28. Larson, M., & Kraus N. C. (1992). Analysis of cross-shore movement of natural longshore bars and material placed to create longshore bars. Coastal Engineering Research Center, Waterways Experiment Station, Corps of Engineers 3909 Halls Ferry Road, Vicksburg, Mississippi 39180-6199, final report, 117 p.
29. Certain, R. (2002). Morphodynamique d'une côte sableuse microtidale à barre, le Golfe du Lyon, Languedoc-Roussillon. Thèse de doctorat, université de Perpignan, 209 p.
30. Yahyaoui, Z. (2016). Modélisation spatio-temporelle de l'évolution morphosédimentaire d'une plage microtidale à deux barres: cas de la plage de Korba, façade orientale de la Tunisie, Cap Bon, Thèse de de Doctorat, Faculté des Sciences de Tunis, 357 p.
31. Barusseau, J. P., Brissaud, L., Drapeau, J., & et Long, B. (1991). Processus hydrodyna-miques et morphosédimentaires de l'environnement des barres d'avant -côte du littoral du Golfe du Lion, Oceano. Acta, vol. spe. no 11, pp. 163–176.
32. Certain, R., & Barusseau, J. P. (2005). Conceptual modelling of sand bars morphody-namics for a microtidal beach (Sete, France). *Bulletin de la Société Géologique de France, 176*(4), 343–354.
33. Certain, R., & Barusseau, J. P. (2004). *Modélisation conceptuel de l'évolution morphody-namique des barres sédimentaires d'une plage microtidale (Sète, France)*, VIIIème Journées Nationales Génie Civil Génie Côtier, Compiègne (pp. 7–9).
34. Akouango, E., & Barusseau, J. P. (1994). Dynamique sédimentaire d'une côte à barres: cas de la côte sableuse du Languedoc-Roussillon. Session II: Dynamique sédimentaire et transports particulaires (pp. 147–154).
35. Rihouey, D. (2004). Analyse statistique de l'évolution morphodynamique des plages sa-bleuses: Application aux sites d'étude du programme national d'environnement côtier et aux plages d'anglet. PhD thesis, Université de Pau et des pays de l'Adour, 350 p.
36. Barusseau, J. P., Radulescu, M., Descamps, C., Akouango, E., & Gerbe, A. (1994). Morphosedimentary pluriannual changes on a barred coast Gulf of Lions, Mediterranean sea, France. *Marine Geology, 122*, 47–62.
37. Jaud, M., Delacourt, C., Allemand P., Grandjean P., Ammann J., Cancouët, R., et al. (2014). DRELIO: Un drone hélicoptère pour le suivi des zones littorales. Revue Paralia (Vol. 7, p. s02.1s02.12). https://doi.org/10.5150/revue-paralia.2014.s02.

Part III
Management and Development of the Territory

Using Aerial Photography for Semi-automatic Extraction of Road Network at a Scale of 1:25000

Karim Mansouri, Noamen Rebai, Sahar Gaaloul and Murad Salhi

Abstract

Through the application of new methods of digital restitution, photogrammetry applied to the urban domain has always contributed to the improvement of the knowledge and control of urban development especially in relation to rapidly changing spaces. In this study, we seek to develop a method of extracting road network from aerial photos at a scale of 1:25000, in the area of El Marsa, a northern suburb of Tunis, Tunisia. As it is generally known, from the intensity variations on the image, the one can make the segmentation, recognize the objects and the forms, and evaluate their spatial positions as well as the relations uniting them. In this context, the roads seem to be of particular interest in urban areas. Their automatic recognition makes them useful in various tasks, including mapping, recalling multisource images, urban planning and automatic navigation. However, despite the various works carried out on this field, none of the sophisticated road detection techniques is perfect. The difficulty of road extraction is seen while establishing a statistical test to know whether a given pixel of the image participates in the road network or not. In fact, this difficulty is accentuated in the urban environment as it is characterized by a strong heterogeneity of attributes. Compared to other cartographic themes such as buildings, contour lines, and vegetation, manual road entry is fast, and the road theme appears to be the easiest one to be automatically restituted. Our major concern, throughout this study, is to elaborate a methodological approach that enables us to replace manual photogrammetric capture by automatic or a semi-automatic captures from the Orthophoto of a 30 cm resolution. In relation to the current state of art, this photogrammetric capture is advantageous in time and material. This work on automatic interpretation of aerial images thus meets the need to optimize the productivity of photogrammetric capture strings without harming the quality of the final product. The problem raised however is related to the automatic extraction of roads in 2D and its transformation into 3D with validation. Our procedure is not limited to image processing, it rather extends to:

- Road extraction from radiometry-based segmentation using the "eCognition" software.
- Making a mask for buildings.

While the first step with "eCognition" allows the transformation of the buildings into vector format and polygon type, the second step with "ArcScene" contributes to the juxtaposition of the roads in a continuous way on the Orthophoto. The last step of our procedure consists in the layering of the roads on a Digital Terrain Model (DTM) using the "LPS" software, with a resolution of 1 m in planimetry. The results obtained are evaluated by quantitative statistical approach. In this perspective, it is important to further develop an automatic chain of quality control and validation of the restituted road network.

Keywords

Road • Marsa city • Segmentation method • Ecognition software • DTM • DSM

K. Mansouri (✉)
ENSI: Ecole Nationale des sciences de l'Informatique Manouba, Manouba, Tunisia
e-mail: mansourikarim11@gmail.com

N. Rebai
LR14ES03 Geotechnical Engineering and Georisk Laboratory, National School of Engineering of Tunis, University of Tunis El Manar, le Belvédère 1002, B.P. 37, Tunis, Tunisia
e-mail: noamen.rebai@enit.utm.tn

S. Gaaloul · M. Salhi
CNCT: Centre National de la Cartographie et de la Télédétection (en Tunisie), Tunis, Tunisie
e-mail: cnct@defense.tn

M. Salhi
e-mail: cnct@defense.tn

N. Rebai and M. Mastere (eds.), *Mapping and Spatial Analysis of Socio-economic and Environmental Indicators for Sustainable Development*, Advances in Science, Technology & Innovation, https://doi.org/10.1007/978-3-030-21166-0_11

1 Introduction

Photogrammetry has continuously grown as a major asset in the field of urban planning, infrastructure, management and monitoring of natural resources.

The evolution of the techniques of aerial photography acquisition permits a fast, numerical, homogeneous and frequent access to the totality of a given territory. This allows to know and control the evolution of this large space.

Applied to the urban domain, photogrammetry can improve the knowledge and control of urban development, especially in rapidly changing areas.

Nowadays, we have mastered the acquisition and digitization of data in raster format, but the automatic extraction of roads and other information contained in these images are not yet perfect and complete [1–5]. It is in the 1970s [6], that the first studies dedicated to the automatic road network extraction from optical images were conducted. This extraction is considered as a multistage process.

The first step is the extraction of the pixels whose radiometry satisfies a model of road forms along the analyzed image.

In the second step, the space operators examine the detected pixels by eliminating the ones that do not belong to the road network and by forming strings from those that have geometric characteristics of a road segment.

This depends on the degree of fragmentation of the detected road segments, the number of those undetected (omissions) as well as the number of the "false" detected ones (commissions), [7].

One can appreciate the quality of the operator applied to the first step which largely determines the accuracy and completeness of the final network extracted from the images.

A classification of the main approaches of road detection from several types of sensors is presented by [8], through which a first class based on the continuation of the detection "road tracking" from "seed points" is identified either manually or automatically.

Both the work of [9] on aerial photographs and those of [10] on IKONOS images aim at identifying linear structures on basis of the analyses of profiles that are perpendicular to one direction. The tracking criteria are essentially geometric and radiometric.

The second class of approaches resides on morphological targeting used for eliminating spectral noises, whose characteristics are similar to road surfaces, from areas prone to roads that have already been identified.

The best example for the first phase of road identification is summarized in [11], which is based on a segmentation process adapted to multispectral images with a high resolution in urban areas. This process is made through classification techniques (supervised or not), filtering or segmentation depending primarily on the image texture.

A number of researchers have developed the third class that relies on algorithms of dynamic programming [4,12–17]).

The iterative process uses the so-called merit analysis until a succession of positions proposes road reconstruction that meets the criteria of proximity, collinearity, and other aspects.

The concept of the fourth class of approaches is introduced by [18] entitled "Snakes: Active contour Models". A variant has been proposed by [19,20] for the extraction at different scales of aerial photographs while exploiting their respective complementary information.

These four main classes of approaches can be generalized by referring to global concepts, namely the successive intervention of local operators applied to the images and spatial operators completing the process. Thus, the local operators are applied to these images in order to extract potential road pixels on the analyzed image. In the same way, from the intensity variations on the image, the one can make the segmentation in order to recognize the objects and forms so as to evaluate their spatial positions and the relations uniting them.

Roads are of particular interest in urban areas and their automatic recognition is useful in a number of areas, including mapping, multisource image registration, urban planning, and automatic navigation.

Compared to other cartographic themes such as buildings, contours, and vegetation, manual road entry is fast, and the road theme appears to be one of the easiest to be automatically interpreted.

Many studies in this context have shown that the difficulty of extracting roads is expressed by establishing a statistical test to know whether a given pixel of the image participates in the road network or not. This difficulty is accentuated in an environment of high heterogeneity. It is therefore difficult to build a method that can perform this extraction. These methods deal with the main characteristics of the roads on the image (geometry, radiometry, etc.) as well as the properties on the network (such as connectivity).

As part of the BDTN25K project of the National Center of Cartography and Remote Sensing (CNCT) of Tunisia, a national topographic database is designed for topographic covers at 1:25000, taking all themes with a geometric precision of the order of the meter in (x, y and z). In this context, we endeavor to replace the manual photogrammetric image capture with a semi-automatic one from the Orthophoto with a 30 cm resolution.

For so doing, we will apply the two steps described to map the roads in 2D, and then represented in 3D by draping on a DEM of a 1m resolution.

The research on the automatic interpretation of aerial images thus answers the need to increase the productivity of photogrammetric capture strings without harming the quality of the final product. In this sense, we have opted for statistical analysis of the performances of our new extraction technique.

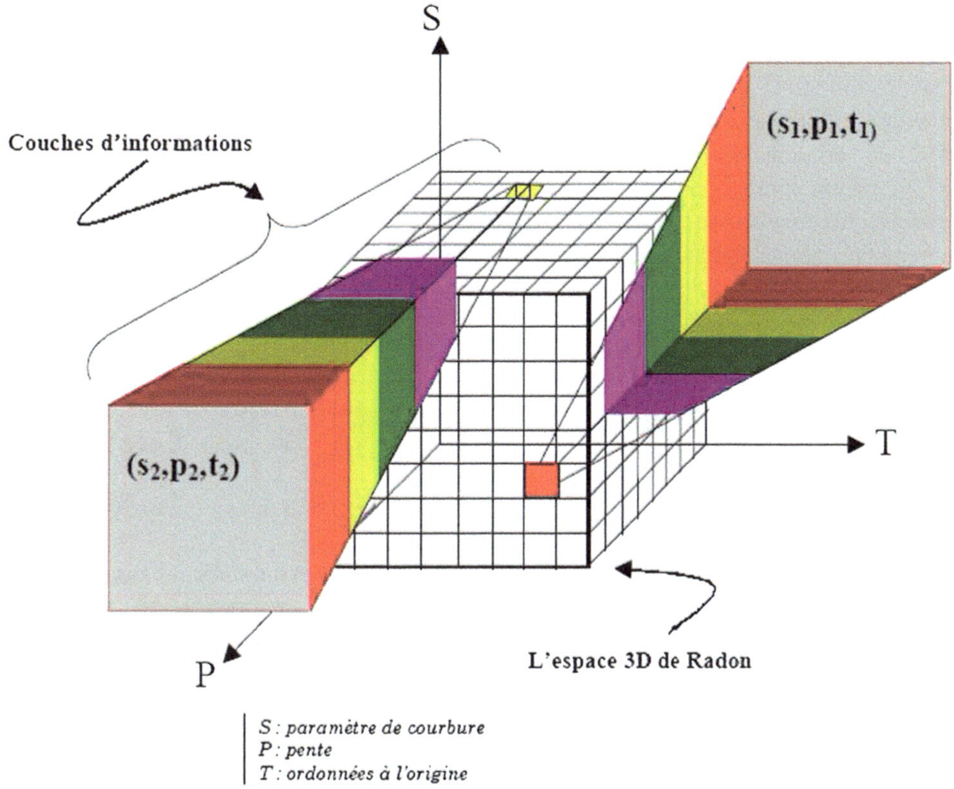

Fig. 1 Schematic presentation of the principle of the M3D method [39]

2 Review on Extraction Methods

The best known methods developed in this area are the local and global methods. In our work, we have used the global segmentation and classification method, which presents a relatively simple and efficient procedure for road extraction.

2.1 Local Methods

They lie in the use of the spectral information of the targeted pixel immediate vicinity. They are mainly based on spatial operators that respond to local variations in intensity [21–23]. Local methods can be grouped into 5 categories, which are as follows, the gradient operator, gradient profile analysis, mathematical morphology, spectral convolution and finally the neural network [24–31].

Two major disadvantages arise when using local methods. First, the image is not considered in its entirety. This leads to discontinuous and incomplete lines in the final result. Second, they cause an inaccuracy and a geometric positioning shift of some extracted elements.

2.2 Global Methods

Global methods have been developed to cope with the observed failure of local methods. It is therefore a matter of converting the image concerned into a new mathematical space and extracting the road network from the transformed forms in this new space. The specificity of global methods (Fig. 1) springs from considering road extraction as a global problem that requires more than local information [4,32–34].

- Combinatorial optimization [12,35]
- Active contours and other methods of expert systems [18,36]
- Stochastic Model and Deterministic Methods [37]
- Profile matching (Kalman filtering) [38]
- Multilayered method
- Three-dimensional multilayered method (M3D) [39]
- Segmentation and classification method [40].

Segmentation is the division of an image into a set of regions according to predefined criteria.

Different approaches can be adopted. First, there is the marginal approach of combining the merged images in a single image after segmenting each spectral component. Second, there is the scalar approach which consists in segmenting the different spectral components after merging them into a single component as an image in the gray level. Third, there is the vector approach in which vector information is used for each pixel [41].

Based on notions of differences and similarities perceived by the human visual system, we distinguish the region approach and the contour approach. The latter is used to identify the changes between regions. The outline of objects is the borders of two or more objects with different gray levels.

According to Croquerez [42], there are several categories of contour approaches: derivative [43,44], morphological [45], and variational [46,47].

The region approaches focus on the regions' content by grouping the pixels used for verifying common properties. Among the regional approaches there are four main methods:

- Threshold segmentation methods [48,49].
- Methods of regions growth types [50,51].
- Division-fusion methods [52,53].
- Classification methods that can be either unsupervised if there is no prior knowledge of the data, or supervised if a pre-classification learning step is required.
  Classification can be developed depending on many factors: the study theme, the user's needs, the study area's characteristics, and the satellite data availability and characteristics.
  There are different classification methods [21,54,55]:
- Classification by the parallelepiped method (hypercubes).
- Classification by distance minimization.
  - Classification by the nearest neighbor K method (KNN).
  - K-means method.
  - Classification by ISODATA.
  - Classification by Fuzzy-C-means.
- Classification by the likelihood method.
- Contextual classification by Markov fields.
- Vectors Supported Machines (SVM).
- Neural networks.

Each of the two methods, whether segmentation by region or contour, has certain disadvantages. This explains the reason why the research has been oriented towards cooperative approaches that allow a better consideration of the characteristics of the image's objects.

There are three types of segmentation by cooperation regions contours:

- sequential cooperation
- the cooperation of results
- mutual cooperation.

3 Case Study

3.1 Study Zone

Our study area is part of the city of El Marsa (Fig. 2). It is delimited by the following four points (A, B, C, D) having the coordinates in the UTM system WGS84 (Table 1):

It is worth noting that in the cartographic series 1:25000 of Tunisia, the map contains between 75 and 80 images with a GSD of 50 cm. However, in our case, with a GSD of 10 cm, the map contains 990 images which correspond to 989 couples.

3.2 Materials

We have recourse to a methodological approach revolving around the combination of satellite images and aerial photo restitution treatments in order to obtain a semi-automatic mapping of 3D roads. As part of our study, we have the dataset of the processing chain set up for the project (CNCT, BDTN25K):

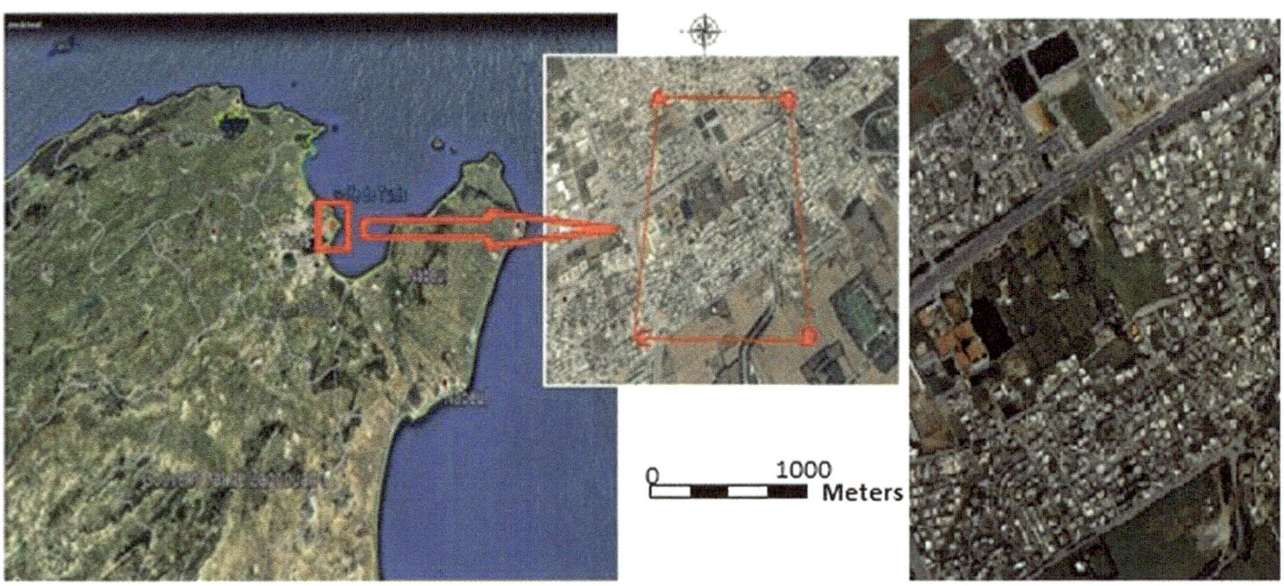

Fig. 2 Urban area of city of El Marsa (North of Tunis-Tunisia)

Table 1 The city of El Marsa delimited by the four points (A, B, C, D)

Coordinates	A	B	C	D
X	617163	618164	617163	618164
Y	4082066	4082066	4080702	4080702

- Multispectral aerial optical images were taken in March 2016 with a DMC camera which is of a large format. Its panchromatic and multispectral focal length is 120 mm. It simultaneously offers a high resolution panchromatic image, the image (R, G, B) and the near infrared spectrum (NIR). For this project, the images are of size 13824 × 7680 pixels with a GSD of 10 cm, taken at an average flight height of about 1000 m, an average terrain altitude of 25 m and an overlap of 80% in-band and 30% interband.
- Coordinates of the images centers are delivered by the GPS embedded in the plane and synchronized with the camera. Indeed, it is from the INS data (Inertial Navigation System) that the computer on the plane triggers the camera. The GPS estimates the resulting delay that will be integrated in the calculation of coordinates.
- Inertial Measurement Unity (IMU) orientations: These are the angles of orientation (omega, phi, and kappa) which describe the attitude of the plane during each photo-taking.

In the context of our study, we will apply the treatment procedure developed by the CNCT (Fig. 3). The work starts with the choice of support points from the canvas for the entire study area. This requires a fieldwork to acquire the coordinates of its points by using:

- RTK GPS.
- Aerial photos in raster format.
- Map of the study area (or old mosaic).
- Material Safety Data Sheets.

After orientation (internal, relative and absolute), aerotriangulation is performed through digital aerial photographs and markup points. This allows us to determine the orientation of the images, create stereoscopic models and correctly read the coordinates indicated in the models.

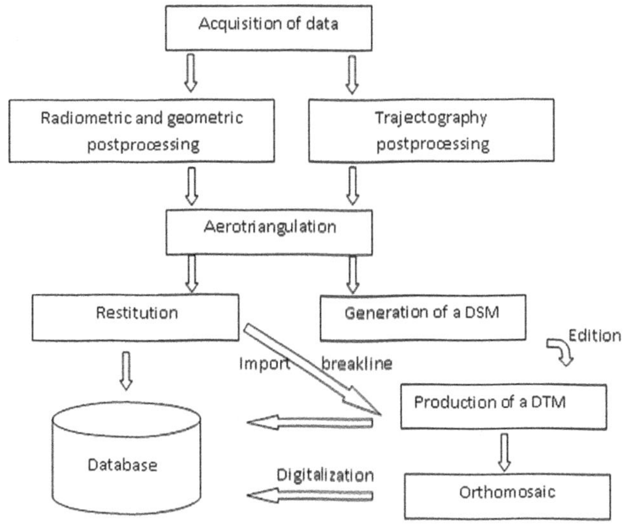

Fig. 3 Photogrammetric processing procedure (CNCT 2010)

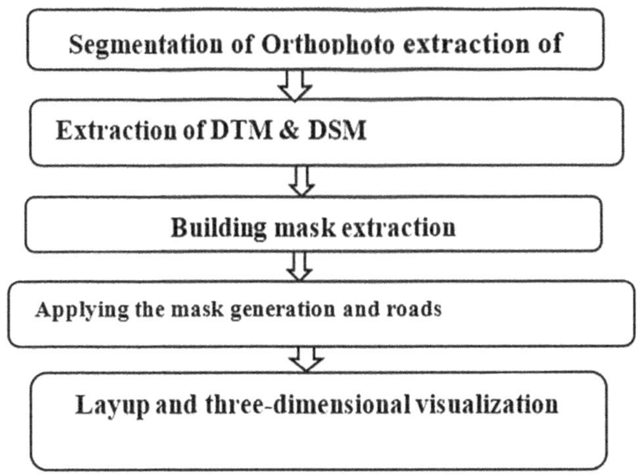

Fig. 4 Flow chart of the methodology adopted for the 3D road extraction

3.3 Methodology

Orthophoto Segmentation and Road Network Extraction (Fig. 4).

Since segmentation is an important step in the entire image analysis process, we have put our focus on it. In fact, segmentation is the partition of an image into a set of regions according to predefined criteria. Many segmentation methods are based on two basic properties of pixels relative to their local vicinity: discontinuity and similarity. While pixel similarity is used by region approach segmentation methods [22], pixel discontinuity is used by edge approach segmentation methods.

In the first approach, we are interested in the boundaries of the regions and in the second, we are concerned with their content. There are several methods for region and contour segmentation. Although they all have certain advantages, they also have a number of disadvantages. Accordingly, several researchers have leaped on cooperative approaches.

The image segmentation by region-contour cooperation can be realized in three types:

- Sequential cooperation: in which one of the segmentation techniques (region or contour) is carried out in the first place; its result is later to be operated by the other techniques.

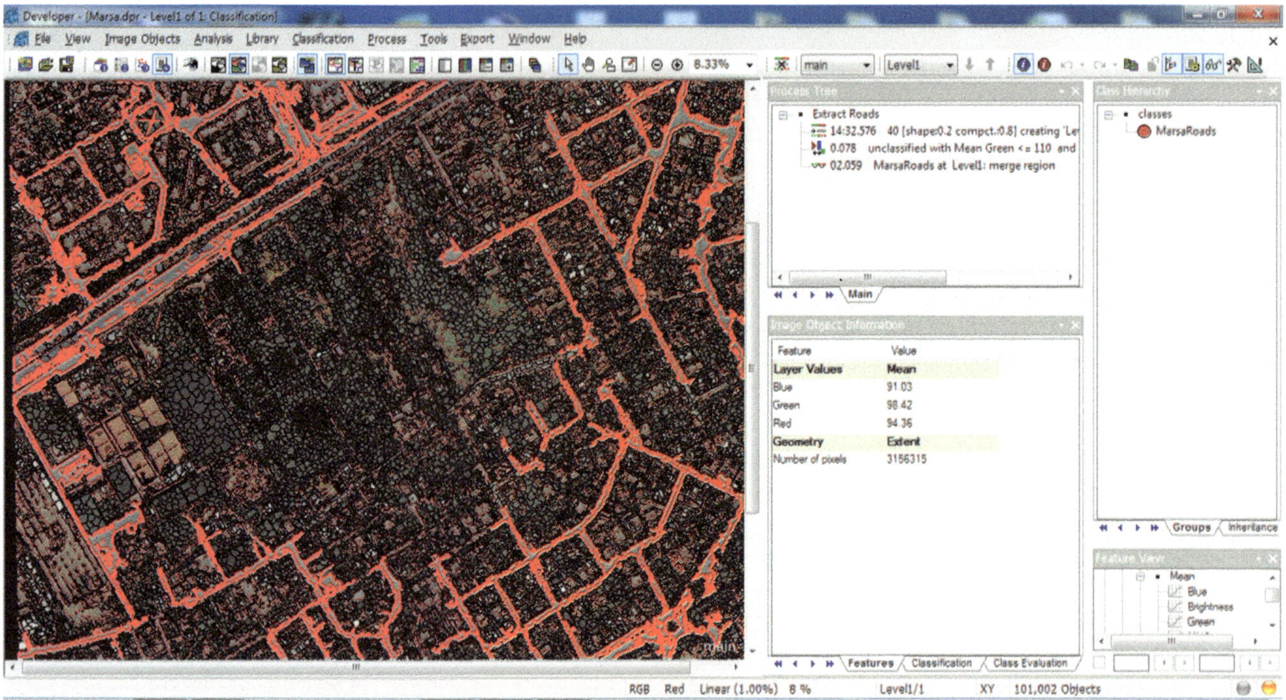

Fig. 5 Result of the road network extraction obtained with the software eCognition V.8.7

- The cooperation of the results: the two types of segmentation will be carried out independently and in parallel; their results will be concerned with the cooperation.
- Mutual cooperation: the two types of segmentation will cooperate with each other during their implementation process.

The segmentation approaches by region-contour cooperation, whether sequential, mutual or results, integrate the two types of information (regions and contours) in order to better consider the characteristics of the objects of the picture. By this cooperation approach, not only homogeneous zones, but also the transitions between them, are respected simultaneously.

In comparison with region or contour segmentation, the result of cooperative segmentation is more faithful to the reality of the image (Fig. 5).

Among the software we used to perform our research are Monteverdi, ENVI and eCognition segmentation. Despite the good results obtained, we prefer eCognition due to the important time saved while studying the objects contained in the prototype. This depends on these objects' density and diversity. In addition, this study is done with the user's intervention which makes our solution less and less automatic. It is crucial to note that this extraction up to this point has been fully automatic.

Realization of the DTM & DSM

After completing the aerotriangulation used for obtaining a block of images from a couple of images of our chosen El Marsa prototype, we have generated the DTM and DSM as well as the Orthophoto. Our DTM is made with a step of 1 m, while our DSM is realized with a step of 0.5 m.

In carrying out its work, the CNCT usually produces DTMs and DSM with a 5 m pitch, which is largely sufficient when producing topographic maps at a scale of 1:25000. However, in our case, the need of refining the DTM and DSM resolution is required in order to be able to apply the mask of the buildings needed for avoiding a deformation in its presentation. This makes a better staggering and a limit to be diffused. In our project, the DTM and the DSM are being generated in image format (with a * .img extension), (Fig. 6).

- Mask buildings extraction

The results obtained so far give us polygons representing roads with some discontinuities due either to the effect of shadow, or to the elevation of some buildings or other obstacles during the aerial photography. The angle of view can cause some road discontinuities.

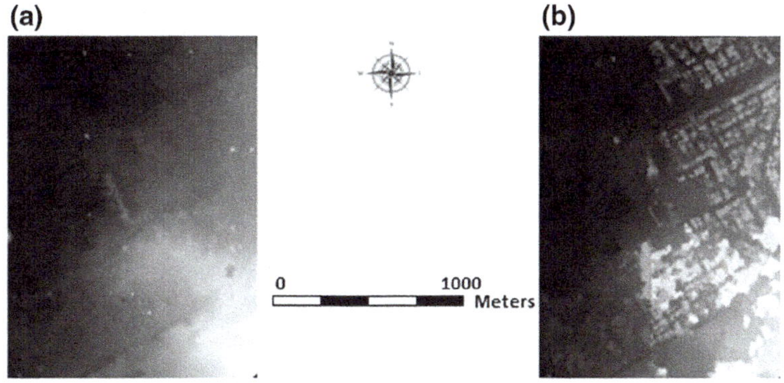

Fig. 6 **a** DTM of the prototype with a pitch of 1 m; **b** DSM of the prototype with a step of 0.5 m

As we will later explain, it is precisely this part that makes our task semi-automatic as we manually intervene with ArcScan. In urban areas, apart from this road discontinuity, it is obvious that when exporting to our polygons in vectors, there is confusion between roads and some buildings, while radiometry cannot solve the problem.

This problem is solved in three different ways that are all inspired by photogrammetry. They are based on the DSM (Digital Surface Model) and the way taken also uses the DTM (Digital Terrain Model), as it will be explained in the next chapter.

The information described by the DSM is the maximum value of height above ground in each mesh. The objective of the three methods is the realization of a mask representing the buildings in our prototype.

The first method under ArcGIS is based on the slope map. Indeed, standards show that roads in general can not vary with slopes more or less higher than 10°. According to the DSM delivered to us by the photogrammetry department within the CNCT with a step of 1 m, we come to realize our slope map using two classes, and we get our mask of anything greater than 10°.

The second method is obtained from the DSM using two filters with "eCognition". Indeed, from the DSM, two filters are produced: a convolution filter and a slope one. Then, we proceed to export the building layer obtained in vector mode and polygon type (Fig. 7).

The third method has been presented in (Fig. 8). We used this method to filter the buildings' layer. For so doing, we have imported, in addition to the automatically imported RGB layers, the DTM and DSM into image format as layers in order to realize the building mask.

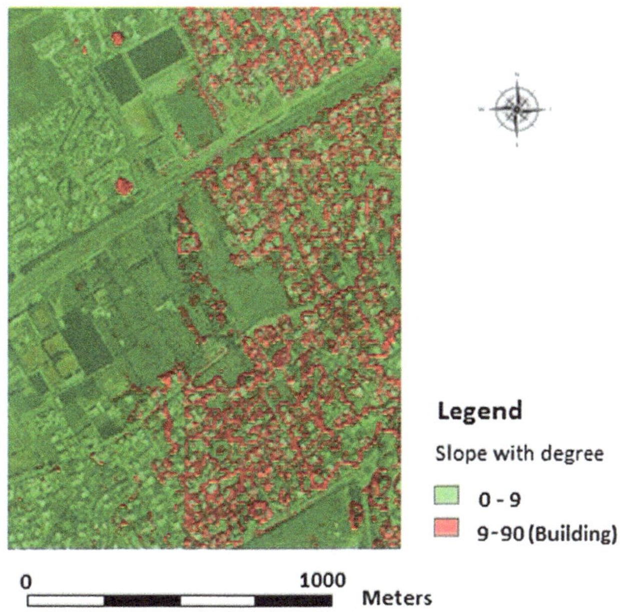

Fig. 7 Slope map obtained from the DSM under ArcGIS 10.4.1

- Applying the mask and generating routes

After obtaining the building mask, the ArcScan tool is used. It is precisely this manual step that makes our method semi-automatic. Indeed, with ArcScan, we will remove the few buildings that have not been removed by the mask, and we will add the missing portions of the roads. Then, from the road polygons, ArcScan will automatically generate the lines n vectors of the roads. Until this point, all the work that has been developed is part of 2D. Indeed, even the use of the DSM and the DTM is limited to the extraction of the mask that includes the buildings, from where the extraction of the road network will permit us to replace the digitization and not the restitution. Although the road layer is one of the simplest and fastest tasks for digitization. This process allow to us to save time and reduce costs in terms of the number of operators and in terms of materials (Fig. 9).

- Drapage and three-dimensional visualization

The third stage of the work lies in the transition from 2D to 3D. For an effective transition, we have chosen to do the draping of the road obtained with our DTM, which is made with a step of 1 m. For the realization of this task, we used the ArcScene software. The figure below shows

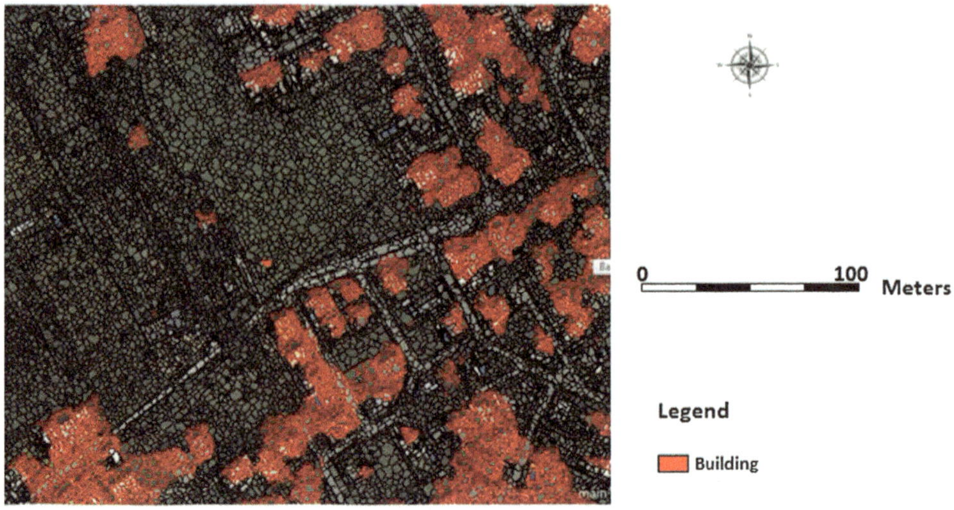

Fig. 8 Making the mask from DTM and DSM

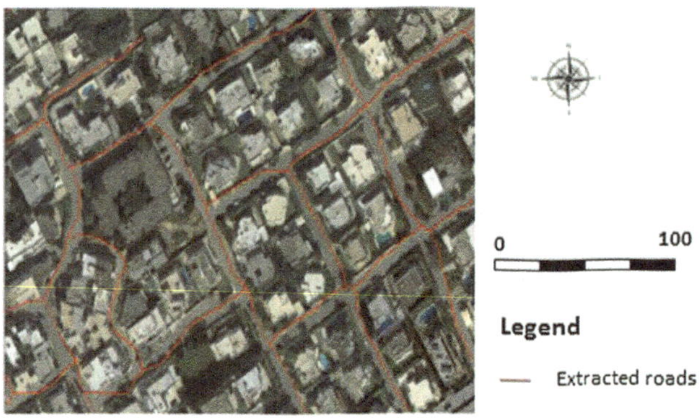

Fig. 9 Generating roads from our extraction method

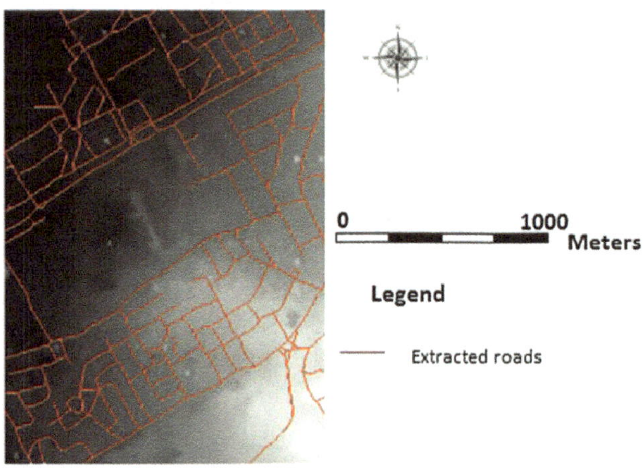

Fig. 10 Displaying roads with DTM under ArcScene for 3D Visualization

the roads as well as the DTM under the ArcScene software (Fig. 10).

4 Results and Discussion

Generally, the assessment of the quality of an extracted road network can be done through either a qualitative or quantitative way. Concerning image processing, the different methods developed have a problem of noises. In our method, the contribution of photogrammetry, using a DTM with a step of 1 m, and DSM with a step of 50 cm, proves to be very useful and effective for the quality of our results. Indeed, if the step of the DSM is greater than 0.5 m, the mask of the buildings will be impossible to realize, because there will be a clear confusion between the roads and the buildings during the visualization. The qualitative evaluation is generally done by a visual comparison of the detected network with a reference network. The quantitative assessment, on the other hand, enables to quantify the quality of detection and therefore it has less subjectivity. Hence, the internal and external validation will be done a priori and posteriori. The evaluation of the road extraction makes possible the evaluation of the performances of the method used, its efficiency as well as its weaknesses. In the case where the one wants to optimize the extraction, this evaluation can give clues on the points to be improved later through developing new ways.

The visual approach facilitates the qualitative assessment of the extraction. On the other hand, the statistical approach gives quantitative measures on the extraction performances. On the other hand, its being global seems disadvantageous as it allows only general measures on the entire road network. Since the visual approach has the disadvantage of being subjective and not rigorous, we have simply adopted the two approaches: the visual approach and the statistical one.

The accuracy obtained is centimeteric. In fact, it is less than 1 m in altimetry and 50 cm in planimetry in most of the prototype, although it sometimes exceeds 1 m.

Although the maximum ΔX is 1.71 m, the maximum ΔY is 1.88 m and the maximum ΔZ is equal to 0.46 m. Our work remains acceptable since we work on a scale of 1:25000. Visual acuity is between 1/10 and 2/10 mm, so that drawing a map scale 1:25000 tolerates an error of 1/10 mm which gives us 2.5 m in reality. In order to acquire the homogeneity of the data and the standardization, we can apply the following formula:

S = 1/X with S = the scale and X = 2 r 103 with r is the pixel resolution in m.

For S = 1:25000, the resolution obtained is therefore 12.5 m [56,57], but in our case the resolution is more efficient and it is 50 cm.

In our case, where the chosen prototype covers a small area, road layer tracking is relatively affordable (Figs. 11, 12, 13 and 14).

The validation approach in our study depends essentially on the scale on which we work. It is of the order of 1:25000, which gives us a margin of error understood according to the standards in cartographies between 1/10th and 2/10th of

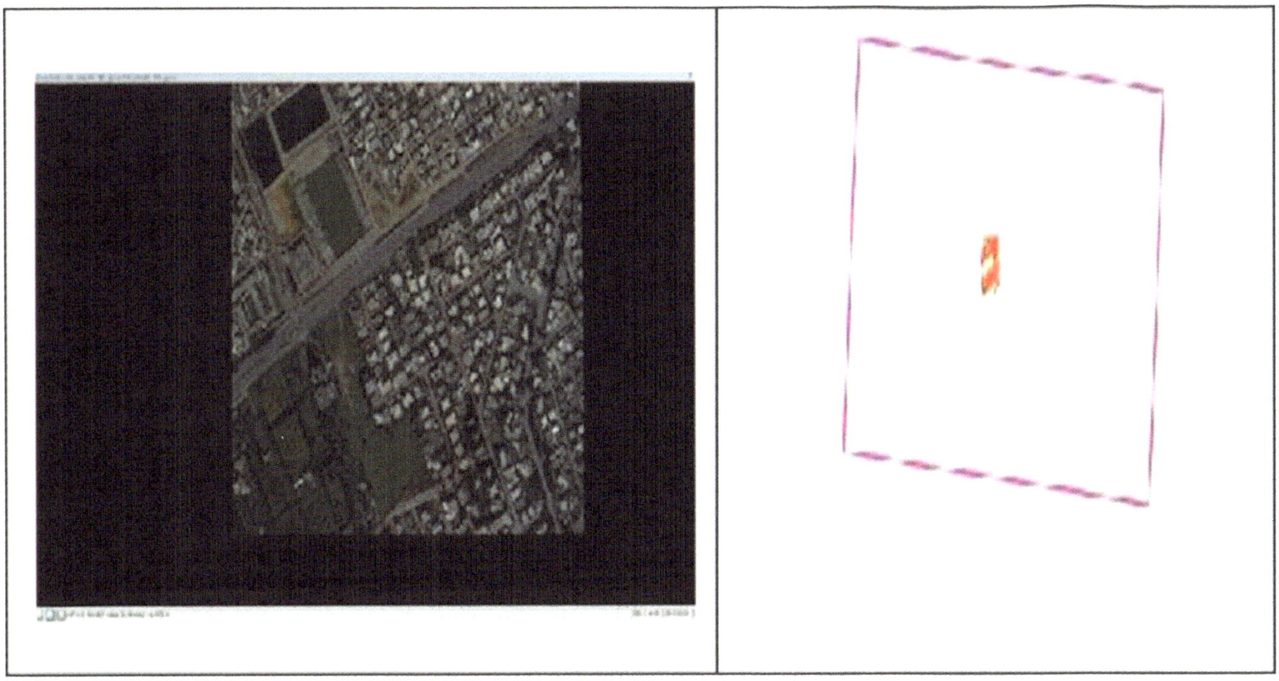

Fig. 11 Arcgis stereo Analyst" interface and the 3D block display of the orthophotos of the study area (.BLK format)

Fig. 12 Presentation of draped roads into stereo analyst

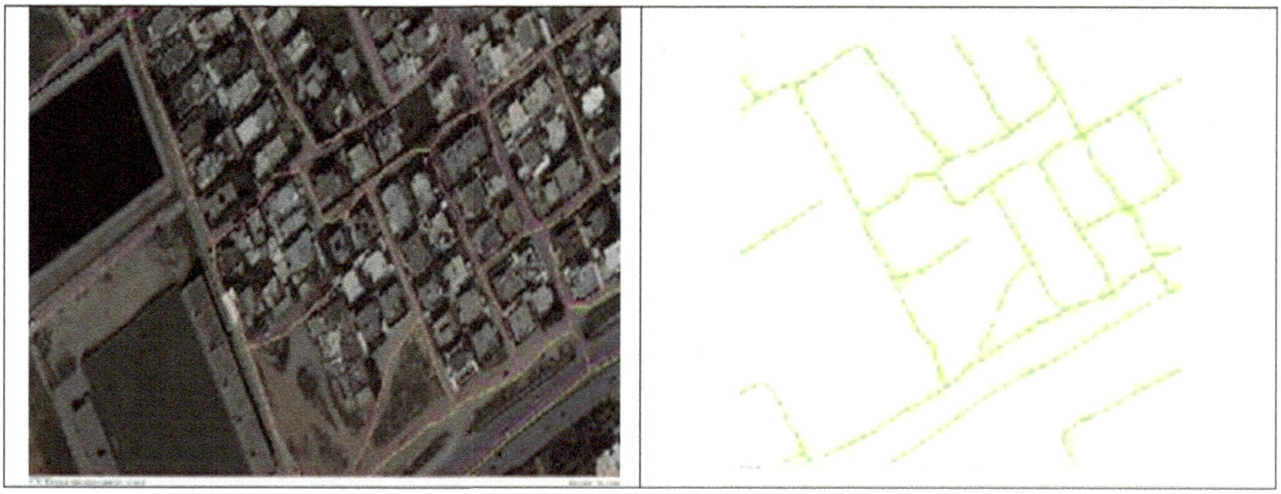

Fig. 13 Comparison of the results with the entities restored

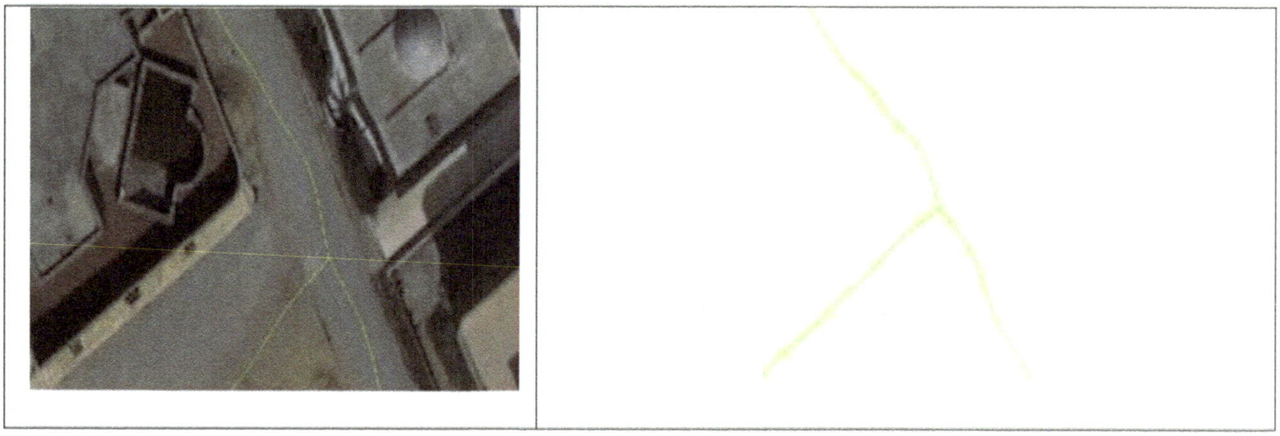

Fig. 14 Verification and validation of the road location in relation to the elevation of the 3D block

millimeters, i.e., between 2.5 m and 5 m. To test the accuracy of the results obtained, we applied a "buffer" of 1 m to our reference road which is in our case a finalized and restored product by the photogrammetry chain. The road restored by our method, as shown in Fig. 15, in no way exceeds the limits of the "buffer", which implies that planimetric errors are largely tolerated.

In addition to the spatial validation that we have carried out, we have applied several statistical methods to estimate the quality of the results obtained. Among the various statistical methods that we have found useful to apply is the "cross validation" method. It estimates the reliability of a model based on a sampling technique [58]. It constitutes a set of statistical parameters enabling the error estimation for each method.

It is based on the analysis of samples of the same points at the different planimetric and altimetric levels. The sampling is based on six points that have been selected from the points that designate the intersections of the roads, as shown in Fig. 16.

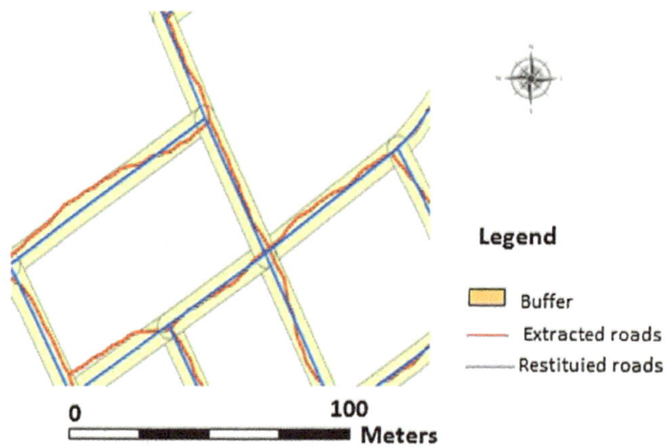

Fig. 15 Estimation of the error by the application of a "buffer" of 1 m

Similarly, we have incorporated many statistical parameters and coefficients to best estimate and perform statistical error analysis.

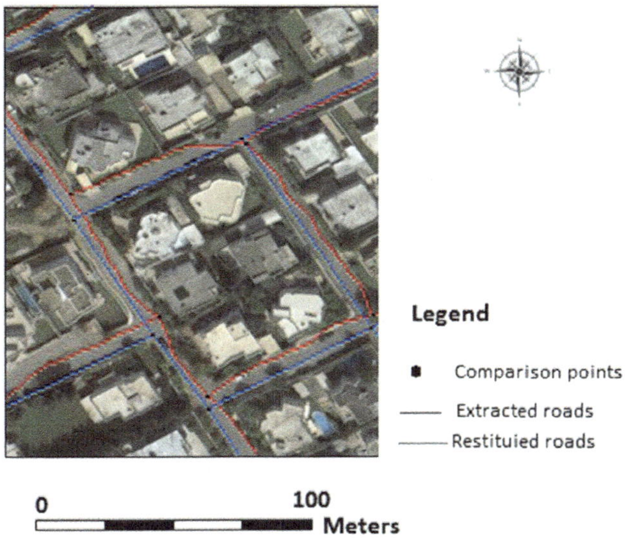

Fig. 16 Estimation of the error by the application of a "buffer" of 1 m

Coefficient of Skewness

This coefficient is sensitive to the distribution of data in the geographical space and allows to measure its asymmetry. For a perfectly symmetric distribution, it is important that this coefficient be equal to zero. This coefficient is calculated by

$$Coef_{Skewness} = \left[\frac{1}{N}\sum_{i=1}^{N}(Zi - Zm)^3\left(\frac{1}{SD}\right)\right] \quad (1)$$

With N being the total number of altitude data, Zi and Zm are respectively the altitudes of a point at a location i = f (x, y) and, the average of the altitudes. SD represents the standard deviation. It is 1.3069 for the restored altitudes (Z) while it is 0.47550981 for the Z calculated by our semi-automatic method.

Coefficient of Kurtosis

The distribution of the data presents a form called "Leptokurtic". This parameter is calculated by the coefficient of "Kurtosis". A value equal to 3 indicates the existence of symmetry in the distribution.

$$Coef_{kurtosis} = \left[\frac{1}{N * SD}\sum_{i=1}^{N}(Zi - Zm)^4\right] - 3 \quad (2)$$

It is important to have a visual overview of the average distribution of the topographic data as well as the distribution of their densities in the geographical space. The values obtained are respectively 1.1818 for the restored Z while it is –0, 44503927 for the Z obtained by the semi-automatic method.

Data Analysis by "Cross Validation"

The "Cross Validation" method is a set of statistical parameters used to estimate the error for all the data.

$$RMSE = \sqrt{\frac{\sum_{i=1}^{N}(Oi - Pi)^2}{n}} \quad (3)$$

With n is the number of data, Oi and Pi are respectively the values restored and the values obtained by the semi-automatic method. In two-dimensional geographic space, the lower the RMSE value, the better is the estimation.

For an effective realization of our Cross Validation method, we have computed statistical indicators applied on a set of three-dimensional coordinates of six samples. The coordinates (XR, YR, ZR) and (XC, YC, ZC) are represented respectively in all the formulas by the restored values (Oi) and the calculated values (Pi). The results are presented in the following (Table 2).

Approach to a Detailed Statistical Analysis

This approach is based on the analysis of the samples extracted from the databases of the values extracted from identical locations.

Each sample contains six values belonging to different altimetric or planimetric levels. Many statistical parameters are applied to analyze all of these samples with respect to the values obtained. These parameters are classified into two types: relative and absolute.

Absolute parameters. It is highly recommended to use at least one absolute error measurement parameter (RMSE or EAM) for a complete and efficient evaluation of the product's model performance. Generally, the frequently used are:
The root mean square error or RMSE which places a particular importance on the peripheral values in the dataset.
The absolute average error or EMA:

$$EMA = n^{-1}\sum_{i=1}^{n}|Oi - Pi| \quad (4)$$

With n is the number of data used (which are six values in our case).

Related Settings

They are considered as nondimensional indices that provide a relative comparison of one model with another [59]. There exist three commonly used parameters:

The coefficient of determination [59]

$$R^2 = \left[\frac{\sum_{i=1}^{n}(Oi - Om)(Pi - Pm)}{\sqrt{\sum_{i=1}^{n}(Oi - Om)^2}\sqrt{\sum_{i=1}^{n}(Pi - Pm)^2}}\right]^2 \quad (5)$$

With Om is the average of the values restored and Pm is the average of the values i extracted (or predicted), n is the number of observations used. R2 is therefore the "Correlation Coefficient" high squared. This measure makes it possible to quantify the strength of the linear link between two variables. R2 is between 0.0 and 1.0 with a high value indicating a high degree of "collinearity".

The coefficient of efficiency [60]

$$E = 1 - \left[\frac{\sum_{i=1}^{n}(Oi - Pi)^2}{\sum_{i=1}^{n}(Oi - Om)^2}\right] \quad (6)$$

A value of 1.0 has perfect interpolation. This parameter is widely used in Earth Sciences.

The agreement index [61]

$$d = 1 - \left[\frac{\sum_{i=1}^{n}(Oi - Pi)^2}{\sum_{i=1}^{n}[|Pi - Pm| + |Oi - Om|]}\right] \quad (7)$$

d varies between 0 and 1. If it is high, it indicates a good agreement between the values restored (Oi) and those extracted (Pi).

While a value of 1.0 indicates perfect agreement, a value of 0.0 denotes complete disagreement. The overall results of this approach are shown in Table 3.

We notice that the RMSE is 0.0868 for the Z, which represents a very good value for the altitudes. This is explained by the fact of working on a DTM with a step of 1m, and an MNS with a step of 0.5 m.

Concerning the planimetry and notwithstanding the fact that the RMSE is 1.133 for the X and Y, we have taken into account the scale factor that is of 1:25000. This result is relatively acceptable.

Table 3 The statistical indicators applied to the restored coordinates (XR, YR, ZR) and the calculated coordinates (XC, YC, ZC)

Coordonnées	(XR, XC)	(YR, YC)	(ZR, ZC)
RMSE	1,1333	1,1337	0,0868
EMA	0,93045	0,84833333	0,238733333
R2	0,999885672	0,999894323	0,95911636
NS	0,999662031	0,99971793	0,92127468
d	0,991562788	0,989035687	0,95230147

The coefficient of determination is greater than 0.9 for X, Y and Z, indicating a high degree of colinearity for both planimetric and altimetry values.

Similarly, the coefficient of effectiveness and the agreement index are very close to 1, which implies that the results obtained are in perfect agreement. All the results of the statistical tests bring us to the conclusion that the elaborated semi-automatic method is largely validated.

5 Conclusion

Road extraction has been treated by several authors who have developed a number of local and global methods.

However, none of them has reached perfect results because of the complexity of the subject. The problem remains open for new efforts to build other extraction algorithms.

In this study, our primary concern is to develop a method of extracting road network from aerial photos at a scale of 1:25000. Via photogrammetry and DSM with a pitch of 50 cm, though previously a DSM with a pitch of 5 m is being used, we are allowed to improve the resolution by factor ten while using aerial photographs.

Table 2 Statistical approach and analysis of restituted data (XR, YR, YR) and calculated data (Xp, Yp, Zp)

Oi&Pi N points	XR	XC	$\|\Delta X\|$	YR	YC	ZR	ZC	$\|\Delta Z\|$
1	617900,4654	617900,8193	0,3539	4081187,5150	4081186,7010	21,4973	21,3550	0,1423
2	617871,5027	617872,4770	0,9743	4081086,4780	4081086,4970	19,6219	19,9889	0,3670
3	617717,1218	617715,4139	1,7079	4081144,7020	4081142,8190	18,4104	18,3714	0,0390
4	617808,7590	617808,4931	0,2659	4081199,2590	4081197,9240	19,9424	20,3994	0,4570
5	617823,8932	617824,7720	0,8788	4081257,9880	4081257,8370	18,9819	19,0857	0,1038
6	617845,1719	617846,5738	1,4019	4081271,4940	4081270,6060	18,5197	18,7530	0,2333
Mean	617827,8190	617828,0915	0,9304	4081191,2393	4081190,3973	19,4956	19,6589	0,2237
Min	617717,1218	617715,4139	0,2659	4081086,4780	4081086,4970	18,4104	18,3714	0,0390
Max	617900,4654	617900,8193	1,7079	4081271,4940	4081270,6060	21,4973	21,3550	0,4570
Variance	4023,8907	4138,4718	0,3211	4823,0094	4822,1688	1,3233	1,2682	0,0260
Mean deviation = emean absolute gap	38,1951	38,7416	0,3694	44,2923	44,3357	0,7357	0,7905	0,1103
Standard deviation	63,4341	64,3310	0,5666	69,4479	69,4418	1,1503	1,1261	0,1613
Standard error (XR, XC)	0,7583							
Standard error (YR, YC)	0,7982							
Standard error (ZR, ZC)	0,2600							

Our first step consists in the extraction of roads from radiometry-based segmentation with the "eCognition" software. Although our Orthophoto has a resolution of 10 cm, the network that has been extracted presents two types of anomalies: road discontinuity and confusion with a number of buildings. Accordingly, our second step is to apply a mask on the buildings. A relatively manual phase based on visual approaches is used with "Arcscan" to eliminate what is not removed by applying our mask, and roads that could not be extracted automatically are added. It should be noted that at this stage, it is useful to automate the visual approach but it remains relatively poorly controlled. The results obtained and validated by both visual and statistical approaches are satisfactory in relation to the scale at which we have worked.

References

1. Singh, P. P., & Garg, R. D. (2013). Automatic road extraction from high resolution satellite image using adaptive global thresholding and morphological operations. *Journal of the Indian Society of Remote Sensing, 41*(3), 631–640.
2. Richards, J. A. (2013). *Remote sensing digital image analysis*. Berlin: Springer.
3. Dosovitskiy, A., Fischer, P., Ilg, E., Hr̈usser, P., Hazırbas, C., Golkov, V., et al. (2015). FlowNet: Learning optical flow with convolutional networks. In *IEEE International Conference on Computer Vision (ICCV)*.
4. Badrinarayanan, V., Kendall, A., & Cipolla, R. (2017). Segnet: A deep convolutional encoder-decoder architecture for scene segmentation. *IEEE Transactions on Pattern Analysis and Machine Intelligence*.
5. Pinheiro, P., & Collobert, R. (2014). Recurrent convolutional neural networks for scenelabeling. In *International Conference on Machine Learning* (pp. 82–90).
6. Bajcsy, R., & Tavakoli, M. (1976). Computer recognition of roads from satellite pictures. *IEEE Transactions on Systems, Man, and Cybernetics, 6*(9), 623–637.
7. Farah, N. (1998). Extraction et évaluation du réseau routier urbain à partir des images satellitaires: développement d'algorithmes. Mémoire (M.Sc) Département de géographie et télédétection Faculté des lettres et sciences humaines Université de Sherbrooke (p. 95).
8. Mena, J. B. (2003). State of the art on automatic road extraction for GIS update: A novel classification. *Pattern Recognition Letters, 24*(16), 3037–3058.
9. Ruskoné, R., & Airault, S. (1997). Toward an automatic extraction of the road network by local interpretation of the scene. In *46th Photogrammetric Week, Stuttgart*.
10. Yoon, H. S., Hackett, J., & Bhattacharya, D. (2002). *Proceedings of the National Academy of Sciences of the USA, 99*, 11724–11729.
11. Faber, A., & Förstner, W. (1999). Scale characteristics of local autocovariances for texture segmentation. *International Archives of Photogrammetry and Remote Sensing, 32*(7–4–3), Valladolid, Spain.
12. Fischler, M., Tenenbaum, J. M., & Wolf, H. C. (1981). Detection of roads and linear structures in low-resolution herid hagery using an mdtisource knowledge integration technique. *Computer Graphics Image Process, 15*, 201–223.
13. Amini, A. A., Weymouth, T. E., & Jain, R. C. (1990). Using dynamic programming for solving variational problems in vision. *IEEE Transactions on Pattern Analysis & Machine Intelligence, 12*, 855–867.
14. Stoica, R., Descombes, X., & Zérubia, J. (2000). A Markov point process for road extraction in remote sensed images. Technical Report 3923, Rapport de recherche de l'INRIA.
15. Dal Poz, A. P., Gyftakis, S., & Agouris, P. (2000). Semiautomatic road extraction: Comparison of methodologies and experiments. In *2000 ASPRS Annual Conference, Washington, DC, USA*.
16. Dal Poz, A. P. & Vale, G. M. (2003). Dynamic programming approach for semi-automated road extraction from medium- and high-resolution images. *ISPRS Archives, 34*(3/W8), 87–91, Munich.
17. Lassalle, P. (2015). Étude du passage à l'échelle des algorithmes de segmentation et de classification en télédétection pour le traitement de volumes massifs de données, Doctorat de l'université de Toulouse.
18. Kass, M., & Witkin, A. (1988). Snakes: Active contours models. *International Journal of Computer Vision, 1*(4), 321–331.
19. Steger, C., Mayer, H., Eckstein, W., Heinrich, E., & Baumgartner, A. (1999). Automatic road extraction based on multi-scale, grouping, and context. *Photogrammetric Engineering and Remote Sensing, 65*(7), 777–785.
20. Laptev, I., Mayer, H., Lindeberg, T., Eckstein, W., Steger, C., & Baumgartner, A. (2000). Automatic extraction of roads from aerial images based on scale space and snakes. *Machine Vision and Applications, 12*, 23–31.
21. Krähenbühl, P., & Koltun, V. (2011). Efficient inference in fully connected CRFs with Gaussian edge potentials. In: *Advances in neural information processing systems*.
22. Chen, L. C., Barron, J. T., Papandreou, G., Murphy, K., & Yuille, A. L. (2016). Semantic image segmentation with task-specific edge detection using
23. Pinheiro, P. O., Lin, T. Y., Collobert, R., & Dollár, P. (2016). Learning to refine object segments. *European Conference on Computer Vision* (pp. 75–91). Berlin: Springer.
24. Wang, J., & Howarth, P. J. (1987). Autornated road network extraction from Landsat TM imagery. *ASPRS-ACSM, Annual Convention, 1*, 429–438.
25. Ton, J., Jain, A. K., Enslin, W. R., & Hudson, W. D. (1989). Automatic road identification and labelling in Landsat 4 TM images. *Photogrammetria, 43*, 276–357.
26. Wang, J., Treitz, P. M., & Howarth, P. J. (1992). Road network detection from SPOT imagery for updating geogaphical information systems in the rural- urban fringe. *International Journal of Geographical Information Systems, 6*(2), 141–157.
27. Wang, T.-Q, Liu, X.-W., & Yin, H.-S, (1994). A new species of Hymenopus (Mantodea Hymenopodidae: Hymenopodinae) from China. *Entomotaxonomia, 16*(2), 79–81.
28. Matheron, G. (1975). Random sets and integral geometry bull. *American Mathematical Society, 81*, 844–847.
29. Serra, J. (1983). *Image analysis and mathematical morphology*. □Cytometry (pp. 184–185). https://doi.org/10.1002/cyto.990040213.
30. O'Brien, D. (1991). Computer assisted feature extraction (InterEx). In *Proceeding of l4th. Canadian Symposium on Rernote Sensing* (pp. 423–427).
31. Wang, D., He, D.-C, Wang, L., & et Morin, D. (1996). L'extraction du réseau routier urbain à partir d'images SPOT HRV. *International Journal of Remote Sensing, 17*(4), 827–833.
32. Wang, J., Song, J., Chen, M., & Yang, Z. (2015). Road network extraction: A neural dynamic framework based on deep learning and a finite state machine. *International Journal of Remote Sensing, 36*, 3144–3169.
33. Khesali, E., Zoej, M. J. V., Mokhtarzade, M., & Dehghani, M. (2016). Semi automatic road extraction by fusion of high resolution optical and radar image. *Journal of the Indian Society of Remote Sensing, 44*(1), 21–29.

34. Reza, H., Riahi, B., Abolfazl, A., Rezaeian, H. (2017). Semi automatic road extraction from digital images. *The Egyptian Journal of Remote Sensing and Space Sciences, 20*(2017), 117–123.
35. Zerubia, J., & et Merlet, N. (1993). Classical Mechanics and Roads Detection in SPOT Images. Rapport de Recherche no 1889, Institut National de Recherche en Informatique et en Automatique. France. 52 p.
36. Cleynenbreugel, J. V., Fierens, F., Suetens, P., & Oosterlink, A. (1990). Delineating road structure on satellite imagery by a GIS-guided technique. *Photogrammetry and Remote Sensing, 56*(6), 893–898.
37. Jedynak, B. (1995). Modèles stochastiques et méthodes déterministes pour extraire les routes des images de la terre vues du ciel". Thèse de Doctorat: Université Paris-Sud, France (p. 186).
38. Vosselman, G., & De Knecht, J. (1995). Road tracking by profile matching and Kalman filtering. In Gruen A., Kuebler O., Agouris P. (Eds.), *Workshop on Automatic Extraction of Man-Made Objects from Aerial and Space Images. Birkhauser, Basel* (pp. 265–274).
39. Tamokoue, H. O. (2011). Etude exploratoire de méthodes d'extraction automatique du réseau routier à partir d'images satellites haute résolution. *ENSG, 2011*, 15–16.
40. Sehad, M. (2014). Segmentation d'images par une approche basée sur des caractéristiques texturales, temporelles et spectrales: Application aux images MSG. Thèse de doctorat en Télédétection: Université Mouloud Mammeri, Tizi Ouzou, Algérie (200 p.).
41. Bouziani, M., et al. (2010). Rule-based classification of very high resolution image. *IEEE Transactions on Geoscience and Remote Sensing, 48*(8), 3198–3211.
42. Croquerez, J. (1995). Analyse d'images - Filtrage Et Segmentation. Edition Masson.
43. Deriche, R. (1987). UsingCanny's Creteria to derive a recursively implemented optimal edge detector. *International Journal of Computer Vision, 1*(2), 167–187.
44. Fortier, M. F. A., Ziou, D., Armenakis, C., & Xang, S. (1999). Nouvelles perspectives en détection de contour: textures et images multispectrales. Vision interface, trios rivière (pp. 19–21).
45. Xue, H., Géraud, T., & Duret-Lutz, A. (2003). Multi-band segmentation using morphological clustering and fusion- application to color image segmentation. In *IEEE International Conference on Image Processing*.
46. Martinez-Uso, A., Plat, F., & Garcia-Sevilla, P. (2005). Multispectral images segmentation by energy minimization for fruit quality estimation. In *Pattern Recognition and Image Analysis* (Vol. 3523, pp. 689–696).
47. Roussen, M., & Deriche, R. (2002). A variational framework for active and adaptive segmentation of vector valued images. In IEEE proceeding of the workshop on motion and video computing.
48. Roula, M. A., Bouridane, A., Kurugollo, F., & Amira, A. (2002). UN supervised segmentation of multispectral images using edge progression and cost function. In *IEEE International Conference of Image Processing* (Vol. 3, pp. 781–784).
49. Cheng, H. D., & Sun, Y. (2000). A hierarchical approach to color image segmentation using homogeneity. *IEEE International Transactions on Image Processing, 9*(12), 2071–2082.
50. Ikonomakis, N., Pataniotis, K. N., & Venetsanopoulos, A. N. (2000). Unsupervised seed determination for a region-based color image segmentation Scheme. In *Proceedings of the international conference on image processing, 10–13 September, Vancouver, BC*, Vol. 1, pp. 537–540.
51. Voisine, N. (2002). *Approche adaptative de coopération hiérarchique de méthodes de segmentation, application aux images multicomposantes*. Ph.D thesis, Université de Rennes 1, 250p.
52. Tremeau, A., Colantoni, P. (2000). Regions adjacency's graph applied to color image segmentation. *IEEE Transactions on Image Processing, 9*(4), 735–744.
53. Li, P., & Xiao, X. (2007). Multispectral image segmentation by a multichannel Watershed-bases approach. *International Journal of Remote Sensing, 28*(19), 4429–4452.
54. Volpi, M., & Tuia, D. (2017). ense semantic labeling of subdecimeter resolution images with convolutional neural networks. *IEEE Transactions on Geoscience and Remote Sensing, 55*, 881–893.
55. Zhang, R., Sun, D., Yu, Y., & Goldberg, M. D. (2012). Mapping night time flood from MODIS observations using support vector machines. *Photogrammetric Engineering and Remote Sensing, 78*, 1151–1161.
56. Girard, M. C., & Girard, C. M. (2003). *Processing of remote sensing data* (487p). The Netherlands: Balkema.
57. Bou Kheir, R., Greve, M. H., Deroin, J. P., & Rebai, N. (2011). Implementing GIS regression trees for generating the spatial distribution of copper in Mediterranean environments: The case study of Lebanon. *International Journal of Environmental Analytical Chemistry*, 20. https://doi.org/10.1080/03067319.2011.603079.
58. Rebai, N., Slama, T., & Turki, M. M. (2007). Evaluation de différentes méthodes d'interpolation spatiale pour laproduction d'un MNT á partir des données topographiques dans un SIG. *Revue XYZ, 110*(1er trimester).
59. Kneale, P., See, L., & Smith, A. (2001). Towards defining measures for neural network forecasting models. In *Proceeding of the 6th International Conference on Geocomputation, University of Queensland, 24–26 September 2001, Brisbane, Australia*, pp. 61–72.
60. Nash, J. E., & Sutcliffe, J. V. (1970). River flow forecasting through conceptual models, I, A discussion of principals. *Journal of Hydrology, 10*(1970), 282–290.
61. Willmott, C. J. (1981). On the validation of models. *Physical Geography, 2*(1981), 148–194.

Methodology of Updating Touristic Map Using Open Source and Open Spatial Data (OSOD). A Case Study of Ben Arous City, Tunisia.

Abdelkader Moussi and Noamen Rebai

Abstract

Landscapes can quickly change in the modern world, such as roads and buildings. There is a very dynamic modern world: cities are expanding, new settlements are emerging, new roads, communication networks, and infrastructure are being built, new areas for the recovery of natural resources are being developed, forests are being cut down and the structure of land use is changing. Since the topographical maps are the basis for each thematic map, official institutions avoid long intervals of time between each revision of the topographical map. Consequently, thematic maps are always subject to updating. Mapping production and publication are linked to numerous aspects, including data availability and legislative requirements. This article examines the potential of open geospatial data for updating touristic map using open source software.

Keywords

Updating map • OSOD • OSM • Raster • Vector • QGIS • Ben Arous - Tunisia

1 Introduction

Most free and open source data and software have become more accessible to geospatial workflows that would otherwise be inaccessible in terms of cost and availability [1]. The availability of commercial data, used in geographic information systems (GIS), usually through a limited number of vendors, which produce highly detailed, accurate and precise information. However, this happens at a price that numerous small and large businesses, private consultants and start-up companies cannot afford.

Open source data, in particular voluntary geographic information (VGI) on OpenStreetMap (OSM) [2–5], represents a community effort to build one of the best touristic maps, and subsequently the best GIS database, available free to the public [6–9].

OpenStreetMap is a web map that any registered user is able to submit geospatial data. Over time, these updates populate the now growing web map that is the OpenStreetMap [10]. Simultaneously, the data that exists on OpenStreetMap can be freely downloaded and used within a GIS for geospatial analysis, mapping and other geospatial tasks [1, 11–16]. Information collected in OpenStreetMap is often referred to as voluntary geographic information (VGI) [17, 18]. With an ever-increasing popularity in recent years, one of the most used, analyzed and referred to VGI platforms is OpenStreetMap (OSM), whose main objective is to produce a freely available geographic database of the whole world [17, 19].

Voluntary Geographic Information (VGI) refers here to a key component of such a geospatial distribution of the phenomenon, including both a range of practices for the production and dissemination of geographic information by volunteers as well as new forms of geospatial data produced and maintained through various mobile devices and interactive online platforms such as OpenStreetMap (OSM) [18].

In this article, we developed a methodology to update a tourist map of Ben Arous (Tunisia) with open geospatial data (OSM) and open source software (QGIS).

2 Open Geospatial Data Solution

Geographic Information Systems (GIS) provide the ability to view, query, and analyze geographic data. Data in a GIS provides a simplified view of the real world. Data modeling represents a set of rules (called entity) to convert the real

A. Moussi (✉)
LR01ES06 Georesources and the Environmental Laboratory, Faculty of Sciences of Tunis, University of Tunis El Manar, 2092 Tunis, Tunisia
e-mail: abdelkader.moussi@fst.utm.tn

N. Rebai
LR14ES03 Geotechnical Engineering and Georisk Laboratory, National School of Engineering of Tunis, University of Tunis El Manar, BP. 37, le Belvédère, 1002 Tunis, Tunisia
e-mail: rebainoamen@gmail.com

N. Rebai and M. Mastere (eds.), *Mapping and Spatial Analysis of Socio-economic and Environmental Indicators for Sustainable Development*, Advances in Science, Technology & Innovation, https://doi.org/10.1007/978-3-030-21166-0_12

Table 1 Public-facing websites to easily find and download open data in a variety of open formats

Open geospatial data	Description	Advantages	Data types
Esri Open Data (https://hub.arcgis.com/pages/open-data)	Since 2017, Esri Open Data is the hidden gold mine of free GIS data. It now allows you, for example, to explore an open data set from many organizations around the world	Obtain highly detailed free open GIS data	Download formats are in spreadsheet, KML, shapefile and API's are OGC WMS, GeoJSON, and GeoService
DIVA-GIS (http://www.diva-gis.org/)	DIVA-GIS has free and quality GIS data sets for all countries in the world. From A to Z, find the most common cultural GIS data such as geographic boundaries, railways, roads, etc.	Search filter by country level data	Country level data for any country in the world: administrative boundaries, roads, railroads, altitude, land cover, population density
FAO GeoNetwork (http://www.fao.org/geonetwork/srv/en/main.home)	FAO GeoNetwork provides satellite imagery and geospatial data to support sustainable development in agriculture, fisheries, and food security	Search wide range of categories and filter by country	Agriculture, fisheries, land resource GIS data
GADM (http://www.gadm.org/)	GADM is a geospatial database of the location of the world's administrative areas (or administrative boundaries) for use in GIS and similar software	Search filter data by country or for the whole world	The data are available as shapefile, ESRI geodatabase, RData, and Google Earth KMZ format
OpenStreetMap (http://www.openstreetmap.org/)	GIS users are harnessing the power of OpenStreetMap (OSM) as a means to crowdsource data from multiple users	Obtain highly detailed free GIS data with different levels of accuracy and completeness	High geospatial resolution cultural vector data. (buildings, land use, railroads, roads, waterways)

world into numerically and logically represented geospatial objects consisting of attributes and geometries. The two main geometric data models are:

- Vector Model: In the vector model, discrete points, lines and/or fields are used for discrete objects with an attribute name or code number.
- Raster Model: The matrix model uses regularly spaced grid cells in a specific order. The pixel is an element of the grid cell that contains a unique value of attributes.

2.1 Vector Data Models

In contrast to the raster data model is the vector data model. In this model, space is not quantified in discrete grid cells like the raster model. Vector data models use points and their associated pairs of X, Y coordinates to represent the vertices of geospatial entities, as if they were drawn by hand on a map [20]. The data attributes of these features are then stored in a separate database management system. The geospatial and attribute information of these models are linked via a simple identification number assigned to each feature in a map. Geographic features in a GIS are often expressed as vectors, considering these features as geometric shapes. Vector data is a representation of the real world using point, linear and surface objects. Vector models are useful for storing data that have discrete boundaries, such as district, provincial and municipal boundaries, parcels of land and streets. These geospatial data are available to the public under free and open licence and exist on many websites (Table 1).

2.2 Raster Data Models

GIS data are organized using a raster or vector data model. Raster data uses pixels to represent a continuous surface. This model is most frequently used to represent surfaces, such as elevations (MNE), by a combination of points, lines, and surface elements. "Raster" refers to a data model based on a regular tessellation of the plane, in which all positional information can be imputed from the sequential position of a record, and is therefore absent from the data structure [21–24]. The raster data model is widely used in applications well beyond geographic information systems (GIS). JPEG, BMP, and TIFF file formats (among others) are based on the raster data model. Each unique color pixels, taken as a whole, combines to form a coherent image. The raster data model consists of rows and columns of pixels of equal size interconnected to

Table 2 Free satellite imagery sources to find and download open data in a variety of open formats

Open geospatial data	Description	Advantages	Data types
USGS earth explorer	Access to Landsat satellite data. 40-years of history of our Earth with consistent spectral bands. Gain full access to NASAs Land Data Products and Services including Hyperions hyperspectral data, MODIS	AVHRR land surface reflectance and disperse Radar data and from no data to hyperspectral data, USGS is the undisputed world champion of free satellite data providers	Download formats are in GeoTIFF data. (https://earthexplorer.usgs.gov/)
Sentinels Scientific Data Hub	Sentinels Scientific Data Hub is the official download headquarters for the European Space Agencys Sentinel satellite data	Simply put, Sentinel satellites give you high quality passive and active data of the entire Earth	The Open AccessHub provides complete, free and open access to Sentinel-1, Sentinel-2, and Sentinel3 user products. (https://scihub.copernicus.eu/dhus/#/home)
NOAA	NOAA uses an online library system called the Comprehensive Large Array-data Stewardship System (CLASS) to store a plethora of environmental data	Because NOAA will take you on a journey to fistfuls of free high quality atmospheric data sets (and more)	Currently, the NOAA National Data Centers support POES, DMSP, GOES, MetOp, Jason2 data, and selected model reanalysis data. (https://www.class.ngdc.noaa.gov/saa/products/welcome)
NASAs Reverb Data Hub	Heres a lot to like about NASAs Reverb Data Hub. Especially, after its new facelift. It has a fresh new look and interface for discovering Earth Science data, NASA Reverb contends	Admittedly, it takes a bit of practice to navigate. There are 30 ways to narrow down your data. Our suggestion is to start with a simple search. Change the time range criteria. Narrow it down, and download your free satellite imagery	Satellite data is incredible: Aqua, Terra, ENVISAT, ALOS, METEOSAT, GOES, ICESAT, GMS, Landsat, RADARSAT, NOAA satellites, GPS (https://search.earthdata.nasa.gov/search).
Earth Observation Link	The Earth Observation Link (EOLi) is the European Space Agencys client for Earth Observation Catalog and Ordering Services	The EOLi is a bit dated. It's a JAVA application that you can download to your PC. It works on any major operating system	You can browse and preview images from Earth Observation data from Envisat, ERS, IKONOS, DMC, SPOT, Kompsat, Proba, IRS, SCISAT.) (https://earth.esa.int/web/guest/eoli)

form a flat surface. These pixels are used as building blocks to create points, lines, areas, networks, and surfaces.

2.2.1 Satellite Imagery

Remote sensing satellite imagery is also becoming increasingly common as satellites equipped with technologically sophisticated sensors are continually sent into space by public agencies and private companies around the world. For free satellite data, there's no better way to do it than to follow (Table 2). Ranked from top to bottom, here are your free satellite image sources.

3 Methods

The Updating map methods of a tourist map, object of our study, continue to progress on the one hand according to the presence of more and more Open Sources, a good control of exchange formats (Raster, vector) and on the other hand the performance of free GIS platform tools. In our study, we present the proposed method and application for update Touristic map of Ben Arous (Tunisia) using OSM data and Open Source GIS software (QGIS) (Fig. 1).

3.1 Available Reference Datasets

Many datasets are used to update touristic map. Touristic map of Ben Arous with scale 1/10000 published in 2000 from Office of Topography and Cartography (Tunisia). Topographic map of LA GOULLETTE NO with the scale of 1/25000 published in 1983 from Office of Topography and Cartography (Tunisia) (Fig. 2).

3.2 Available Open Geopatial Datasets

Open geospatial data can be obtained from the OpenStreetMap through a variety of methods (Table 3). The first way is to download the map data from the OpenStreetMap

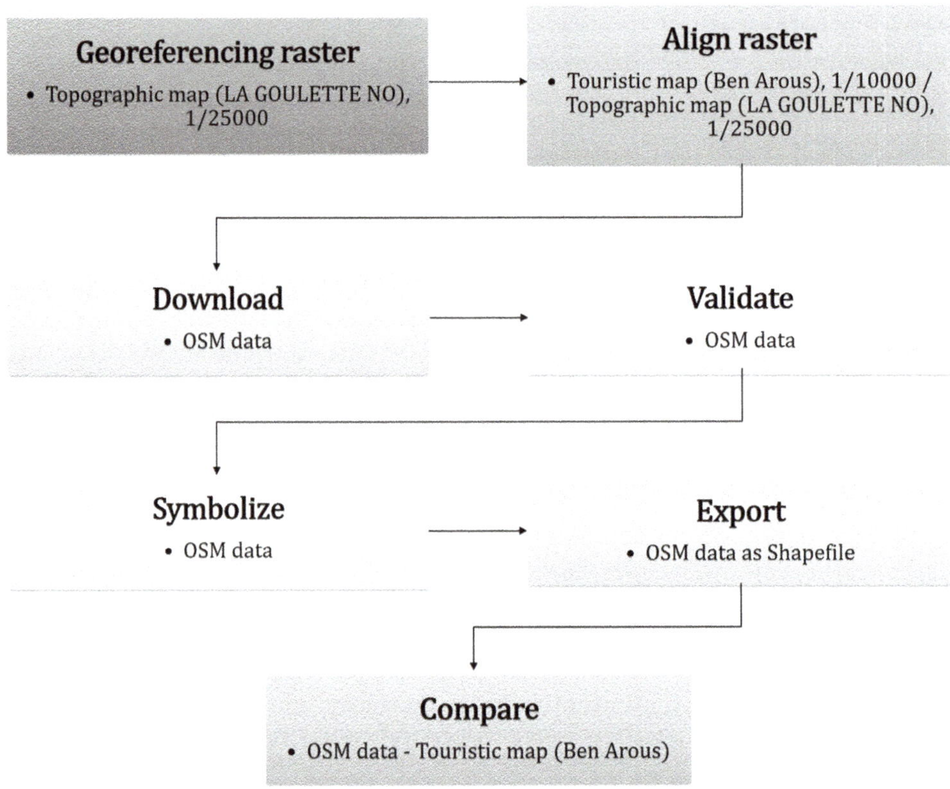

Fig. 1 Methodology for update touristic map of Ben Arous (Tunisia) using OSM data

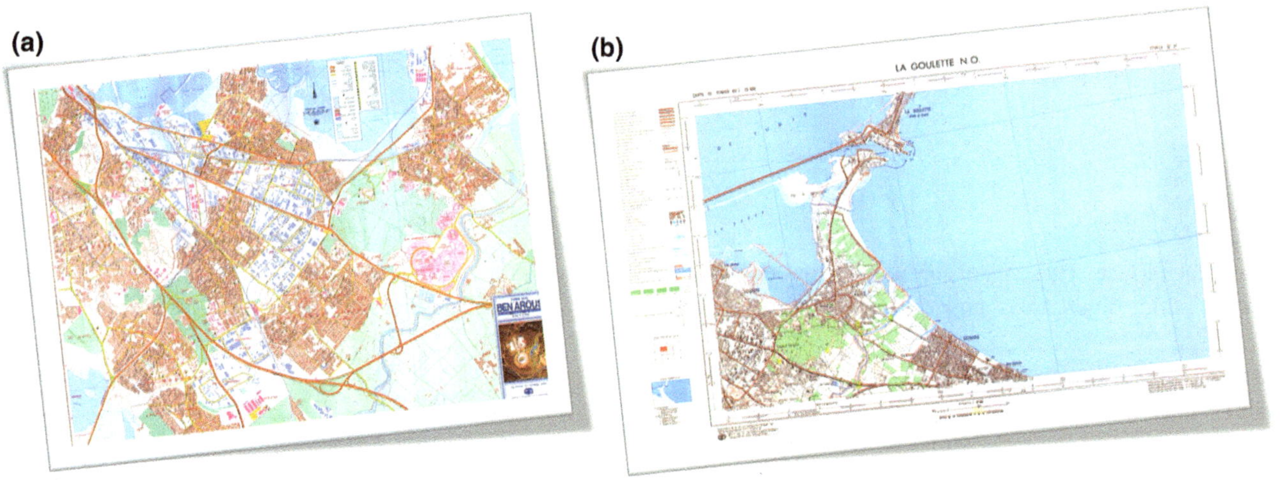

Fig. 2 Reference maps used in this study. **a** Touristic map of Ben Arous with scale 1/10000. **b** Topographic map of La Goullette with the scale of 1/25000

dataset. The whole data set is available in the download area of the OpenStreetMap website. It is also possible to select smaller areas to download.

The latter is to use OpenStreetMap Plugin for the Open Source GIS software QGIS. It adds support for OpenStreetMap raw vector data, bringing it as a layer from an .osm file or through direct download from the OpenStreetMap API. The OSM data for Ben Arous were downloaded in OSM format then exported in shapefile format and finally the desired urban area was clipped.

4 Application and Results

The application of the methodology developed can be summarized in seven steps. This methodology is a new approach that

Table 3 OpenStreetMap data extraction tools

Name	Geographies	File formats	Software
Overpass API	Any, though large files take time	OSM XML, OSM JSON	None required
Geofabrik	Continents	OSM, PBF	Web-based
PlanetOSM	Global	OSM, PBF	Web-based
MapZen metro extracts	Metropolitan Areas	OSM, .PBF, Osm2pgsql SHP, GeoJSON, IMPOSM SHOP, IMPOSP GEOJSON, Water and Coastline SHPs	Web-based
BBBike.org	Bounding box to define	OSM, PBF, SHP, CSV, SVG, Garmin	Web-based
HOT exports	Bounding box to define, HOT countries only	SHP, Garmin, OSM, KMZ, SQlite	Web-based
QuickOSM	Any, though large files take time	OSM	QGIS
QSM downloader	Any, bounding box to define	OSM	QGIS
QGIS vector menu	Any, bounding box or layer to define	OSM, PBF, JSON, SHP	QGIS
Osm2pgsql	Any, though large files take time	PostGIS	PostgreSQL, PostGIS
Imposm	Any, though large files take time	PostGIS	PostgreSQL, PostGIS

essentially allows to edit after the execution of the intermediate phases of the treatment, a map with a complete and complete update (geometric element, symbols, and toponomy).

The interest of this update is essential, on the procedural level based only on free downloadable data and from the processing point of view we use only open source tools where there is an economic gain while maintaining the quality compared to other proprietary data and tools.

The different methods currently used by public institutions specialized in cartography and, which are developed by several authors remain insufficient at several levels (Align raster, Validate OSM data and compare OSM data throw Touristic map of Ben Arous).

We notice on the one hand that for most of the solutions adopted are hybrid, there is not a complete OSOD solution and, on the other hand, validation and quality control are deficient, especially at the level of calibration of the new cartographic elements with respect to the source map.

Our results obtained from our conceptual approach of free data mapping update and the use of free software show that the methodology is generally applicable with quality georeferencing and symbolization automation. A validation test is performed knowing that an improvement is possible in the comparison of raster formats by taking into account the colors for the extraction in a simple way the new objects that are updated.

4.1 Georeferencing Topographic Map (LA GOULETTE NO), 1/25000

Most GIS projects require georeferencing some raster data. In this study, the georeferencing in QGIS of the topographic 1/25000 scale map of LA GOULETTE NO is done via the "Georeferencer GDAL" plugin. This is a core plugin—meaning it is already part of your QGIS installation. We choose the raster coordinate reference system (CRS) Carthage/Lambert Nord Tunisie. This is to specify the projection and datum of control points.

We choose the Transformation type as (In our study we choose Polynomial 2 used for the input raster that needs to be bent or curved and requires six GCPs) First-order polynomial transformation (affine) preserves collinearity and allows scaling, translation and rotation only. The image aligned with an acceptable RMS error equals to 0 (RMSE < 1). RMSE is a metric that indicates the quality of the transformation. It will change depending on the Transformation type value chosen.

The general rule of thumb is that the RMSE should not be larger than half the pixel size of the raster in map units. However, it is only an indication. Another indication is how well the georeferenced imagery aligns with other datasets (Fig. 3).

4.2 Align Raster: Touristic Map (BEN AROUS), 1/10000 / Topographic Map (LA GOULETTE NO), 1/25000

Georeferencing Touristic 1/10000 scale map of BEN AROUS, using second raster dataset, the topographic 1/25000 scale map of LA GOULETTE NO. Control points may be entering from the map canvas as here reference is raster dataset which is present on the map canvas. The remaining ground control points until you have entered eight GCPs.

We choose appropriate values for the Transformation type, Resampling method fields. This setting will determine how

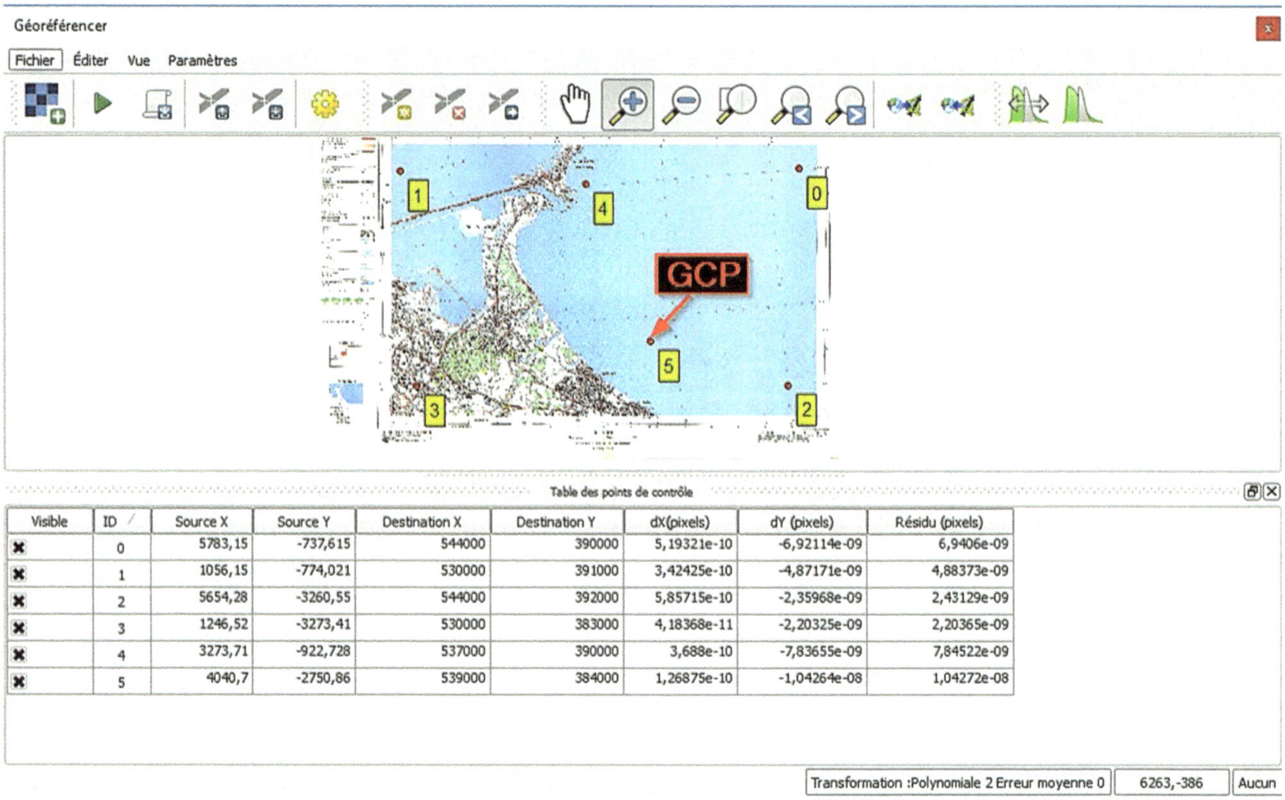

Fig. 3 Georeferencing, into QGIS, of the topographic 1/25000 of LA GOULETTE NO

the ground control points are used to transform the image from source to destination coordinate space. In our study we choose Polynomial 2 used for the input raster that needs to be bent or curved and requires six GCPs.

The following screenshot shows the image in the Georeferencer window with six GCPs entered. Their location is identified by numbered boxes within the image window. The **To** and **From** coordinates are display in the GCP table window along with the RMSE values: 0 m (Fig. 4).

The calibration result of both the topographic and the touristic maps clearly shows the quality of georeferencing (Fig. 5). The result map will then be used to download the open data.

4.3 Download OSM Data Using 'QGIS V2.18.14-Las Palmas'

Recently, the concept of voluntary geographic information (VGI) has emerged from new Web 2.0 technologies. The OpenStreetMap project is actually the most significant example of a system based on VGI [25]. It aims to produce free vector geographic databases from the contributions of Internet users.

4.3.1 QuickOSM Plugin

QuickOSM is a plugin for GIS OpenSource software QGIS. It allows to quickly obtain data OpenStreetMap (OSM) through web service Overpass API. QuickOSM allows to query the OpenStreetMap database, extract data, transform GIS data and displayed in the QGIS. It was developed by Etienne Trimaille during his internship at 3Liz. QuickOSM features are available so be grouped as a single window are so dissociated in the form of panels that can be integrated (Fig. 6) or not with the QGIS interface (menu View/Panels).

4.3.2 OpenStreetMap Plugin

The QGIS OpenStreetMap Plugin is a plugin for the desktop GIS application QGIS. It adds support for OpenStreetMap raw vector data, bringing it in as a layer either from .osm XML file or by direct download from the OpenStreetMapAPI (Fig. 7). It also permits editing and upload back the OSM server.

4.4 Validating OSM Data

Validating data is an important part of OpenStreetMap [3]: having experienced volunteers check OSM data to make sure it is complete, accurate and thorough [26–31].

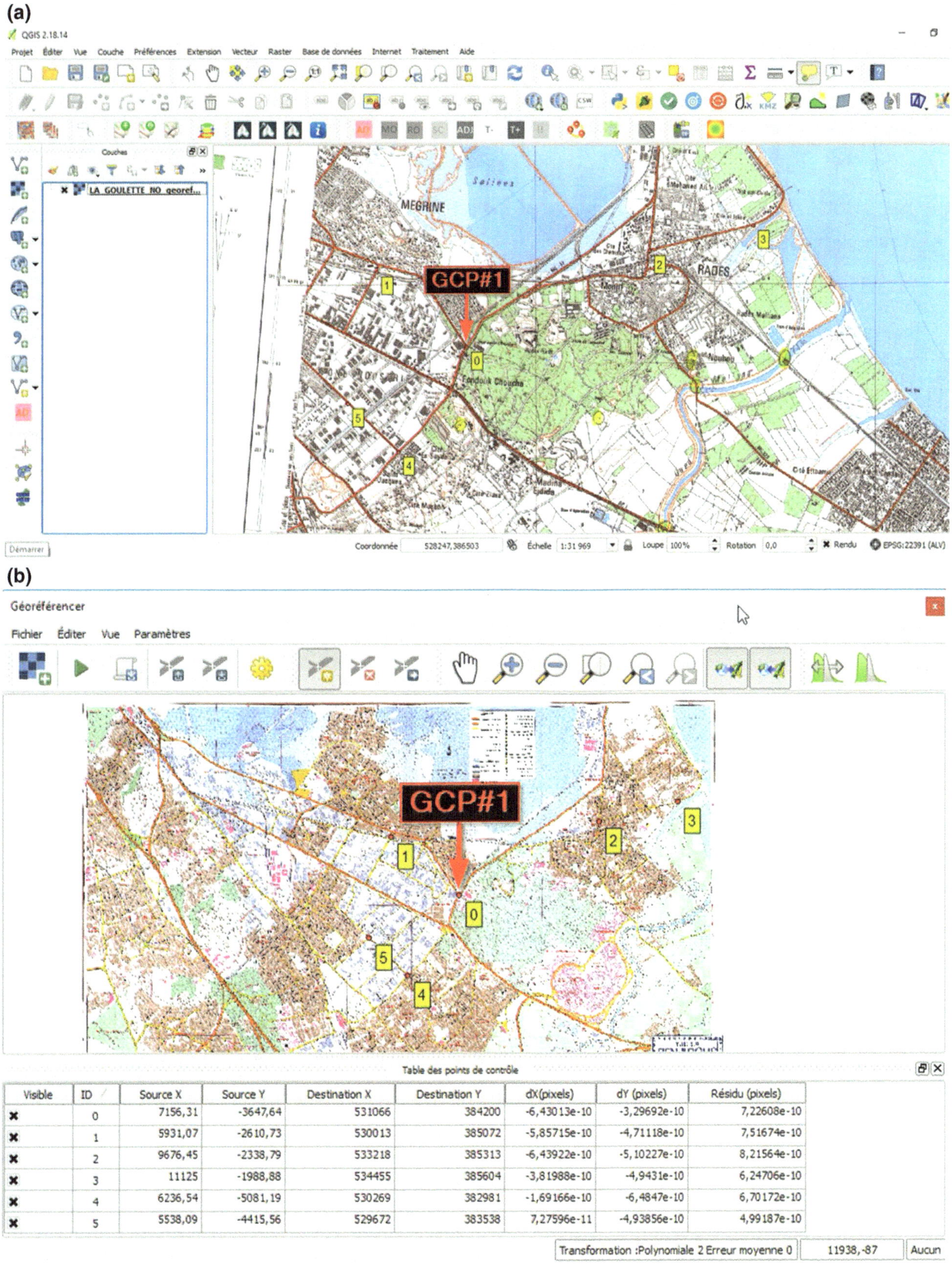

Fig. 4 Align, into QGIS, **a** of the topographic 1/25000 map of LA GOULETTE NO and **b** Georeferencer window with six GCPs

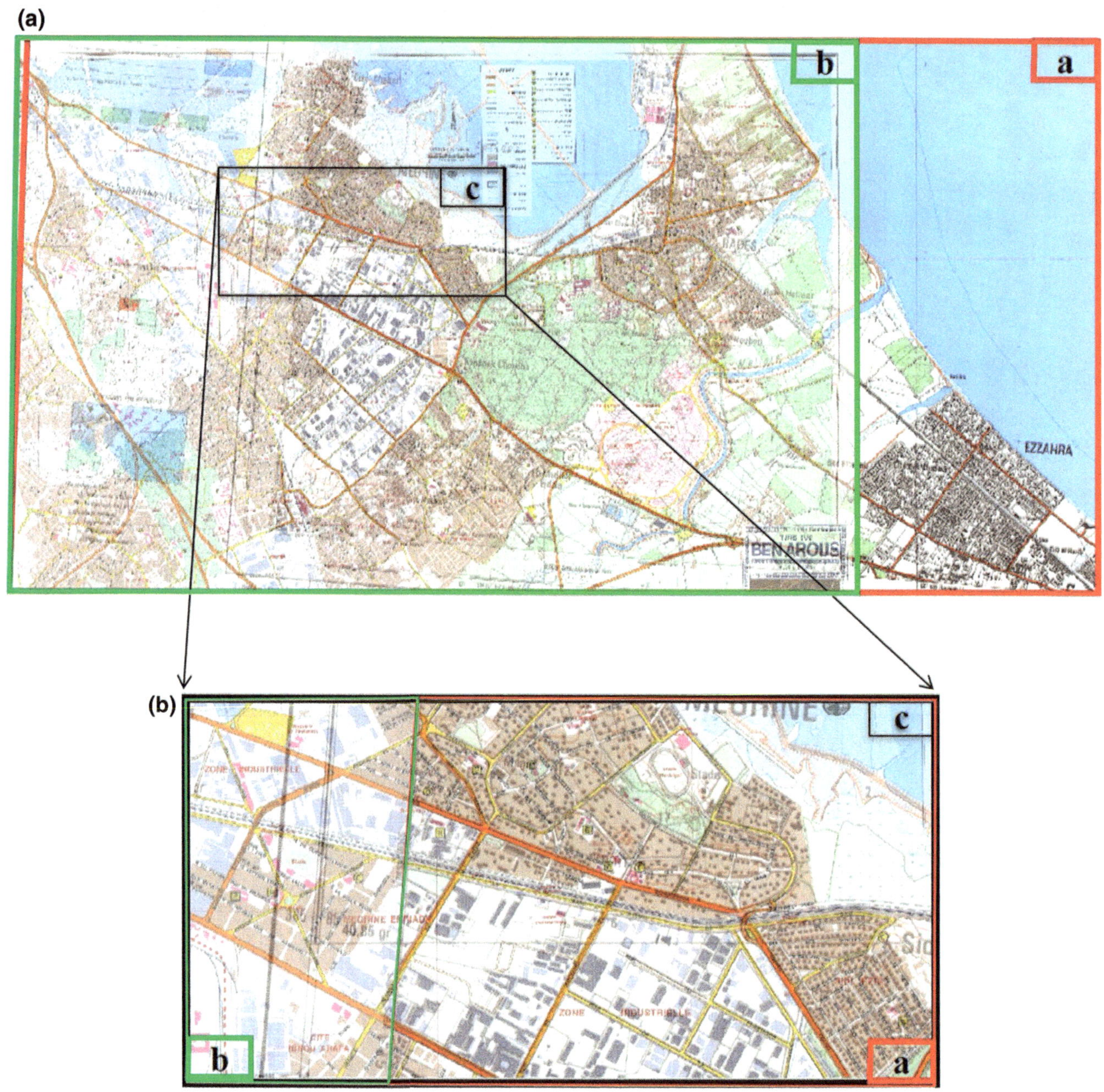

Fig. 5 Align, into QGIS, **a** of Touristic 1/10000 map [b] with topographic 1/25000 map of LA GOULETTE NO [a] and **b** zoom on zone [c]

The OSM data has precious resources, including the mapping data and especially the volunteers who spend time updating and improving it. The validation process tries to ensure that these two resources are maintained at a high level, which is to always be improving the quality of the data [25].

Many people whose first attempts at mapping are the result. There is much to learn by examining input data that has already been validated.

JOSM Validation and Digitization Tool (Update)

The JOSM Validator is a core feature of JOSM which checks and fixes invalid data. JOSM (Java OpenStreetMap Editor) is a desktop application originally developed by Immanuel Scholz and currently maintained by Dirk Stöcker [32]. Its homepage is located at https://josm.openstreetmap.de/.

Although it has a relatively steep learning curve, JOSM is popular among experienced editors thanks to its plugins and

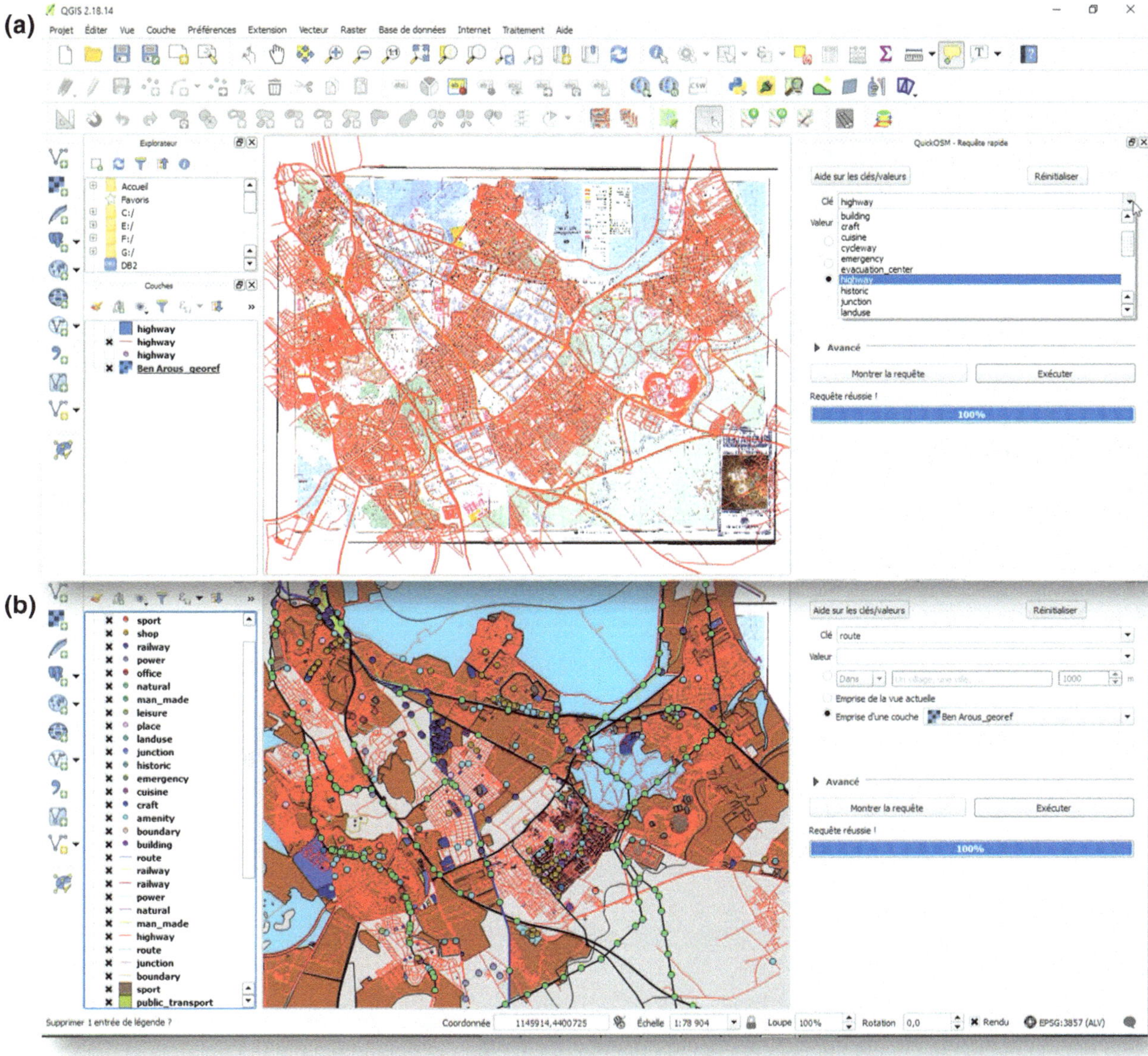

Fig. 6 Download, open and view **a** Highway OSM data of Ben Arous and **b** all OSM data using the QuickOSM plugin into QGIS software

stability [33]. JOSM is a feature-rich editor with an interface that may seem complex at first use. Take a look at the comparison of publishers if you're still not sure which publisher to choose. Using the JOSM tool, the OSM data for the Ben Arous area was validated (Fig. 8).

Edition with JOSM

Editing OpenStreetMap with JOSM is similar to editing with the iD editor we saw previously [34]. However, JOSM is a desktop application, it works a little differently. The process of editing and adding to OpenStreetMap with JOSM will always be as follows:

1. Download the current map data from OSM;
2. Edit it using Satellite Imagery, GPS, Field Papers;
3. Save changes to OpenStreetMap.

The first step in the editing process has therefore been completed: the download. We have prepared JOSM with satellite imagery as a reference. The next step is to edit the map and add new elements. Depending on the area you have chosen to download (Ben Arous), there may be much or very little existing map data. But note that this is the same kind of data we saw earlier: points, lines, and shapes that represent real places. Use the techniques you have already learned to add a few points to the map of places you know. If you see errors, try to correct them (Fig. 8).

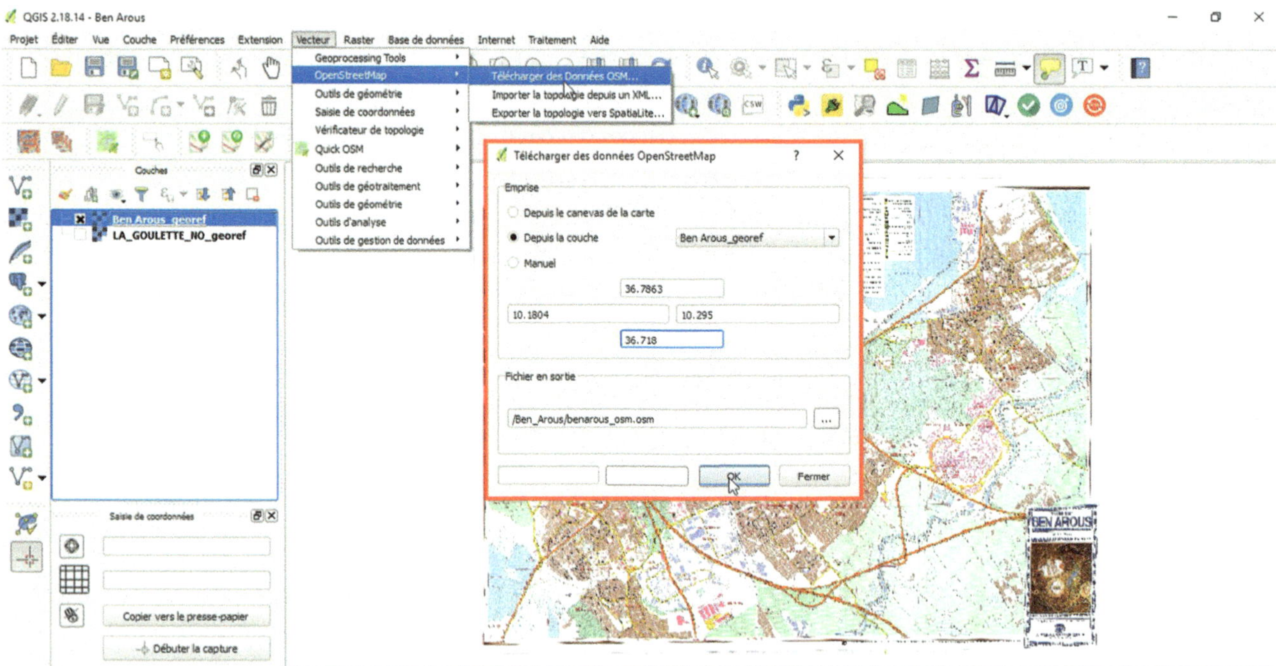

Fig. 7 Download OSM data using the OpenStreetMap plugin

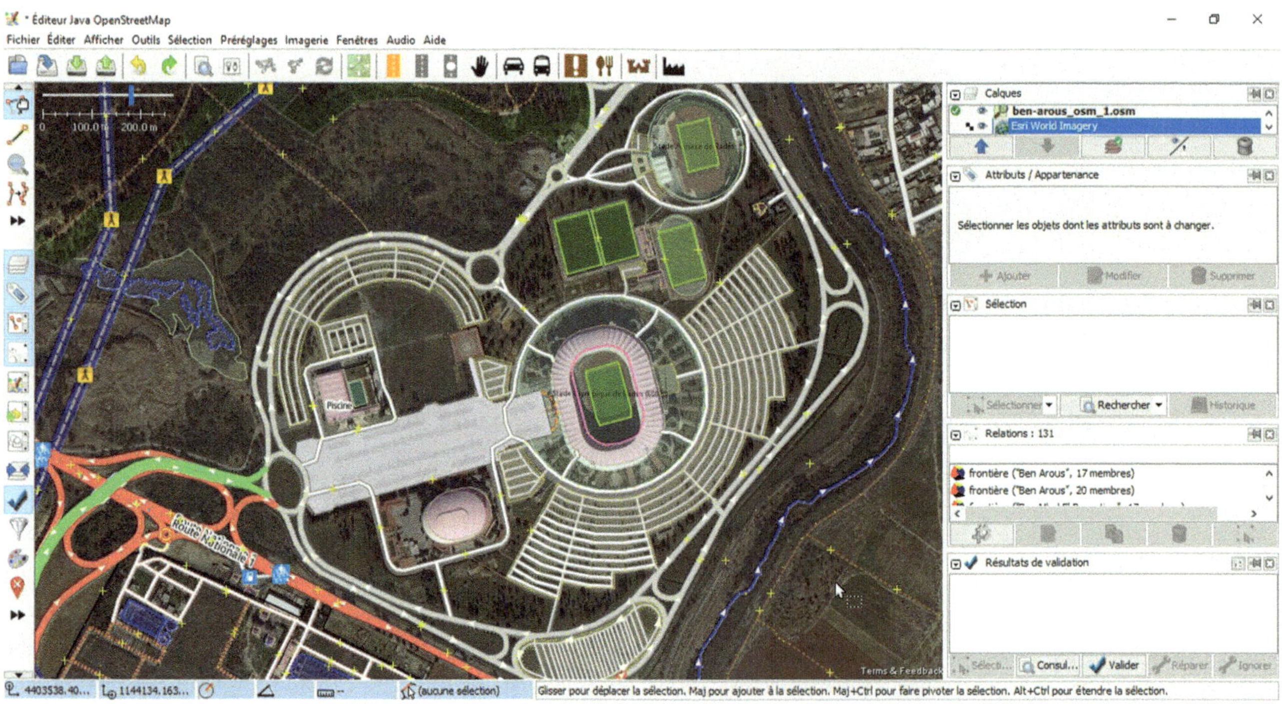

Fig. 8 Validation, digitization and downloading OSM data Symbolize OSM data of Ben Arous into the JOSM Tool

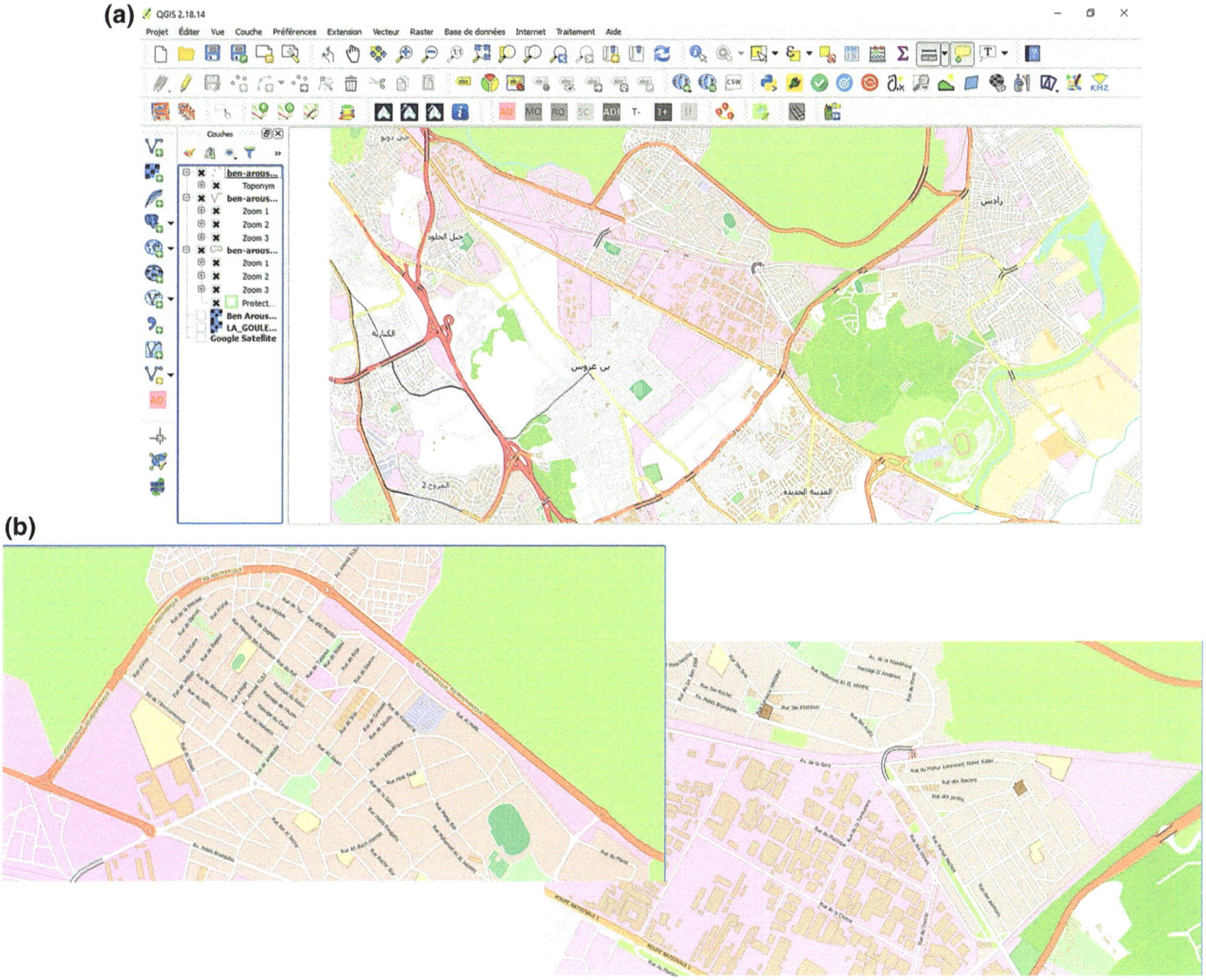

Fig. 9 Symbolize **a** OSM data of Ben Arous into QGIS software and **b** zoom on Symbolized OSM data

4.5 Symbolize OSM Data

We tried to symbolize all of the downloaded OSM data of Ben Arous exactly like it is displayed on Openstreetmap.org. We used additional stylesheets symbols (e.g., Symbolize Points, Symbolize Lines, Symbolize Polygons) in QGIS (Fig. 9) (available at: https://github.com/yannos/Beautiful_OSM_in_QGIS).

4.6 Export OSM Data as Shapefile

The result and the layers are only temporary! Exporting any GIS data layer that appears in the layer list is as simple as right-clicking and save as (Fig. 10). You will find options under format for Shapefiles, GeoJSON, KML, GML, MapInfo, AutoCAD DXF, and others. Through this process, up-to-date OSM data can be easily obtained and integrated into the QIS. Once you have layers like this in QGIS, it is possible to save them as shapefiles, execute filters and queries, and so forth.

4.7 Compare OSM Data—Touristic Map

On the other hand, in order to determine VGI (Volunteered Geographical Information) [25, 26, 35] quality in quantitative terms, a comparison with official data of a higher quality (data produced by accepted quality standards) is required. In this study, the reference topographic map at a scale of 1:25000 that was produced from the topographic map of LA GOULETTE NO and the Touristic 1/10000 scale map of BEN AROUS.

The comparison result of the downloaded, validated, and symbolized OSM data of the Ben Arous area shows semilar

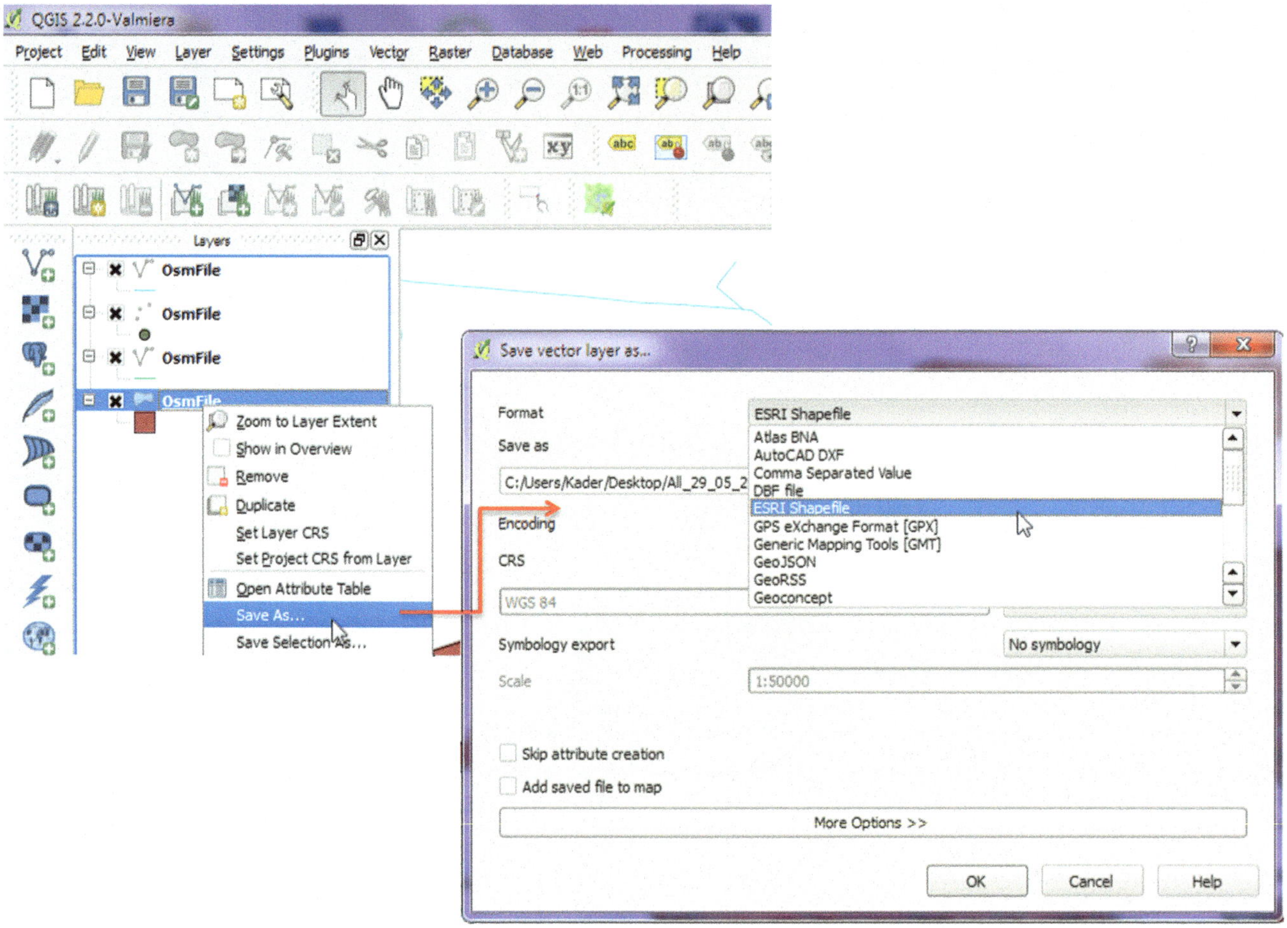

Fig. 10 Exporting OSM data to any GIS data layer (Shapefiles, GeoJSON, KML, GML, MapInfo, AutoCAD DXF, and others) into QGIS

areas with the same roads, buildings, and localities (Figs. 11 and 12) and new roads and constructions that have been added in almost 18 years (Fig. 11).

5 Conclusion

Open geospatial data is raw information accessible in an open and free format without any restrictions of a technical, legal or financial nature. More than a simple legal obligation, Open geospatial data is a great opportunity in terms of openness and innovation. Indeed, governments and communities are already engaged in an Open geospatial data process showing a better flow of information to the outside but also within their structure.

The results obtained from our methodological approach of free updating of cartographic data and the use of open source software require a good control of raster data georeferencing (topographic map) and validation of OSM data. The comparison of raster formats (OSM and touristic map) by considering colors for the simple extraction of new objects that are updated is essential to better determine the urban dynamics in such region.

As we know today, geography is at the heart of a more robust and sustainable future. Creating responsible products and solutions stimulates passion for improving quality of life around the world. Open data can produce an interesting commercial effect such as local advertising or application development.

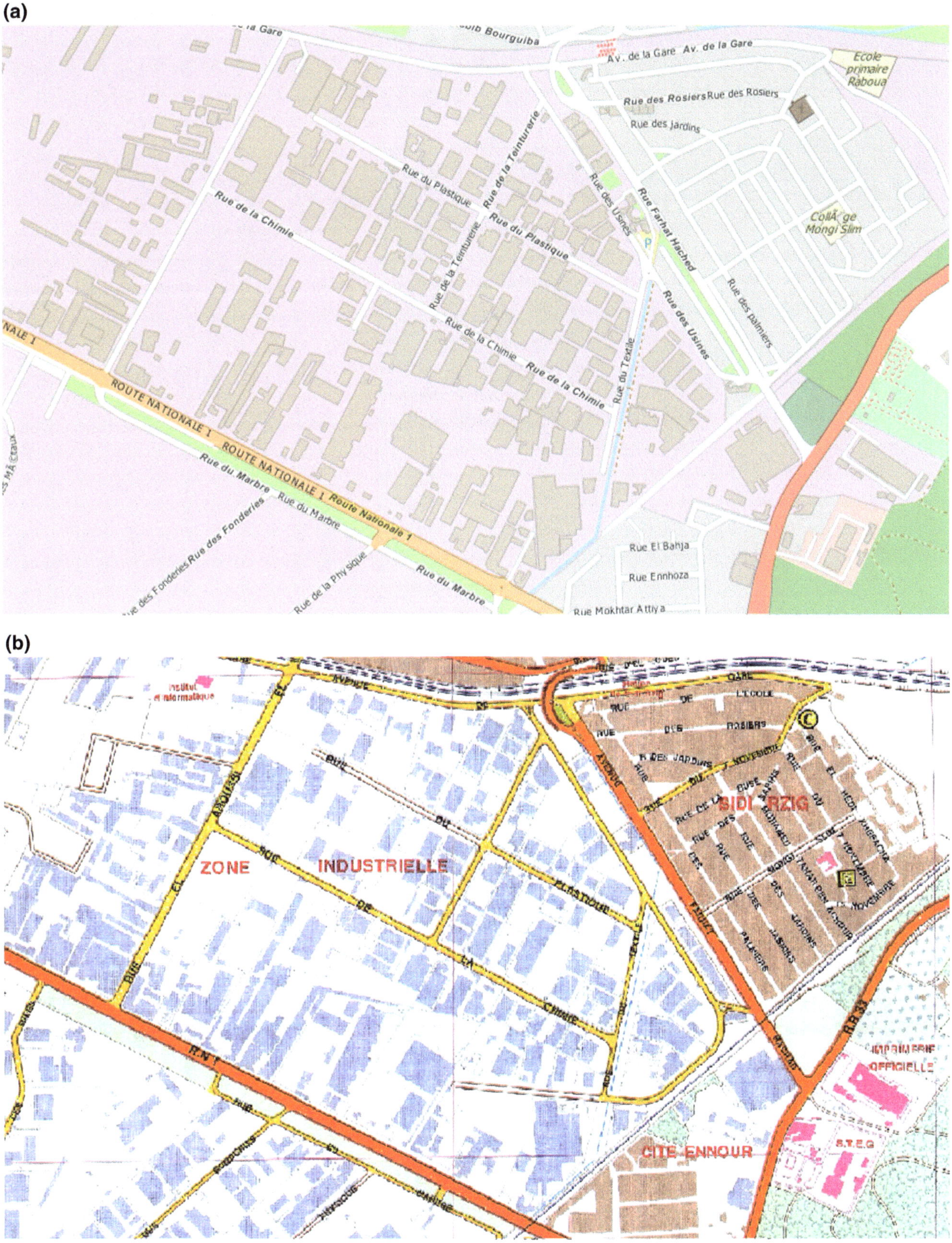

Fig. 11 Compare **a** OSM data—**b** Touristic map of Ben Arous (industrial zone of SIDI RZIG)

(a)

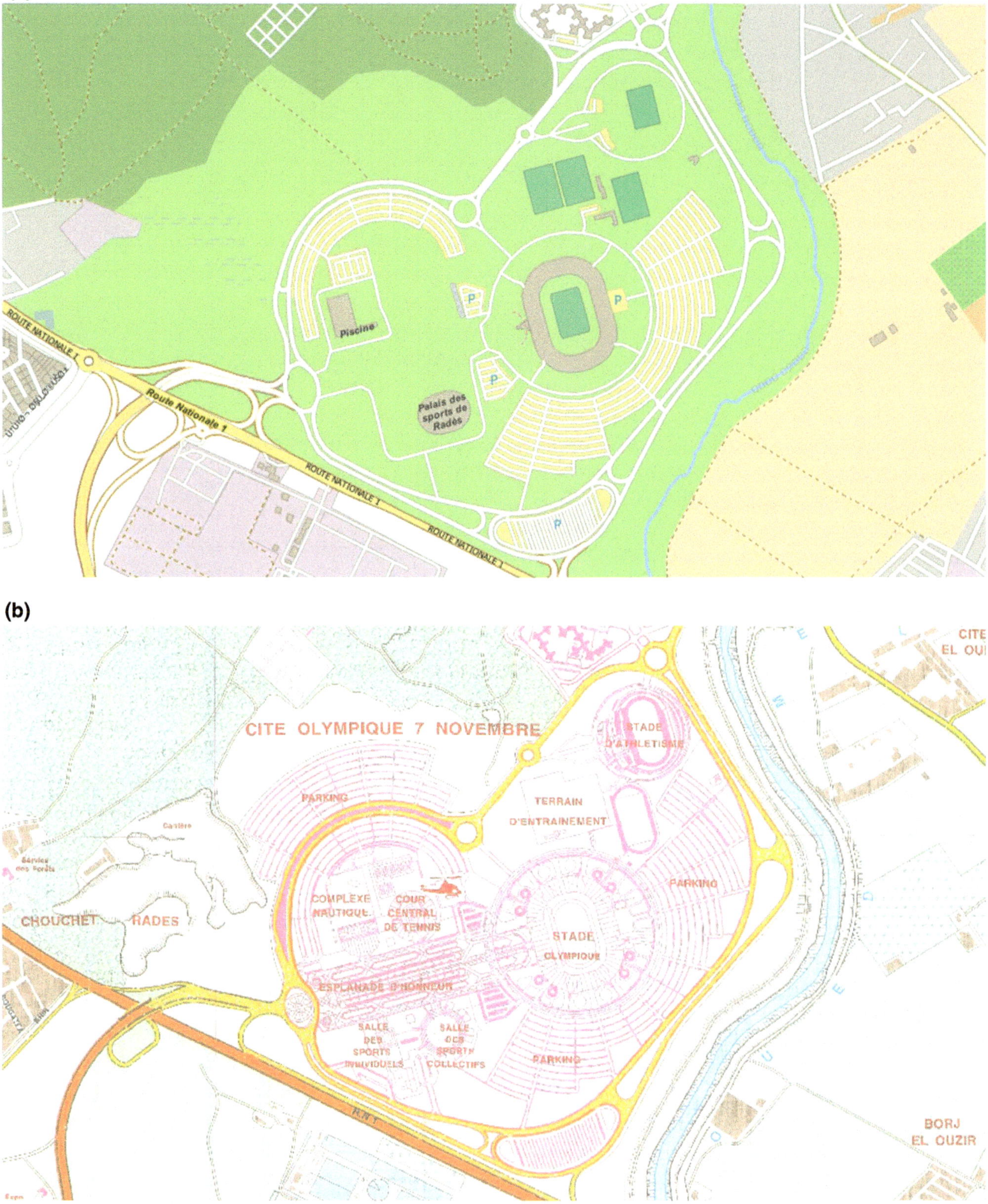

Fig. 12 Compare **a** OSM data—**b** Touristic map of Ben Arous (Olympic city November 7th)

References

1. Markieta, M. (2012). Using OpenStreetMap data with open-source GIS. *Cartographic Perspectives*, 91–104.
2. Haklay, M., & Weber, P. (2008). OpenStreetMap: User-generated street maps. *IEEE Pervasive Computing*, *7*(4), 12–18.
3. Bennett, J. (2010). *OpenStreetMap*. Birmingham: Packt Publishing Ltd.
4. Ramm, F., Topf, J., & Chilton, S. (2011). *OpenStreetMap: Using and enhancing the free map of the world*. Cambridge: UIT Cambridge.
5. OpenStreetMap Wiki. (2013).
6. Neis, P., & Zielstra, D. (2014). Recent developments and future trends in volunteered geographic information research: The case of openstreetmap. *Future Internet*, *6*(1), 76–106.
7. Kunze, C., & Hecht, R. (2015). Semantic enrichment of building data with volunteered geographic information to improve mappings of dwelling units and population. *Computers, Environment and Urban Systems*, *53*, 4–18.
8. Zhang, H., & Malczewski, J. (2017). Quality evaluation of volunteered geographic information. *Volunteered geographic information and the future of geospatial data* (p. 19).
9. Senaratne, H., Mobasheri, A., Ali, A. L., Capineri, C., & Haklay, M. (2017). A review of volunteered geographic information quality assessment methods. *International Journal of Geographical Information Science*, *31*(1), 139–167.
10. Arsanjani, J. J., Zipf, A., Mooney, P., & Helbich, M. (2015). OpenStreetMap in GIScience.
11. Longley, P. (2005). *Geographic information systems and science*. New York: Wiley.
12. Bernhardsen, T. (2002). *Geographic information systems: An introduction*. New York: Wiley.
13. McCurley, K. S. (2001). Geospatial mapping and navigation of the web. In *Proceedings of the 10th International Conference on World Wide Web* (pp. 221–229). ACM.
14. Burrough, P. A., McDonnell, R. A., & Lloyd, C. D. (2015). *Principles of geographical information systems*. Oxford: Oxford University Press.
15. Longley, P. A., Goodchild, M. F., Maguire, D. J., & Rhind, D. W. (2015). *Geographic information science and systems*. New York: Wiley.
16. Thomas, R. K. (2018). Data sources for demography. In *Concepts, methods and practical applications in applied demography* (pp. 31–51). Berlin: Springer.
17. Craun, K. J., & Chih-Hung, M. (2017). A comparison of volunteered geographic information (VGI) collected in rural areas to VGI collected in urban and suburban areas of the united states. *Citizen empowered mapping* (pp. 173–197). Berlin: Springer.
18. Lin, W. (2018). Volunteered geographic information constructions in a contested terrain: A case of OpenStreetMap in China. *Geoforum*, *89*, 73–82.
19. Ballatore, A., Bertolotto, M., & Wilson, D. C. (2013). Geographic knowledge extraction and semantic similarity in OpenStreetMap. *Knowledge and Information Systems*, *37*(1), 61–81.
20. Aronoff, S. (1989). Geographic information systems: A management perspective.
21. Gregory, J. W. (1908). *Geography*. London: Blackie and Son.
22. Goodchild, M. F. (1992). Geographical data modeling. *Computers & Geosciences*, *18*(4), 401–408.
23. DeMers, M. N. (2002). *GIS modeling in raster*. Number 910.285 D4G5. New York: Wiley.
24. Chang, K.-T. (2015). *Introduction to geographic information systems*. New York: McGraw-Hill Science/Engineering/Math.
25. Girres, J.-F., & Touya, G. (2010). Quality assessment of the French OpenStreetMap dataset. *Transactions in GIS*, *14*(4), 435–459.
26. Wang, M., Li, Q., Hu, Q., & Zhou, M. (2013). Quality analysis of open street map data. *ISPRS-International Archives of the Photogrammetry, Remote Sensing and Spatial Information Sciences*, *1*, 155–158.
27. Fan, H., Zipf, A., Qing, F., & Neis, P. (2014). Quality assessment for building footprints data on OpenStreetMap. *International Journal of Geographical Information Science*, *28*(4), 700–719.
28. Du, H., Nguyen, H. H., Alechina, N., Logan, B., Jackson, M., & Goodwin, J. (2015). Using qualitative spatial logic for validating crowd-sourced geospatial data. In *AAAI* (pp. 3948–3953).
29. Zofka, M. R., Klemm, S., Kuhnt, F., Schamm, T., & Marius Zöllner, J. (2016). Testing and validating high level components for automated driving: Simulation framework for traffic scenarios. In *Intelligent Vehicles Symposium (IV), 2016 IEEE* (pp. 144–150). IEEE.
30. Da Costa, J. N. (2016). Towards building data semantic similarity analysis: OpenStreetMap and the polish database of topographic objects. In *Geodetic Congress (Geomatics), Baltic* (pp. 269–275). IEEE.
31. Fonte, C. C., Patriarca, J. A., Minghini, M., Antoniou, V., See, L., & Brovelli, M. A. (2017). Using OpenStreetMap to create land use and land cover maps. *IIASA*
32. Scholz, I., & Community, O. (2013). JOSM (Java OpenStreetMap Editor).
33. Vandecasteele, A., & Devillers, R. (2015). Improving volunteered geographic information quality using a tag recommender system: The case of OpenStreetMap. In *OpenStreetMap in GIScience* (pp. 59–80). Berlin: Springer.
34. Piñol, P., López, O., Martínez, M., Oliver, J., & Malumbres, M. P. (2012). Modeling video streaming over VANETs. In *Proceedings of the 7th ACM Workshop on Performance Monitoring and Measurement of Heterogeneous Wireless and Wired Networks* (pp. 7–14). ACM.
35. Mehdipoor, H., Zurita-Milla, R., Rosemartin, A., Gerst, K. L., & Weltzin, J. F. (2015). Developing a workflow to identify inconsistencies in volunteered geographic information: A phenological case study. *PloS One*, *10*(10), e0140811.

Open Remote Sensing Image Classification Using NDVI and Thermal Bands

Mohamed Mastere, Abdelkrim Achbun and Bouchta El Fellah

Abstract

In this paper the authors try to use time series of NDVI index, interpretation of thermal bands of images acquired by the Landsat 5 and 7, and object-based classification, to reveals the geo-biophysical causes observed that explains changes in vegetation greenness and globally in land cover evolution in the study area defined between Wadi Law and the northern Mediterranean coast of the Tinjitane peninsula (North of Morocco). The study area is a vast mountainous field, with rivers and agricultural patterns, forests and artificial surfaces. The combination of multi-sources data was crucial when establishing a classification within heterogeneous data in complexes landscapes. The object-based classification helped to best categorize the land cover classes in the study area. We observed clearly a global decreasing of forest areas between 1973 and 2015, and in parallel an increasing of the surface of Agriculture and natural vegetation class. The post-classification change vector detection is an optimal method to describe the evolution of the land cover in the context of this study.

Keywords

Land cover • Change analysis • Object based • NDVI • Thermal bands

M. Mastere (✉) · B. El Fellah
Geopac Research Center, Scientific Institute, Mohammed V University of Rabat, Avenue Ibn Battouta, Agdal, Rabat, Morocco
e-mail: mohamed.mastere@gmail.com

A. Achbun
Laboratoire de géomatique, Institut National d'Aménagement et d'Urbanisme, Avenue Allal El Fassi BP 6215 Rabat-Instituts, Rabat, Morocco

Laboratoire de Géomorphologie et de Cartographie, Institut Scientifique de Rabat, Université Mohammed V, Rabat, Morocco

1 Introduction

LULC assessment is one of the most important parameters to meaningfully plan for land resources management. The knowledge of spatial land cover information is essential for proper management, planning and monitoring of natural resources [1]. The obtainment and updating of information about the current condition and the continuous dynamic changes of our earth's surface in remote high-mountain regions is a task where remote sensing technologies can best display their advantages.

For instance, in agricultural land use, which is systematically linked through seasonal-temporal interactions, the land might be associated with a sequence of covers through a year, as a vegetated land (vegetation), barren land (soil) and inundated land (water) [2].

The complicated change of land cover could be categorized into three types and mechanisms: (1) seasonal change, driven by annual temperature and rainfall interactions on vegetation phenology; (2) gradual change, caused by inter-annual climate variability or land management; and (3) abrupt change, caused by disturbances such as deforestation, urbanization, floods, and fires.

Satellite remote sensing imagery is a viable source of gathering quality land cover information at local, regional and global scales [3,4]. Advances in remote sensing technology enable land scientists to identify ongoing land cover change processes and their locations [5]. Moreover, remote sensing data are on the whole, useful for land cover change detection and mapping in mountainous regions [2]. The area of our study is monotonous zone, the study of land cover change there within a remote sensing process, may be influenced by a large number of factors, since: the presence of shadows caused by high altitude of the terrain, the cloud cover, deep narrow valleys and ravines, low sun angles, steep slopes and differential vegetation cover. For that reason, classification is only on the basis of spectral data from a remote sensing sensor alone may not be sufficient to gather effective land cover informa-

N. Rebai and M. Mastere (eds.), *Mapping and Spatial Analysis of Socio-economic and Environmental Indicators for Sustainable Development*, Advances in Science, Technology & Innovation, https://doi.org/10.1007/978-3-030-21166-0_13

tion [6]. Moreover, Normalized Difference Vegetation Index (NDVI) is a general biophysical parameter that correlates with photosynthetic activity of vegetation and provides an indication of the 'greenness' of the vegetation [7]. The NDVI data layer is defined as: $NDVI = (NIR - R)/(NIR + R)$. Lyon et al. (1998) concluded that the NDVI differencing technique demonstrated the best vegetation change detection. We used during this work the NDVI time-series images to best facilitate the interpretation of the classes when classifying the images; it's helped us to extract the forest class that was used as an environmental indicator of the land cover change and the anthropogenic disturbance.

There are so many methods to do it but the common change detection methods include the comparison of land cover classifications, multi-date classification, band arithmetic, simple rationing, vegetation index differencing and change vector analysis [8]. In general, multi date remote sensing data can be used for detection of the LULC changes [9]. Medium-resolution sensors are intended to provide appropriate scales of information for a wide-variety of Earth-resource applications [10].

The used methodological process is a combination of the NDVI, thermal bands interpretation, and a post-classification change detection process; this is the most obvious method of change detection which requires the comparison of independently produced classified images. By properly coding the classification results for times t1, and t2, the analyst can produce change maps which show a complete matrix of changes. In addition, selective grouping of classification results allows the analyst to observe any subset of changes which may be of interest [11]. More precisely, the chance detection was studied after the realization of an Object-Based Classification and Change vector (of classified images) and was analyzed. The vector describing the direction and magnitude of change from the first to the second date is a spectral change vector [11], the direction of the vector contains information about the type of change, i.e. clear cut or regrowth [12].

In this paper authors try to use time series of NDVI index, interpretation of thermal bands of the Landsat 5 and 7, images and object-based classification to reveals the geo-biophysical causes observed that explain changes in vegetation greenness and globally in land cover in the study area defined between Wadi Law and the northern coast of Tinjitane peninsula in the north of Morocco. The study area is a mountainous zone with rivers, agricultural patterns and artificial surfaces.

2 Materials and Methods

2.1 Study Area

The study area is a mountainous zone intermitted by rivers, ravines, plains, artificial surfaces, and large forests. It's located in the Rifean mountain chain between Wadi (river) Law at the East and the northern coast of Tinjitane peninsula, in the Nord of Morocco (Fig. 1).

The study area covers 126,622 ha. From geological point of view, the study area belongs to the northern Morocco. This area is part of peri-mediterranean Alpine chains, united under the name of betico-Rifo-telleien [13,14]. It consists essentially of Mesozoic Cenozoic age and land. The area belongs specially to the internal Rifain field and particularly the unity of Ghomarides; This unit consists of layers within the highest element of internal Rif which contains epi-metamorphic, schist detrital and carbonate Silurian to Carboniferous age formations [15].

2.2 Data

Four images were downloaded from the platform: Earth Science Data Interface (ESDI) at the Global Land Cover Facility (http://glcfapp.glcf.umd.edu:8080/esdi/). Images acquired from different sensors (MSS, TM and ETM+) of the Landsat satellites. Table 1 and Fig. 2 present the acquisition dates and the resolutions of the images used. Images are projected in the system UTM, WGS84, zone 3. Moreover, Quick looks of Google Earth have been used in the phase of understanding the land cover classes and image interpretation, by using their high spatial precision.

2.3 Methodology

The methodological process (Fig. 3) applied in this study, began by a pre-processing of the data acquired from different sensors. The Landsat (MSS, TM and ETM) images projected on the UTM system were firstly layer stacked to prepare composed multiband images. Then a subset extraction was applied on each image on the base of the study area defined.

In order to make easy the interpretation of different classes, an elaboration of NDVI maps was deed for each date and an object based classification of the NDVI maps was deed (Fig. 4), this to facilitate the recognition of the vegetation areas and to revels the probable contrast caused by a season effect that appears on the images. In parallel, an analysis of thermal bands of Landsat 7 and 8 was performed to enhance a comprehension of the land cover change effects, especially at the seasonal scale. The cited pre-processing helped to initiate the processing by the Object Based Classification. A post-classification change detection process; this is the most obvious method of change detection which requires the comparison of independently produced classified images [11]. Each image has been firstly segmented on eCognition® 8.7 software and prepared for classification on the base of the objects selected under segmentation process. The vector shapes of

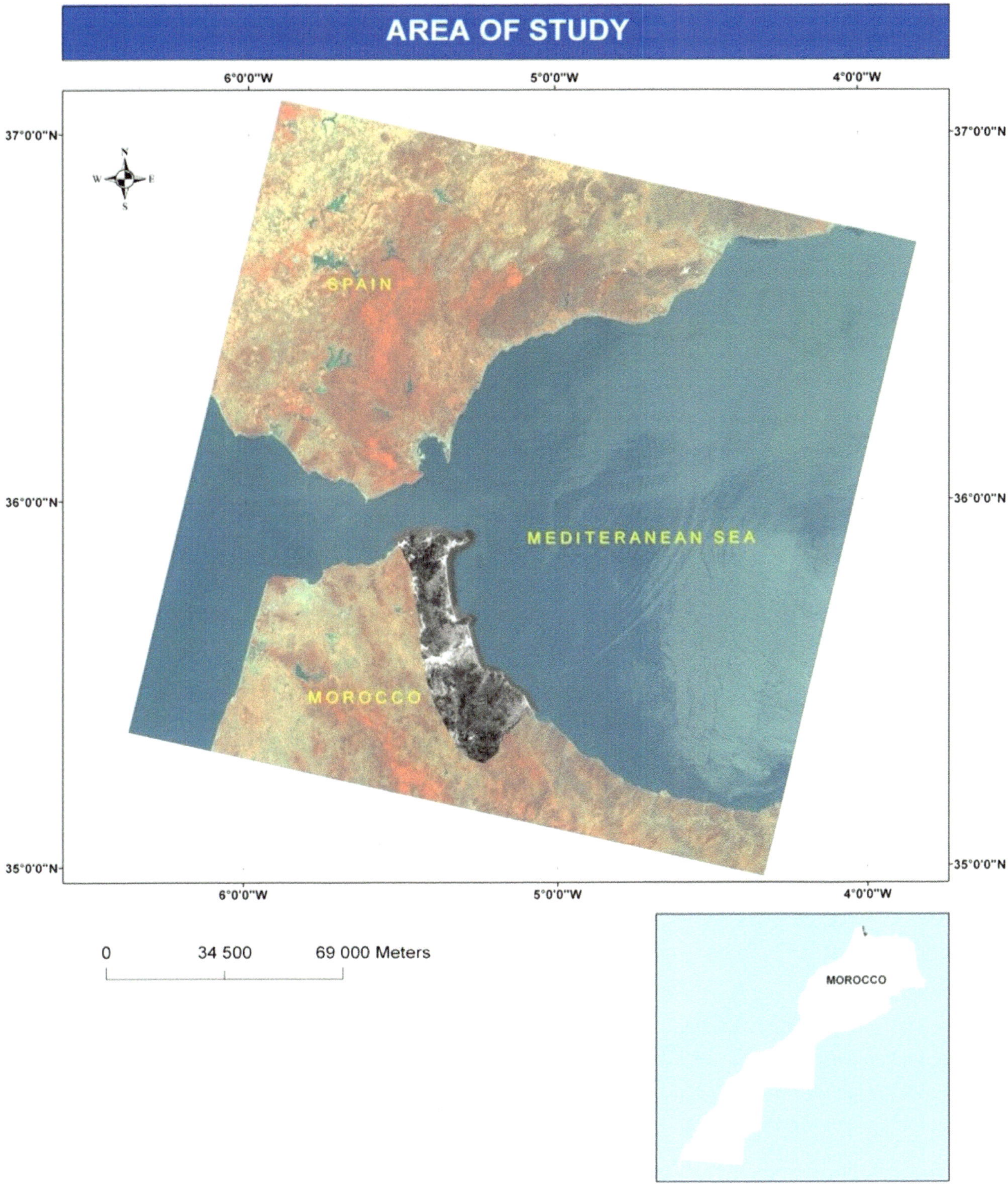

Fig. 1 Geographic situation of the study area

Table 1 Image characteristics

Date of acquisition of the image	Resolution (m)	Sensor
A: MSS 14/11/1973	60	Landsat 1, MSS
B: TM 06/03/1988	28.5	Landsat 5, TM
C: ETM 20/08/1999	28.5	Landsat 7, ETM+
D: ETM 07/07/2015	15	Landsat 8, Oli Tirs

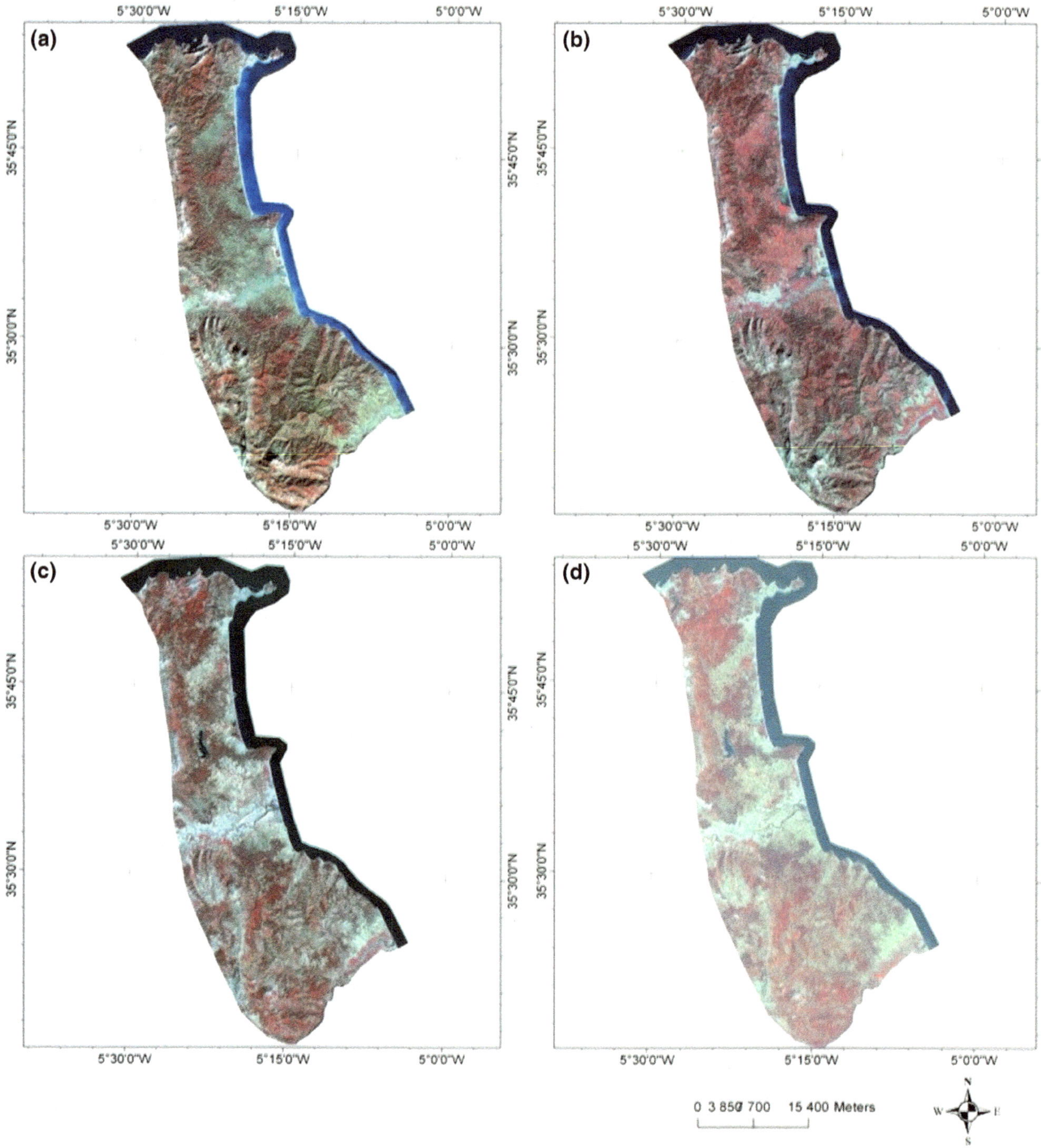

Fig. 2 Subsets of the images: **a** MSS 1973, **b** TM 1988, **c** ETM 1999 and **d** ETM 2015

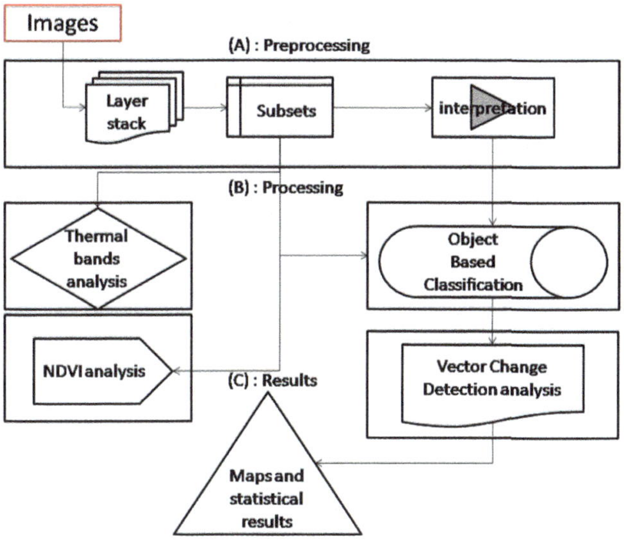

Fig. 3 Methodological work flow

Object Oriented classification of each date, were exported to the ArcGis® software on vector format, this to prepare the classified maps. Moreover, a vector change detection analysis and mapping were accomplished on ArcGis® and change detection map of the period between 1973 (date of acquisition MSS image) and 1999 (date of acquisition ETM image) and between 1999 and 2015 was prepared and statistical results were extracted. The described methodological work flow was tested on a heterogeneous data in the aim to concept a global methodology adapted to similar situations.

3 Results and Discussion

3.1 Image Interpretation, NDVI, and Thermal Bands Analysis

The interpretation step still very important in a semi-automatic process of classification. We used the quick looks to recognize land cover classes in addition to field knowledge. This process produced the following four classes: 1-Water, 2-Agriculture and natural vegetation, 3-Barren soil, 4-artificial surfaces, and 5-forests. The second classes are composed; this to allow unifying the key of interpretation for the four heterogeneous spatial resolution images of views. Especially, we noted for the MSS image having low spatial resolution (60 m) many interferences, which did not allow us to identify more specific classes. Moreover (Fig. 2), constraints; as the shadow of high mountains noted particularly during the image interpretation of MSS image and confusion observed between the Barren soil and artificial fields marked this recognition stage. This kind of thematic generalization [16, 17], allows us later to process the change detection change on a common base. The image interpretation was conducted just before the segmentation process of the images directly on the software eCognition® 8.7 and the NDVI maps, thermals band maps and graphs are analyzed in the aim to perform the object-based classification.

Maps of NDVI extracted from images have facilitated the identification of the vegetation related classes, including the class of forest. During the process of interpreting the land cover classes, the NDVI maps attends the recognition of zones and allow reducing the likely confusion between forest classes and agriculture and natural vegetation; compared to other classes.

We also used thermal bands of satellite images acquired from Landsat 7 and 8 to measure the surface temperature at the study area. Thus, we screened 20 sample points on each image (TM 1988 ETM + 1999 and ETM + 2015) to measure the average surface temperature for these images. This is done to monitor the effect of this environmental indicator on image interpretation and also to assess the contrast and the seasonal effect which appears due to the difference noted in the image acquisition dates. Table 3 and Fig. 5 summarize the results of this analysis. Thus, for the ETM + 2015 image acquired on 07/07/2015 in summer the average surface temperature (estimated by spectral value) is equal to 195, 12 and gave evidence of the effect of the season, and as a result; the areas of the following classes: Agriculture and Natural vegetation, water has been reduced in comparison to those derived from ETM image acquired on 06/03/1999 in the spring. The same applies to the lowest surface temperature recorded for the image MSS 1973 acquired at 07/11/1973 in winter; agricultural land not yet sown and natural vegetation was in his last phases of phrenology, which resulted in increasing the interpreted area of Barren soil and Artificial field (urban areas) (Fig. 6). The variation of the surface temperature was more important for the ETM 2015 image (Table 2).

The use of additional data represented by the NDVI and spectral values of thermal bands, attended the interpretation classes of the land cover and allowed relatively to quantify the impact of environmental factors and the effect of the season on the analysis of the satellite images before starting the process of object-oriented classification.

3.2 Object Based Classification

On eCognition 8.7 the process of segmentation has been started on each of the four images. The segmentation algorithms are used to subdivide entire images at a pixel level (Fig. 7), or specific image objects from other domains into smaller image objects (Definiens Developer, 2011).

The extraction of meaningful image objects needs to take into account the scale of the problem to be solved. Being as an optimization procedure which, for a given number of image objects, minimizes the average heterogeneity and maximizes

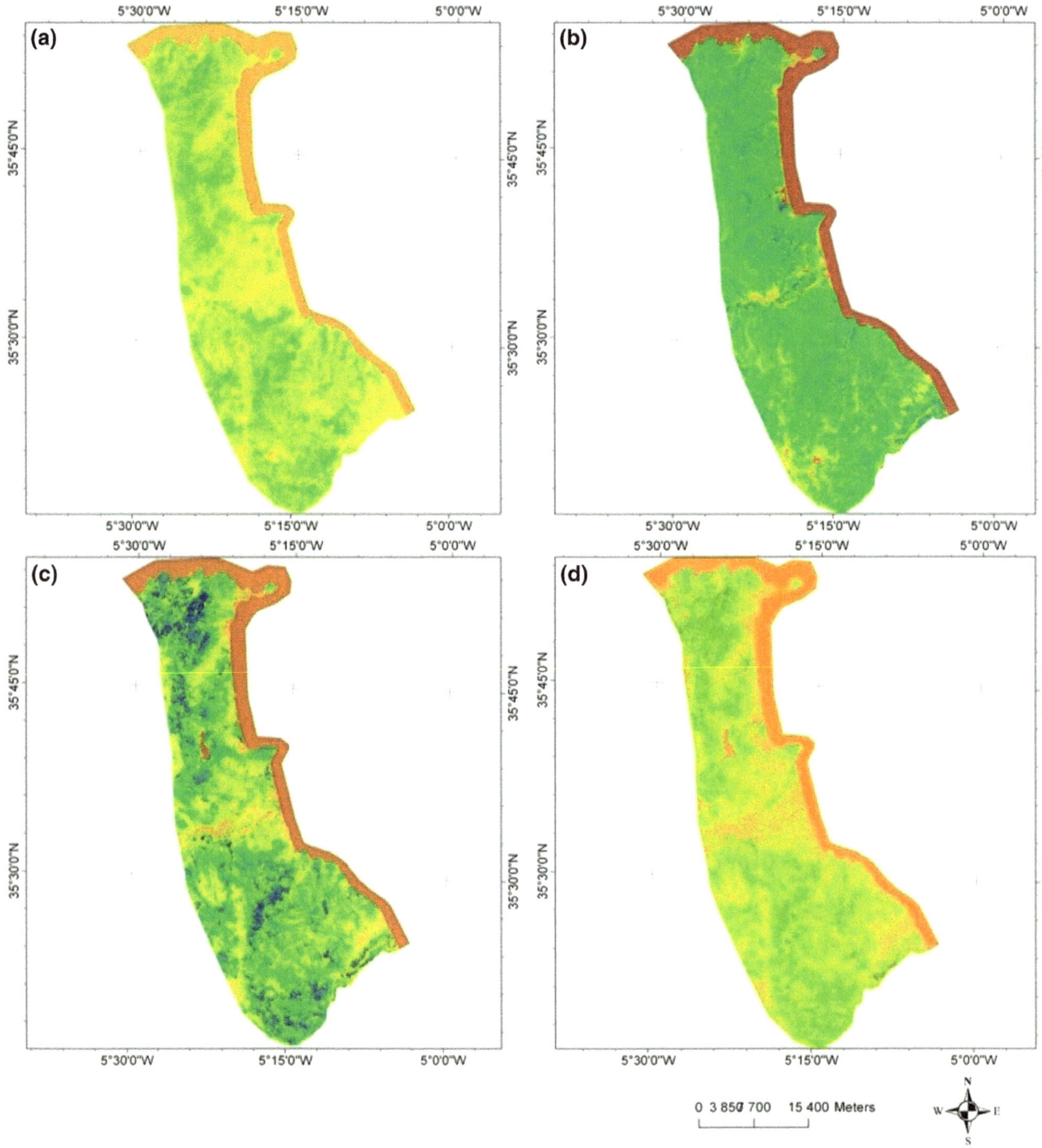

Fig. 4 NDVI index. **a** MSS 1973, **b** TM 1988, **c** ETM 1999 and **d** ETM 2015

their respective homogeneity; in this circumstance, homogeneity is used as a synonym for minimized heterogeneity. Internally, three criteria are computed: color, smoothness, and compactness Definiens Developer (2011). We used the multi-resolution algorithm that minimizes the average heterogeneity of image objects for a given resolution of image; this algorithm merges pixels or existing image objects. Two principal variables are controlled by this algorithm: scale (5-255) and color level (0, 1-0, 9). The Scale Parameter is an abstract term that determines the maximum allowed heterogeneity for the resulting image objects.

A hierarchical process of Object Based classification was conducted on the base of the segmentation results. The vector shape of classification was exported and mapping was deed

Table 2 Image characteristics

Images acquisition date	Spectral mean values of a 20 sample points tacked on thermal bands	Season
TM 06/03/1988	106,00	Spring
ETM 07/07/2015	195,12	Summer
ETM 20/08/1999	155,65	End of the summer

Table 3 Land covers change between 1973 and 1999

Changed classes		
Class name 1973	Class name 1999	Area
Forests	To artificial surfaces	7
Forests	To agriculture and natural vegetation	7
Agriculture and natural vegetation	To artificial surfaces	24
Agriculture and natural vegetation	To agriculture and natural vegetation	24
Water	To barren soil	65
Water	To agriculture and natural vegetation	164
Water	To artificial surfaces	220
Artificial surfaces	To water	261
Water	To forests	274
Barren soil	To water	308
Forests	To water	411
Agriculture and natural vegetation	To water	431
Barren soil	To forests	512
Artificial surfaces	To barren soil	872
Barren soil	To agriculture and natural vegetation	1035
Artificial surfaces	To forests	1065
Agriculture and natural vegetation	To barren soil	1654
Forests	To barren soil	1784
Barren soil	To artificial surfaces	1980
Artificial surfaces	To agriculture and natural vegetation	2129
Agriculture and natural vegetation	To forests	6458
Forests	To artificial surfaces	6553
Agriculture and natural vegetation	To artificial surfaces	9899
Forests	To agriculture and natural vegetation	14,015
Unchanged classes		
Agriculture and natural vegetation		15,723
Water		17,469
Forests		38,661
Artificial surfaces		3623
Barren soil		994
Total		126,622 ha

on ArcGIS® software. Figure 8 presents the resulting maps of the Object Based classification for the four dates. We observed clearly a global decreasing of forest areas from 1973 to 2015, and in parallel an increasing of the surface of Agriculture and natural vegetation class. The demographic creasing and anthropogenic disturbance are the most probable factors of this evolution.

3.3 Post-classification Change Detection

In order to assess trends in the change of land use between 1973 and 2015, we applied a change detection process through the post-classification analysis. According to Singh 1989 [11], this method requires the comparison of independently produced classified images. By properly coding the classifica-

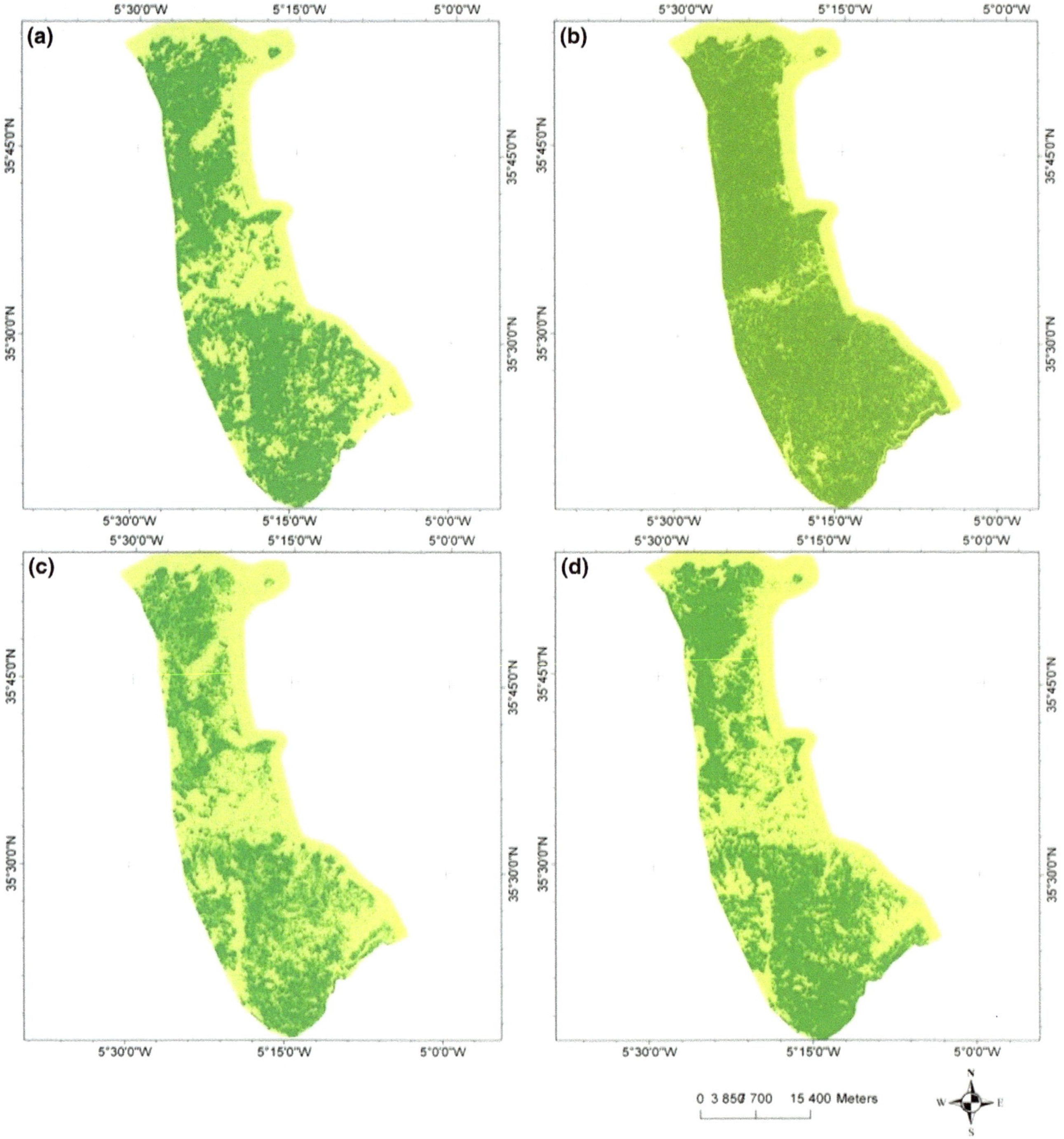

Fig. 5 Object based NDVI maps. **a** MSS 1973, **b** TM 1988, **c** ETM 1999 and **d** ETM 2015

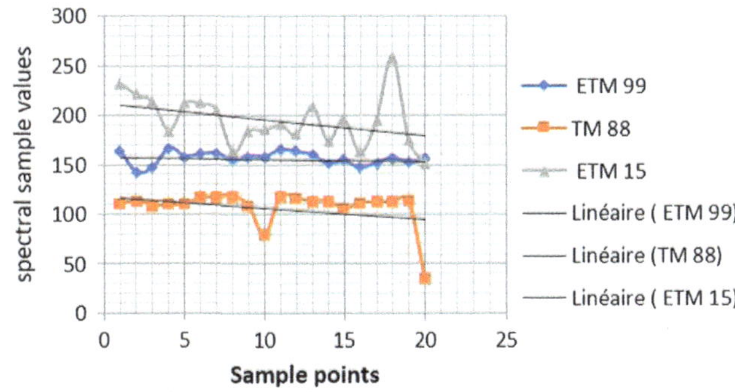

Fig. 6 Diagram of sample points of spectral values established on thermal bands

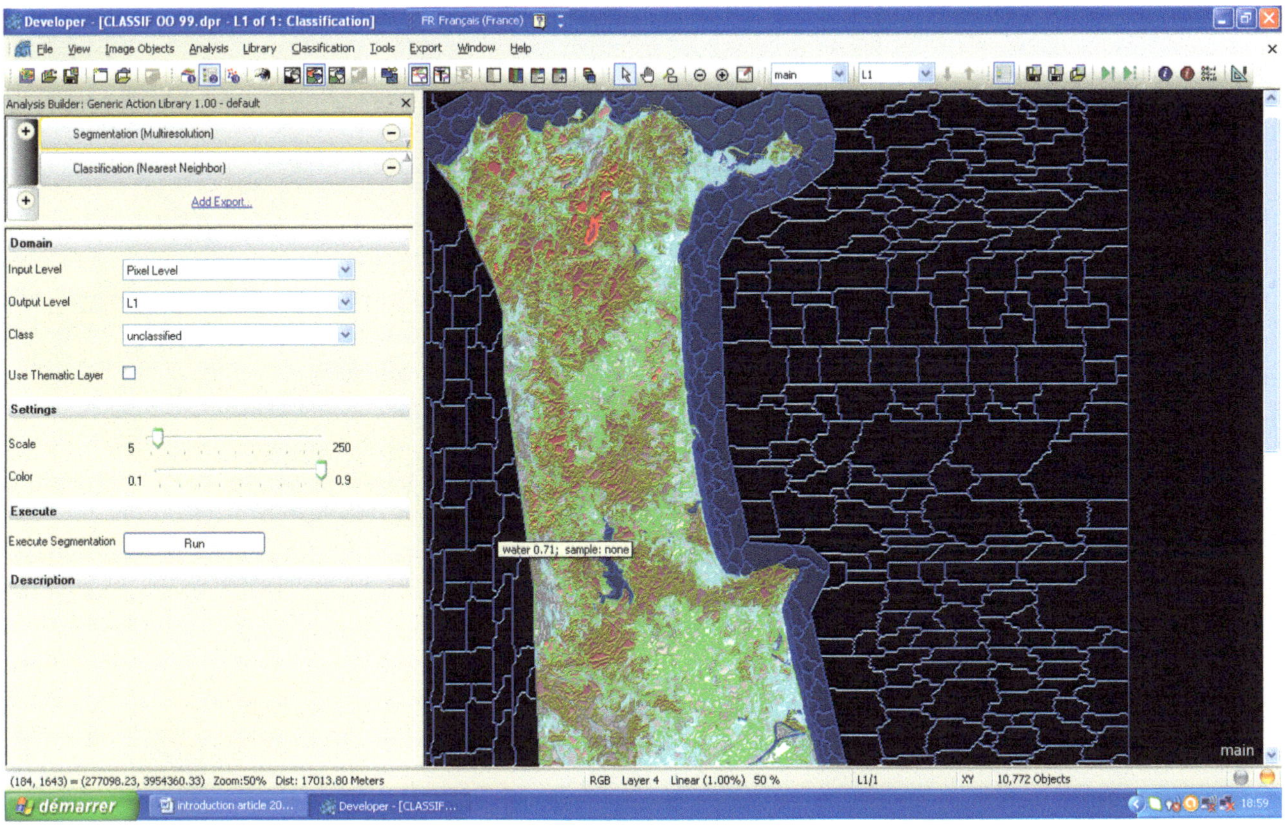

Fig. 7 Example of segmentation process

tion results for times t1, and t2, more precisely we used a change vector analysis. The vector describing the direction and magnitude of change from the first to the second date is a spectral change vector. We tried to experiment a change detection analysis of an object-based classification vector between two periods (1973–2015) and (1999–2015), this by using the Intersect tool implemented on ArcGIS® software.

Concerning the unchanged classes (Figs. 9a–c and 10), we noted the decease of the forest surfaces reserves and the barren soil surfaces between the two periods (1973–1999) and (1999–2015), this due to the extension of the agricultural fields and the artificial surfaces represented by the infrastructures and the urban extension, the installation of new dams and retained explains the increase in inland water areas. Figure 9c shows the annual rate of change of land cover. We note a net increase in the rate of change of land cover in the study site between the two above mentioned periods. Thus about 1695 (ha) of land per year where converted between 1973 and 1999 against 3135 ha per year between 1999 and 2015, which shows a large dynamic of development and installation of infrastructures during the second period (Fig. 9c).

About the changed classes, major changes in both periods studied, concerned the conversion of a large proportion of forest area into agricultural land (14,015 ha), followed by the increase in artificial surfaces despite agricultural areas (Tables 3 and 4).

Fig. 8 Object based map of the images (**a** MSS 1973, **b** ETM 1999, **c** ETM 2015)

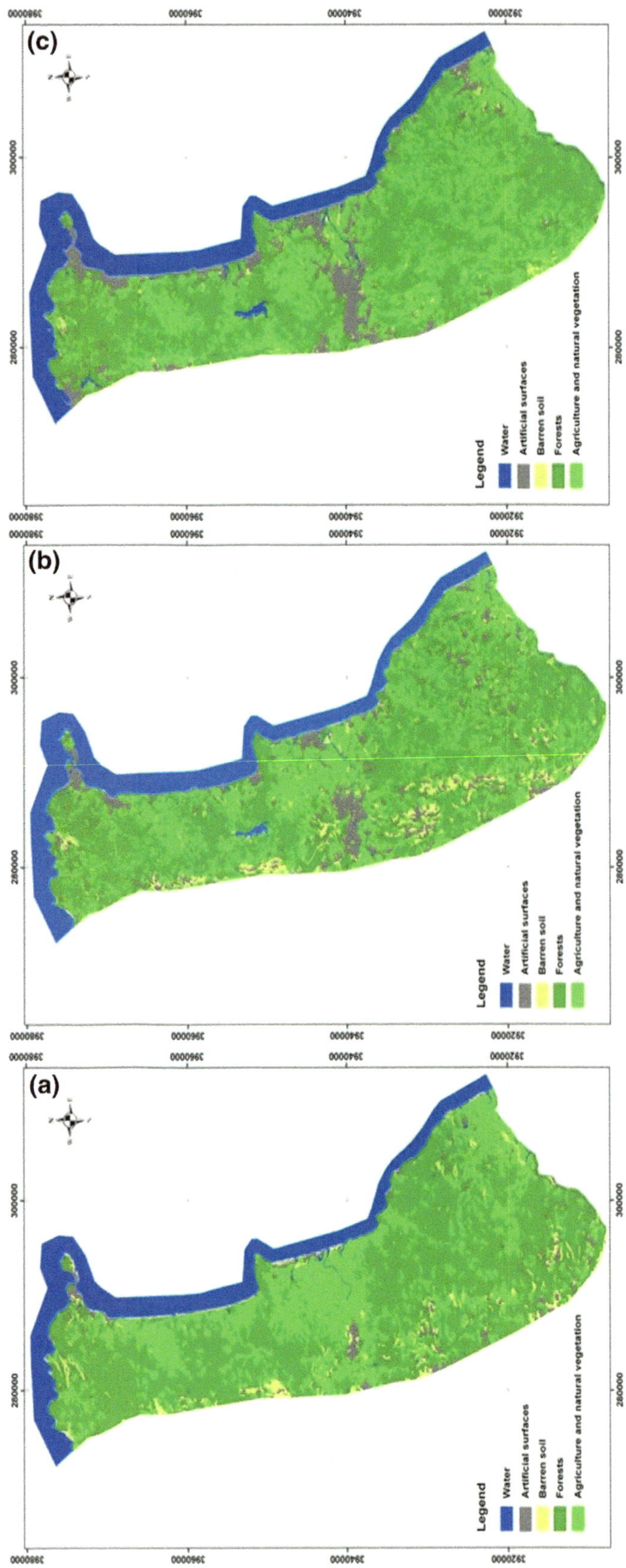

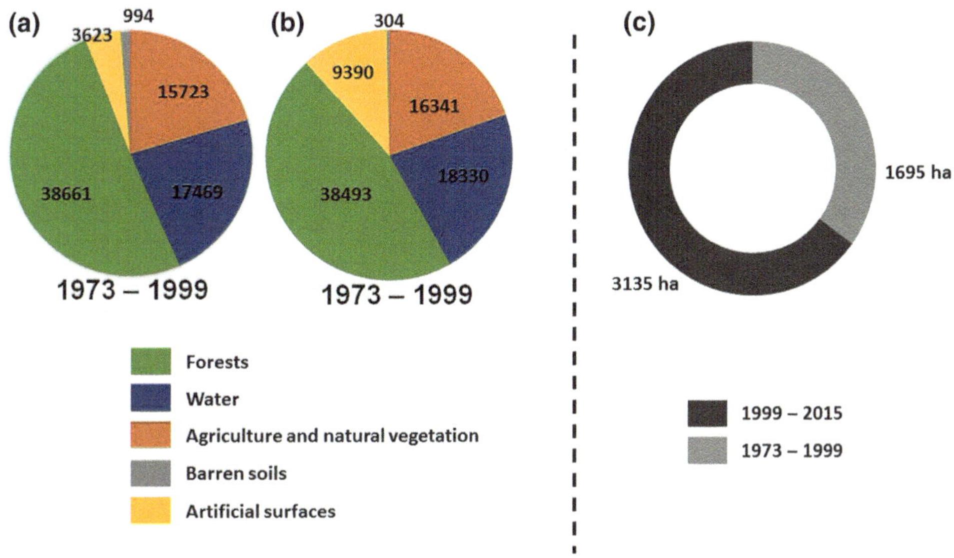

Fig. 9 Unchanged class rates between **a** (1973–2015) and **b** (1999–2015); **c** Unchanged class rate per year of land cover

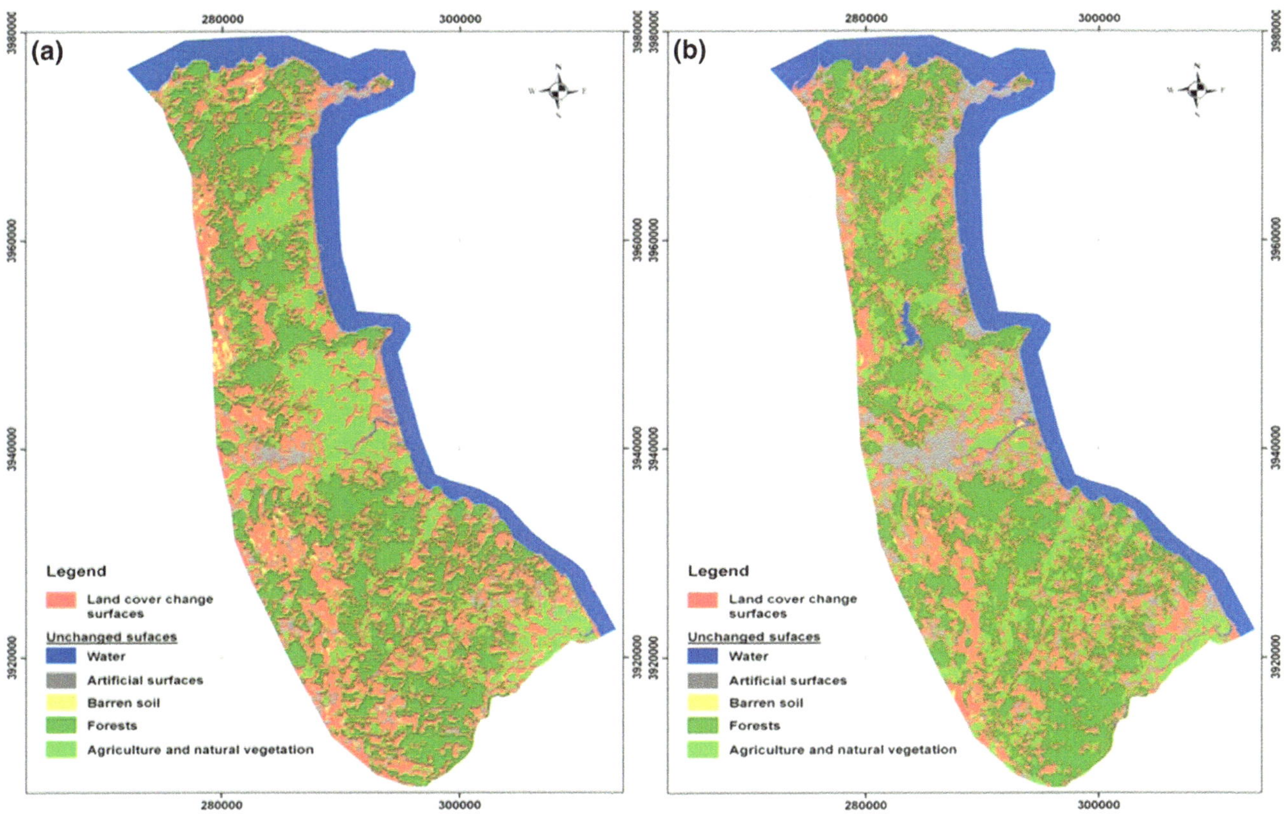

Fig. 10 Maps of land cover change (**a** 1973–1999; **b** 1973–2015)

Table 4 Land covers change between 1999 and 2015

Changed classes		
Class name 1999	Class name 2015	Area
Water	To barren soil	1
Artificial surfaces	To artificial surfaces	1
Agriculture and natural vegetation	To artificial surfaces	1
Artificial surfaces	To agricultural and natural vegetation	14
Agriculture and natural vegetation	To agricultural and natural vegetation	14
Artificial surfaces	To forests	16
Agriculture and natural vegetation	To forests	16
Forests	To barren soil	32
Barren soil	To water	62
Water	To agricultural and natural vegetation	104
Water	To forests	105
Artificial surfaces	To water	252
Agriculture and natural vegetation	To water	278
Water	To artificial surfaces	340
Agriculture and natural vegetation	To barren soil	407
Forests	To water	488
Artificial surfaces	To barren soil	558
Barren soil	To artificial surfaces	873
Barren soil	To forests	1386
Forests	To artificial surfaces	1731
Barren soil	To agricultural and natural vegetation	2744
Artificial surfaces	To forests	3186
Forests	To agricultural and natural vegetation	6226
Agriculture and natural vegetation	To artificial surfaces	6430
Artificial surfaces	To agricultural and natural vegetation	8889
Agriculture and natural vegetation	To forests	9610
Unchanged classes		
Agriculture and natural vegetation	Agricultural and natural vegetation	16,341
Water	Water	18,330
Forests	Forests	38,493
Artificial surfaces	Artificial surfaces	9390
Barren soil		304
Total		126,622

4 Conclusions

During the interpretation process, the use of additional data represented by the NDVI and spectral values of thermal bands, attended the interpretation classes of the land cover and allowed relatively to quantify the impact of environmental factors and the effect of the season on the analysis of the satellite images before starting the process of object-oriented classification. The combination of multi-sources of information is crucial when establishing a classification within heterogeneous data and in complexes landscapes.

The object-based classification helped to best categorize the land cover classes in the study area. We observed clearly a global decreasing of forest areas from 1973 to 2015, and in parallel an increasing of the surface of Agriculture and natural vegetation class. The demographic creasing and anthropogenic disturbance due to the implantation of infrastructure and the urban extension are the most probable factors of this evolution. The post-classification change vector detection is an optimal method to describe the evolution of the land cover in the context of this study.

Homage: The authors would like to pay a tribute to Abdelkrim ACHBUN the second author of this chapter, who died in a tragic accident just before the appearance of this paper. He was in the last year of his PhD preparation at Mohammed V university, Rabat Morocco. He received his first Master degree in geo-sciences and the second in remote sensing and GIS from the Regional African Center of Space

Sciences and Technologies respectively in 2005 and 2010. He was also an urban planner at the urban agency of Khemisset (Morocco). May God rest his soul in peace.

References

1. Zhu, A. X. (1997). Measuring uncertainty in class assignment for natural resource maps under fuzzy logic. *Photogrammetric Engineering and Remote Sensing*, *63*(10), 1195–1202.
2. Yudi, S., & Kunihiko, Y. (2012). Change detection in land-use and land cover dynamics at a regional scale from MODIS time-series imagery. *ISPRS Annals of the Photogrammetry, Remote Sensing and Spatial Information Sciences*, *I-7*, 243. XXII ISPRS Congress, 25 August–01 September 2012, Melbourne, Australia.
3. Csaplovics, E. (1998). High resolution space imagery for regional environmental monitoring status quo and future trends. *International Archives of Photogrammetry and Remote Sensing*, *32*(7), 211–216.
4. Foody, G. M., Campbell, N. A., Trodd, N. M., & Wood, T. F. (1992). Derivation and applications of probabilistic measures of class membership from maximum likelihood classification. *Photogrammetric Engineering and Remote Sensing*, *58*, 1334–1335.
5. Herold, M., Woodcock, C., Di Gregorio, A., Mayaux, P., Belward, A., Latham, J., et al. (2006). A joint initiative for harmonization and validation of land cover datasets. *IEEE Transactions on Geoscience and Remote Sensing*, *44*, 1719–1727.
6. Arora, M., & Kand Mathur, S. (2001). Multi-source classification using artificial neural network in a rugged terrain. *Geocarto International*, *16*(3), 37–44.
7. Sellers, P. J. (1985). Canopy reflectance, photosynthesis, and transpiration. *International Journal of Remote Sensing*, *6*, 1335–1372.
8. Jomaa, I., & Kheir, R. B. (2003). Multitemporal unsupervised classification and NDVI to monitor Land cover change in Lebanon (1987–1997).
9. Lu, D., Mausel, P., Brondizio, E., & Moran, E. (2004). Change detection techniques. *International Journal of Remote Sensing*, *25*, 2365–2407.
10. Rogan, J., & Chen, D. (2004). Remote sensing technology for mapping and monitoring land-cover and land-use change. *Progress in Planning*, *61*, 301–325.
11. Singh, A. (1989). Digital change detection techniques using remotely-sensed data. *International Journal of Remote Sensing*, *10*, 989–1003.
12. Malilaw, A. (1980). Change vector analysis: An approach for detecting forest changes with Landsat. In *Proceedings of the 6th Annual Symposium on Machine Processing of Remotely Sensed Data held at Purdue University in 1980 (Indiana: Purdue University)* (pp. 326–335).
13. Faillot, P. (1937). Essai sur la géologie du Rif septentrional. *Notes et mémoires du Service Géologique du Maroc*, *40*, 533.
14. Durand-Delga, M. (1963). Essai sur les structures des domaines émergés autour de la Méditerranée occidentale. *Geologische Rundschau Stuttgart*, *5-3-2*, 534–535.
15. Achbun, A., Mastere, M., & El Fellah, B. (2015). Up-scaling approach for a multi-series object based classification using CORINE typology. *European Journal of Scientific Research*, *132*(2), 107–117.
16. Achbun, A., Mansour, M., Layelman, M., & Smiej, F. (2011). Etude comparative de la classification orientée objet d'une image spot5 pour la cartographie de l'occupation du sol via eCognition® 9", Géoobservateur, 19.
17. Diallo, Y., Hu, G., & Wen, X. (2010). Assessment of land use cover changes using NDVI and DEM in Puer and Simao Counties, Yunnan Province, China. *Report and Opinion*, *2*(9).

Analysis of Noisy Satellite Image Using Statistical Approach

Salma El Fellah, Salwa Lagdali, Mohammed Rziza and Younes El Fellah

Abstract

Technological advances and increasing availability of satellite sensors acquire more information about the earth and offer the potential for more accurate land cover classifications and pattern analysis. However, this type of image (satellite image) is rich and various in content, however it suffers from noise that affects the image in the acquisition. Therefore, there is a requirement of an effective and efficient method for features extraction from the noisy image. This paper presents an approach for satellite image segmentation that automatically segments image using a supervised learning algorithm into urban and nonurban area. We have applied a statistical feature including local feature computed by using the probability distribution of the phase congruency computed (El Fellah S, El haziti M, Rziza M, et Mastere M, A hybrid feature extraction for satellite image segmentation using statistical global and local feature, 2016, [1]). The results provided, demonstrate a good detection of urban area with high accuracy in absence of noise. However when noise is added to images, the classification results deteriorate. Hence, to improve these results we propose a novel features based on higher order spectra known by their robustness against noise.

Keywords

Computer vision • Segmentation • Classification • Satellite image • Statistical feature • Phase gradient • Higher order statistics

S. El Fellah (✉) · S. Lagdali · M. Rziza
Faculty of Sciences, Mohammed V University in Rabat, Rabat, Morocco
e-mail: elfellah.s@gmail.com

Y. El Fellah
Energy and Farm Machinery Department, Rural Engineering, Institut Agronomique et Veterinaire Hassan II, Rabat, Morocco

1 Introduction

The analysis of satellite image has long attracted the attention of the remote sensing community and the results are the basis for many environmental and socioeconomic applications [2]. Each image contains a lot of information inside it and can have a number of objects with characteristics related to nature, shape, color, density, texture, or structure. It is very difficult for any human to go through each image, extract, and store useful patterns. An automatic mechanism is needed to extract objects from the image to be analyzed for classification, detection, etc. The problem of urban object recognition on satellite images is rather complicated because of the huge amount of variability in the shape and layout of an urban area, in addition to that, the occlusion effects, illumination, view angle, scaling, are uncontrolled [3]. Therefore, more robust methods are necessary for good detection of the objects in remotely sensed images. Generally, most of existing approaches are based on frequency features and use images in gray levels. Texture-based methods partition an image into several homogenous regions in terms of texture similarity. Most of the work has concentrated on pixel-based techniques, [4]. The result of pixel-level segmentation is a thematic map in which each pixel is assigned a predefined label from a finite set. However, remote sensing images are often multispectral and of high resolution which makes its detailed semantic segmentation excessively computationally demanding task. This is the reason why some researchers decided to classify image blocks instead of individual pixels [5]. We also adopt this approach by automatically dividing the image into a single sub-image (block), and then evaluate classifiers based on support vector machines, which have shown good results in image classification. The process of generating descriptions represents the visual content of images. In this paper, we focus on the emerging image classification method that uses statistical and bi spectrum feature; in order to model local feature and analyze the effect of noise on the classification accuracy. The rest of the paper is organized as follows. Section 2 presents a review of some works related to our work, Sect. 3 describes

N. Rebai and M. Mastere (eds.), *Mapping and Spatial Analysis of Socio-economic and Environmental Indicators for Sustainable Development*, Advances in Science, Technology & Innovation, https://doi.org/10.1007/978-3-030-21166-0_14

the general framework of the proposed approach, while Sect. 4 shows the experimental results and finally Sect. 5 concludes the paper.

2 Related Work

There are many techniques for classification of satellite images. We briefly review here some of the methods that are related to our work: Mehralian and Palhang [6] separate urban terrains from nonurban terrains using a supervised learning algorithm. Extracted feature for image description is based on principal components analysis of gradient distribution. Ilea et al. [7] considered the adaptive integration of the color and texture attributes and proposed a color-texture-based approach for SAR image segmentation. Fauquer et al. [8] classify aerial images based on color, texture, and structure features; the authors tested their algorithm on a dataset of 1040 aerial images from 8 categories. Ma and Manjunath [9] use Gabor descriptors for representing aerial images. Their work is centered on efficient content-based retrieval from the database of aerial images and they did not try to automatically classify images to semantic categories. In this work, we use local features, believing that is beneficial on identifying image and more suitable to represent complex and noisy scenes and events categories, our feature vector extracted will be the result of combination of the statistical local feature.

3 Proposed Approach of Urban Terrain Recognition

In this paper, a new approach for satellite images segmentation is presented. The method comprises three major steps:

- Firstly, we start by splitting the image into blocks with size of (20 × 20).
- The second step consists in feature extraction, we calculate the computation of phase congruency map of each sub-image (block) then, the statistical local features (mean, variance and skewness) are calculated from the probability distribution of the phase congruency. These features are combined to construct a feature vector for each block. These vectors are used to characterize each sub-image.
- Finally, these vectors are used as training and testing where Gaussian noise is added to the test images. Classification is performed using SVM to distinguish urban classes from nonurban classes. The tests show that the proposed method can segment images with high accuracy in absence of noise but when the images are corrupted by noise the accuracy becomes progressively worse.

In what follows, it is assumed that satellite images are being analyzed for segmentation to urban and nonurban terrain. Below, we describe each of these steps.

3.1 Partitioning

For evaluation of the classifiers we used 800 × 800 pixel (RGB) image taken from Google Earth related to Larache city, Morocco, satellite images are sometimes very large and handling. We split this image into smaller blocks of 20 × 20 pixels, with an overlap of 4 pixels at the borders. So we have in total 4493 blocks in our experiments. We classified all images into 2 categories, namely: urban and nonurban. Examples of sub-images from each class are shown in Fig. 1.

3.2 Feature Extraction

1. **Phase congruency histogram features**

To describe each sub-image, the statistical features of the 2D phase congruency histogram applied on each block and obtained values as principals used for features, so we have a 1D feature vector for each block, which will be used in training process. The statistical features [10] provide information about the properties of the probability distribution. We use statistical features of the phase congruency histogram (PCH) as mean, variance, and skewness that are computed by using the probability distribution of the different levels in the histograms of PCH. Let be a discrete random variable that represents different levels in a map and let be the respective probability density function. A histogram is an estimation of the probability of occurrence of values as measured by We content with three statistical feature of histogram:

Mean (m): computes the average value. It is the standardized first central moment of the probability distribution in image.

$$m = \sum_{i=0}^{L-1} \lambda_i p(\lambda_i). \tag{1}$$

Variance (σ): It's second central moment of the probability distribution, the expected value of the squared deviation from the mean.

$$\sigma = \sum_{i=0}^{L-1} P(\lambda_i)\lambda_i^2 - m^2. \tag{2}$$

Skewness (k): computes the symmetry of distribution. S gives zero value for a symmetric histogram about the mean and otherwise gives either positive or negative value depending on whether histogram has been skewed right or left to the mean.

Fig. 1 Some sample blocks from satellite images; first row: Nonurban areas, second row urban areas

$$k = \sum_{i=0}^{L-1} (\gamma_i - m)^3 p(\gamma_i). \quad (3)$$

2. **Bispectral features**

Bispectrum is the third order spectrum known by its ability to nullify Gaussian noise where the bispectrum of Gaussian noise is zero [11,12]. Mathematically the bispectrum $B(f_1, f_2)$ of 1D signal x is expressed as:

$$B(f_1, f_2) = X(f_1).X(f_2). * X(f_1 + f_2)$$

where $X(f_1)$ and $X(f_2)$ are respectively the fourier transform of at frequencies f_1 and f_2 and $*X(f_1 + f_2)$ the conjugate. The bispectrum is a complex value and it can be expressed as:

$$B(f_1, f_2) = |B(f_1, f_2)|\, exp(i\pi)$$

where$|B(f_1, f_2)|$ is the magnitude and π is the phase of bispectrum.

Features extracted from bispectrum are very robust against noise where every Gaussian noise added to the image is eliminated. For this reason, we are interested in the bispectrum magnitude features.

The mean of the bispectrum magnitude is choosed as a feature vector for classifying satellite images:

$$MB = Mean(B(f_1, f_2))$$

3.3 Classification

After the calculation of these statistical features for each PCH, the feature vectors of each block are constructed as:

$$fPC = \{mPCH, \lambda PCH, kPCH\}$$

The feature vectors of all the blocks images including urban and nonurban sub-image are constructed and stored to create a feature database.

After feature extraction, we use this vector feature to train and test SVM. Given a set of training examples, each marked as belonging to one of two categories. We used half of the images for training and the other half for testing.

4 Experimental Result of the Proposed Method

4.1 Classification Using Phase Congruency Features

We experiment the approach on two samples of satellite image, taken both from Google Earth, related to Larache city, Morocco [1]. There sizes are 800 × 800 pixel (RGB) image. We have split (with an overlap of 4 pixels at the borders) these images into smaller sub-images of 20 × 20 pixels. A total of 4493 sub-images occurs.

We characterize each sub-image by 1D statistical feature vector, which is based on the statistical local features (mean, variance, and skewness) calculated from the probability dis-

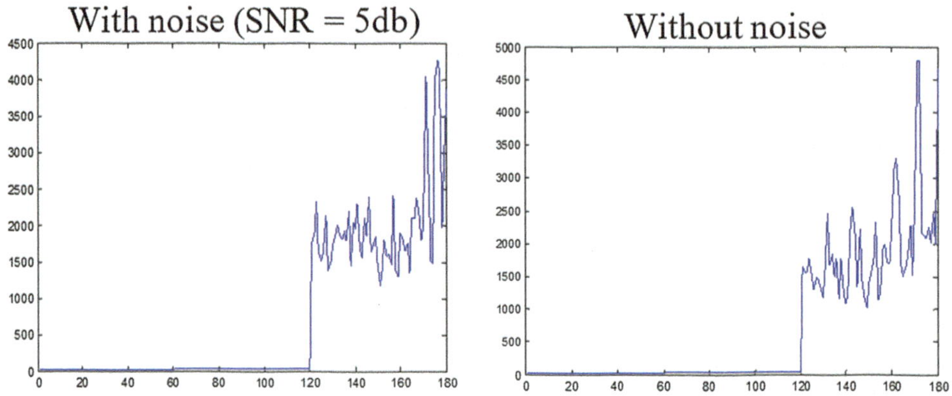

Fig. 2 Local feature plot of a satellite image with a level noise (5 dB) and without noise

Table 1 Block detection results for noisy test data using local feature with size of sub-image [20 × 20]

Metric	Without noise	SNR = 5 dB	SNR = 10 dB	SNR = 20 dB
Accuracy	0.94	0.74	0.85	0.93
Precision	0.96	0.76	0.87	0.95

tribution of the phase congruency. In order to evaluate the effect of noise on the classification of satellite images, we add Gaussian noise with different noise levels (SNR = 20, 10 and 5 dB) on the test images (Fig. 2). The results of the classification experiment with and without noise are reported in Table 1. Figure shows the plot of local feature of a satellite image block (see Fig. 3) with a very high-level noise (5 dB) and without noise, it can be observed that some picks have despaired when adding noise; which means that the feature has been affected by the noise.

We use half of the features for training and the other half for testing, we label the training data manually with 1 and −1, where label 1 refers to urban category and the −1 refers to nonurban category, and the obtained model will be tested on the test data. To evaluate the robustness of the features against noise, we add Gaussian noise with different levels of noise on the test data. In the sections below the results of the test will be discussed.

To examine the ability of proposed approach, we have used accuracy and precision statistical measures:

From Table 1 we observe that in the absence of noise, the phase congruency features give good results (94%). However, when the images are corrupted by Gaussian noise, the correct accuracy becomes to deteriorate from 93% at SNR = 20 dB to 74% at SNR = 5 dB. This degradation yields us to think about exploiting Higher order spectra in the feature extraction procedure. Higher order spectra and especially the third order namely the bispectrum are known by their ability to nullify Gaussian noise and therefore they manage well in high-level noise scenes as in the case of satellite images.

4.2 Classification Using Bispectral Feature

Figure 4 shows the mean magnitude of bispectrum of a Satellite image without noise and under a very high-level noise (5 dB). It can be observed that the feature is not affected by the noise. Hence it can be used as a feature for classifying satellite images.

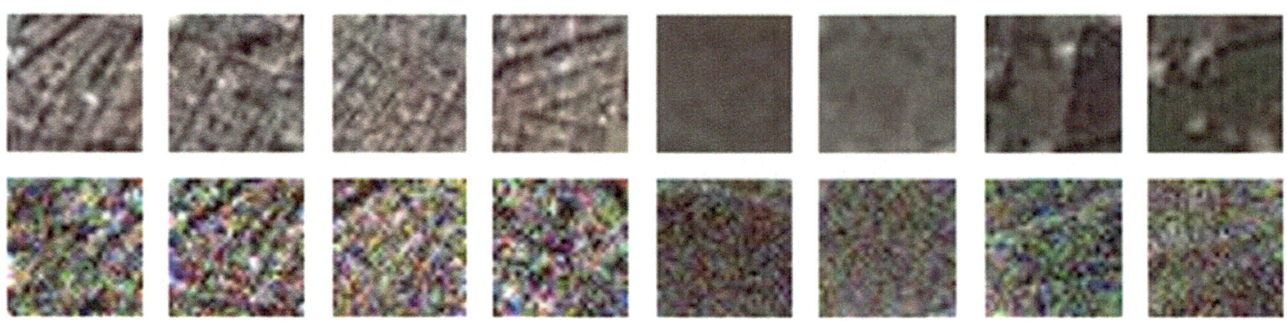

Fig. 3 Satellite image block without noise (first line) and with a level noise (5 dB) (second line)

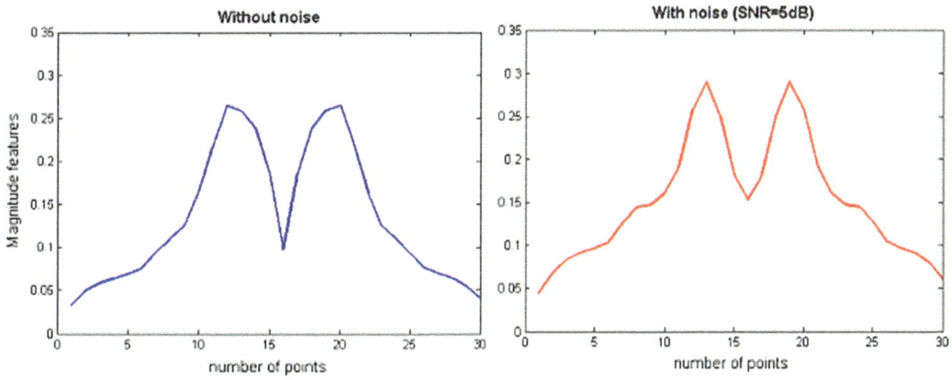

Fig. 4 Mean magnitude of bispectrum of a satellite image with a level noise (5 dB) and without noise

Table 2 Block detection results for noisy test data using bispectrum mean feature

Metric	Without noise	SNR = 5 dB	SNR = 10 dB	SNR = 20 dB
Accuracy	0.95	0.76	0.87	0.94
Precision	0.96	0.96	0.88.2	0.94.3

Table 2 shows the classification results using the bispectrum mean features. As can be seen, the features derived from bispectrum are robust against noise due to the invariance to noise property of higher order spectra.

5 Conclusions

In this work, we have focused on the type of Feature Extraction Technique, and we have proposed a statistical feature including local features in which we compute the probability of distribution of 2D phase congruency. We distinguish urban from nonurban terrain, the algorithm makes decision about image block (not a pixel) in both size (20 × 20), so each block is described by 1D vector features, then SVM is used for classification. The results of the approach yield good performance in absence of noise. However, when the images are corrupted by Gaussian noise the accuracy deteriorate which yields as to analyze bispectrum features that do not change with noise.

References

1. El Fellah, S., El haziti, M., Rziza, M., & et Mastere, M. (2016, April). *A hybrid feature extraction for satellite image segmentation using statistical global and local feature*. Volume 380 of the book series lecture notes in electrical engineering (LNEE) (pp. 247–255). https://doi.org/10.1007/978-3-319-30301-7_26.
2. Blaschke, T. (2010). Object based image analysis for remote sensing. *ISPRS Journal of Photogrammetry and Remote Sensing*, *65*(1), 2–16.
3. Sirmacek, B., & Unsalan, C. (2011). A probabilistic approach to detect urban regions from remotely sensed images based on combination of local features. In *5th RAST2011 Conference*.
4. Szummer, M., & Picard, R. W. (1998, January). Indoor-outdoor image classification. In *Proceedings of the IEEE ICCV Workshop Bombay, India* (pp. 42–51).
5. Pagare, R., & Shinde, A. (2012, January). A study on image annotation techniques. *International Journal of Computer Applications*, *37*(6), 42–45.
6. Mehralian, S., & Palhang, M. (2013). Principal components of gradient distribution for aerial images segmentation. In *11th Intelligent Systems Conference*.
7. Ilea, D. E., & Whelan, P. F. (2011). Image segmentation based on the integration of color texture descriptors - A review. *Pattern Recognition*, *44*, 2479–2501.
8. Fauqueur, J., Kingsbury, G., & Anderson, R. (2005). Semantic discriminant mapping for classification and browsing of remote sensing textures and objects. In *Proceedings of IEEE ICIP*.
9. Ma, W. Y., & Manjunath, B. S. (1998). A texture thesaurus for browsing large aerial photographs. *Journal of the American Society for Information Science*, *49*(7), 633–648
10. ShamikTiwari, V. P. Shukla, S. R. Biradar, & Singh, A. K. (1993). A blind blur detection scheme using statistical features of phase congruency and gradient magnitude. *Advances in Electrical Engineering* Article ID521027, 10 pages, 2014. Lang. Syst. *15*, 5 (1993), 795–825. https://doi.org/10.1145/161468.16147.
11. Nikias, C. L., & Mendel, J. M. (1993). Signal processing with higher-order spectra. *IEEE Signal Processing Magazine*, *10*(3), 10–37.
12. Petropulu, A. (1999). Higher-order spectral analysis. In Madisetti V. K. & Williams D. B. (eds.), *Digital signal processing hand-book*. Chapman & Hall/CRCnetBASE.